Springer Proceedings in Physics 77

Springer Proceedings in Physics

Managing Editor: H. K. V. Lotsch

46 *Cellular Automata and Modeling of Complex Physical Systems*
Editors: P. Manneville, N. Boccara, G. Y. Vichniac, and R. Bidaux

47 *Number Theory and Physics*
Editors: J.-M. Luck, P. Moussa, and M. Waldschmidt

48 *Many-Atom Interactions in Solids*
Editors: R. M. Nieminen, M. J. Puska, and M. J. Manninen

49 *Ultrafast Phenomena in Spectroscopy*
Editors: E. Klose and B. Wilhelmi

50 *Magnetic Properties of Low-Dimensional Systems II: New Developments*
Editors: L. M. Falicov, F. Mejía-Lira, and J. L. Morán-López

51 *The Physics and Chemistry of Organic Superconductors*
Editors: G. Saito and S. Kagoshima

52 *Dynamics and Patterns in Complex Fluids: New Aspects of the Physics–Chemistry Interface*
Editors: A. Onuki and K. Kawasaki

53 *Computer Simulation Studies in Condensed-Matter Physics III*
Editors: D. P. Landau, K. K. Mon, and H.-B. Schüttler

54 *Polycrystalline Semiconductors II*
Editors: J. H. Werner and H. P. Strunk

55 *Nonlinear Dynamics and Quantum Phenomena in Optical Systems*
Editors: R. Vilaseca and R. Corbalán

56 *Amorphous and Crystalline Silicon Carbide III, and Other Group IV-IV Materials*
Editors: G. L. Harris, M. G. Spencer, and C. Y. Yang

57 *Evolutionary Trends in the Physical Sciences*
Editors: M. Suzuki and R. Kubo

58 *New Trends in Nuclear Collective Dynamics*
Editors: Y. Abe, H. Horiuchi, and K. Matsuyanagi

59 *Exotic Atoms in Condensed Matter*
Editors: G. Benedek and H. Schneuwly

60 *The Physics and Chemistry of Oxide Superconductors*
Editors: Y. Iye and H. Yasuoka

61 *Surface X-Ray and Neutron Scattering*
Editors: H. Zabel and I. K. Robinson

62 *Surface Science: Lectures on Basic Concepts and Applications*
Editors: F. A. Ponce and M. Cardona

63 *Coherent Raman Spectroscopy: Recent Advances*
Editors: G. Marowsky and V. V. Smirnov

64 *Superconducting Devices and Their Applications*
Editors: H. Koch and H. Lübbig

65 *Present and Future of High-Energy Physics*
Editors. K.-I. Aoki and M. Kobayashi

66 *The Structure and Conformation of Amphiphilic Membranes*
Editors: R. Lipowsky, D. Richter, and K. Kremer

67 *Nonlinearity with Disorder*
Editors: F. Abdullaev, A. R. Bishop, and S. Pnevmatikos

68 *Time-Resolved Vibrational Spectroscopy V*
Editor: H. Takahashi

69 *Evolution of Dynamical Structures in Complex Systems*
Editors: R. Friedrich and A. Wunderlin

70 *Computational Approaches in Condensed-Matter Physics*
Editors: S. Miyashita, M. Imada, and H. Takayama

71 *Amorphous and Crystalline Silicon Carbide IV*
Editors: C. Y. Yang, M. M. Rahman, and G. L. Harris

72 *Computer Simulation Studies in Condensed-Matter Physics IV*
Editors: D. P. Landau, K. K. Mon, and H.-B. Schüttler

73 *Surface Science: Principles and Applications*
Editors: R. F. Howe, R. N. Lamb, and K. Wandelt

74 *Time-Resolved Vibrational Spectroscopy VI*
Editors: A. Lau, F. Siebert, and W. Werncke

75 *Computer Simulation Studies in Condensed-Matter Physics V*
Editors: D. P. Landau, K. K. Mon, and H.-B. Schüttler

76 *Computer Simulation Studies in Condensed-Matter Physics VI*
Editors: D. P. Landau, K. K. Mon, and H.-B. Schüttler

77 *Quantum Optics VI*
Editors: D. F. Walls and J. D. Harvey

Volumes 1–45 are listed at the end of the book

D. F. Walls J. D. Harvey (Eds.)

Quantum Optics VI

Proceedings of the Sixth International Symposium
on Quantum Optics
Rotorua, New Zealand, January 24-28, 1994

With 130 Figures

Springer-Verlag
Berlin Heidelberg New York
London Paris Tokyo
Hong Kong Barcelona
Budapest

Professor Dr. Dan F. Walls
Professor Dr. John D. Harvey

University of Auckland, Department of Theoretical Physics
Private Bag 92019
Auckland, New Zealand

ISBN-13: 978-3-642-79103-1 e-ISBN-13: 978-3-642-79101-7
DOI: 10.1007/978-3-642-79101-7

Library of Congress Cataloging-in-Publication Data. International Symposium on Quantum Optics (6th: 1994: Rotorua, N. Z.) Quantum optics VI: proceedings of the Sixth International Symposium on Quantum Optics, Rotorua, New Zealand, January 24-28, 1994 / D. F. Walls, J. D. Harvey, eds. p. cm. – (Springer proceedings in physics; v. 77)

Quantum optics–Congresses. 2. Interferometry–Congresses. 3. Light, Wave theory of–Congresses. I. Walls, D. F. II. Harvey, J. D. III. Title. IV. Series. QC446.15.I6567 1994 535–dc20 94-25883

Softcover reprint of the hardcover 1st edition 1994

SPIN 10427149 54/3140 - 5 4 3 2 1 0 - Printed on acid-free paper

Hannes Risken (1934–1994)

We wish to dedicate these Proceedings to the memory of Hannes Risken who died shortly after the Symposium.

Hannes was an active participant at previous Symposia and only his ill health prevented him from attending this Symposium.

His contribution which appeared as a poster is represented in the Proceedings.

Preface

The New Zealand Quantum Optics meetings have now become established internationally as an important forum for discussions of advances in theoretical and experimental studies of the interaction of light with atoms. From the first meeting in 1973 to the sixth in 1994, the conference has grown steadily in both the number of participants and the number of talks.

This volume contains detailed notes provided by the invited speakers at, or shortly after, the meeting. Whilst earlier meetings were held every three years, the anticipated symposium in 1992 was delayed by two years to avoid a clash with an Australian meeting and to ensure that the 6th and 7th meetings were spaced as far as possible from the IQEC 96 in Sydney. This increased time between the most recent meetings has allowed the field to evolve more substantially than at previous meetings, and this is apparent in the shift of emphasis towards atom optics as well as quantum optics. In order to accommodate an increased number of speakers, the format was changed slightly from that in 1989 to include a larger number of invited talks of 20 or 30 minutes duration, but it was still possible to avoid parallel sessions. In addition, 40 posters were presented divided into two separate sessions.

Just under 100 participants (with over half of these from outside Australasia) gathered together in Rotorua in January 1994 to discuss various aspects of quantum optics. This field has a very strong participation from Australia and New Zealand, which is reflected in the number of contributions from "local" Australasian speakers. A large number of graduate students were also able to benefit from the opportunity to attend an international meeting in New Zealand.

The conference was again held in the Kingsgate Hotel at Rotorua with all sessions conducted in the mornings and evenings, keeping the afternoons free for informal discussions and exploration of the local environment.

The editors would like to thank the participants for providing their notes for publication promptly after the meeting, and also the following organisations who provided financial and other support:

University of Auckland,
International Union of Pure and Applied Physics,
The Royal Society,
US Army Research Office, Far East,
Coherent Scientific Pty, Ltd.,
The British Council.

We look forward to another stimulating meeting in 1998.

Auckland
April 1994

Daniel F. Walls
John D. Harvey

Contents

Part IV New Theoretical Techniques

Part V Applications in Laser Physics

Part I

Quantum Optics Experiments

An Axially Symmetric Imaging System for Ultra-Cold Neutral Atoms

H. Takuma[1], *T. Nayuki*[1], *A. Shindoh*[1], *J. Kawanaka*[1], *K. Shimizu*[1], *and F. Shimizu*[2,1]

[1]Institute for Laser Science, UEC, 1-5-1, Chofugaoka, Chofu, Tokyo 182, Japan

[2]Department of Applied Physics, University of Tokyo, Bunkyo-ku, Tokyo 113, Japan

1. Introduction

The quantum mechanical observation problems have had hypothesis which could not be observed directly until recent progress in laser technology brought such problems in the light of experimental reality in various cases. Atom interferometry is one of such examples[1]. By the use of recently developed laser cooling and trapping techniques, such experiments have been carried out at a long de Broglie wavelength range, allowing the observation of interference fringes in various types of interferometer. Thus it has been demonstrated beautifully that the quantum mechanical wave function of an atom behaves exactly as predicted by quantum mechanics[1-5].

Such experimental successes on atom optics are interested not only as a demonstration of the quantum mechanical observation, but also as a new technique having potential applicability in number of practical problems. As an example, high accuracy measurement of the earth gravitational field was shown to be possible by atom interferometer using stimulated Raman process[6]. It is also expected that a gyroscope based on the atom interferometer will provide a sensitivity much higher than that has been established by laser gyroscopes[7].

The use of atom optics is not limited in interferometry. Since the behavior of the wave function of a particle moving in the free space is analogous with the optical waves, it is a quite natural attempt to develop atom optics components corresponding to various optical components such as imaging lenses. First image formation in the atom optics was realized by the use of Fresnel zone plate made for X-ray optics[8]. However, it was only one dimensional image and the intensity was extremely weak because a Fresnel zone plate has extremely low transmission coefficient for ultra-cold atoms.

Thus an imaging lens for neutral atoms having axial symmetry and high transmission coefficient has been looked for in order to realize a two dimensional imaging system for atom optics. The purpose of the present paper is to propose such a scheme of atom lens for the first time and report our recent preliminary experiment, which demonstrated the feasibility of such a system.

2. Principle

The wave function of a neutral atom which is freely moving in the potential field represented by $V(r)$ is described by the Schrödinger equation

$$\nabla^2\phi + \frac{2m}{\hbar}(E - V(\mathbf{r}))\phi = 0 \qquad (1)$$

By substituting

$$k^2 = \frac{2m}{\hbar}(E - V(\mathbf{r})),$$

Springer Proceedings in Physics, Vol. 77 Editors: D.F. Walls · J.D. Harvey

Quantum Optics VI

we have a common type of wave equation

$$\nabla^2\phi + k^2\phi = 0. \tag{2}$$

This is an identical expression with all kinds of classical waves including optical waves. In the optical waves, wave number k is inversely proportional to the index of refraction n, which is given as the ratio of the wave number in the optical medium to that in the vacuum:

$$n = \frac{k}{k_0}. \tag{3}$$

Therefore the propagation of the wave function of an atom is equivalent to the propagation of optical wave in a space having the index of refraction represented by

$$n = \left[1 - \frac{V(\mathbf{r})}{E}\right]^{1/2}. \tag{4}$$

In order to make a lens for atoms, we must have a potential distribution analogous to the distribution of the index of refraction in an optical lens. For neutral atoms, potential formation in free space may be possible by applying magnetic and/or electric field. Because the potential gradient should be large enough to deflect the atoms, the potential formation by the second order Stark effect seems to be the most reasonable way.

We have shown in our former work on double-slit atom interferometry that a moderately strong inhomogenous electric field gives a good amount of deflection and phase-shift in wave function when the de Broglie wavelength is on the order of 10 nm. The effect of non resonant electromagnetic field on neutral atoms is normally through the second-order Stark effect which is expressed as

$$V = -\alpha F^2, \tag{5}$$

where α is a positive constant for a stable state, and F the electric field. Therefore the atom is deflected toward stronger field, and the focusing lens should have a field maximum on the axis, which is impossible to realize by dc field, making it difficult to produce an electrostatic-static positive lens for atoms.

However, it is possible to make a potential maximum in one dimensional arrangement, and atoms should experience axially symmetric focusing field if the dc field is rotated around the axis so that slowly dropping atoms experience axially symmetric focusing field in average. Such a device was shown to work as a focuser in the previous work[9].

However, such a system is not perfectly analogous to the optical lens, and more versatile focusing and imaging scheme has been looked for keenly as a most important component in atom optics. We propose here more idealistic arrangement, which is provided by the use of microwave electric field in a cylindrical cavity in the lowest TM_{01} mode, which is resonant to the cut off frequency of the circular wave guide in TM_{01} mode. In this case, the electric field is parallel to the axis of the cylindrical cavity, and the intensity shows axially symmetric distribution which is expressed as a Bessel function $A_0 J_0(k_0 r)$, where r is the radius in the cylindrical coordinate, and k_0 the wave number of the microwave which is tuned at resonant frequency of the TM_{01} cut-off mode resonator. The boundary condition is $E = 0$ on the inner wall of the resonator, therefore the inner diameter of the cavity, which is represented by R, is given by

$$R = \frac{2.405}{k_0},$$

taking into account of that the first root of $J_0(x) = 0$ is $x = 2.405$. Namely, the microwave frequency determines the radius of the cavity cross section. Although the potential distribution given by Bessel function is not identical with the index of refraction distribution in the optical lens, the two systems behaves in exactly the same way for paraxial rays. The focal length of the microwave atom lens is determined by the potential distribution $V(k_0 r)$, the kinetic energy of the atom, input power and the quality factor of the resonator. Calculation based on the classical approximation of the atomic motion in the gravitational and microwave field in the present arrangement gives the focal length, which is designated as f, as

$$f = 5.76\frac{mv^2c^4\varepsilon_0\pi}{\alpha\omega^3QP}, \tag{6}$$

where Q and P stand for the loaded quality factor of the resonator and the power dissipated in the resonator, respectively.

The most critical question is whether the coefficient α is large enough to obtain reasonably short focal length with practically obtainable input power and quality factor of the resonator. Deflection and phase shift were investigated on neon metastable state in our work on double-slit interferometry, and good amount of shift of the interference fringe was observed by applying dc voltage of only a few hundred volt on the pin electrode placed in front of the double-slit [10]. This shows that such a focusing device can be obtained by feeding a few watts of microwave power into a resonator having a quality factor on the order of several thousand. Equation (3) gives a focal length of about 20 cm for Q=10,000 and P=3 W, by the use of α obtained in our double-slit experiment[9].

3. An Experimental Demonstration

An experimental demonstration of such a lens system for atoms was carried out by using a setup shown schematically in Fig. 1. An ultra-cold magneto-optical trap of metastable state neon atoms is prepared and a train of pumping pulses of 599 nm is focused in the trap to release the metastable atoms located in the beam waist. About one third of atoms in the

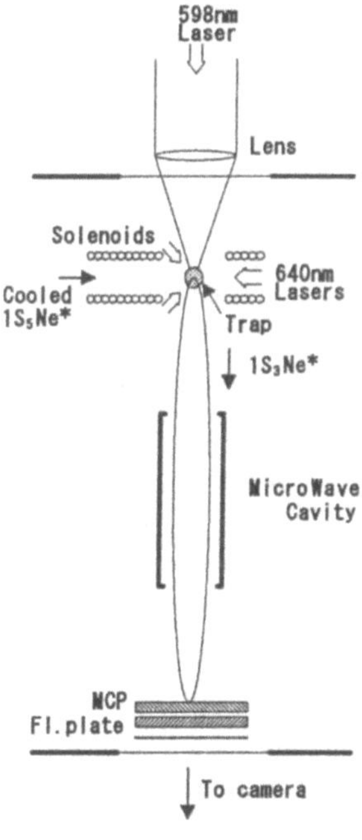

Fig. 1 A schematic diagram of the experimental setup. Trapped ultra-cold atoms are reliesed from the trap by the irradiation of 598 nm laser pulse focused in the center of the trap. A resonant microwave field in TM01 mode in the cylindrical microwave cavity provides a focusing field for neon atoms in the metastable state. The atoms are detected by a two-dimensional detector which consists of multichannel plate and fluorescent plate.

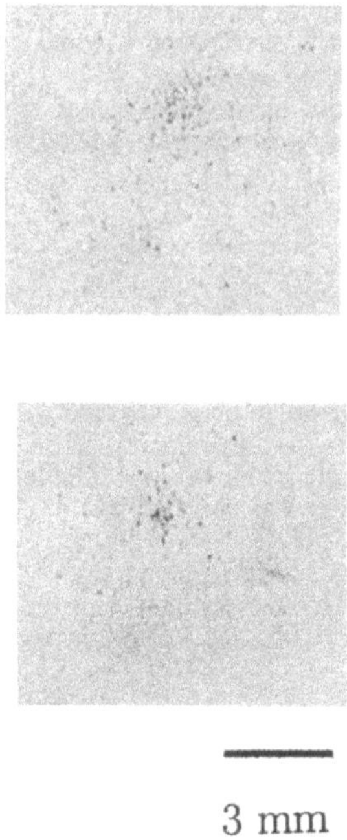

Fig.2 Distribution of the matastable state neon atoms detected by the tow dimensional detector. Upper picture: without microwave field Lower picture: with 3 W microwave input.

focal region is transferred to the J=0 state, released from the trap and start to fall freely. Those atoms having sufficiently small transverse momentum pass through a 13.0 cm long cylindrical resonator the upper edge of which is about 39 cm bellow the trap, deflected by the microwave field, and detected by a two-dimensional metastable state detector, which consists of a pair of micro-channel plate and a fluorescent plate, placed about 20 cm below the lower edge of the resonator. The inner wall of the resonator is uniformly machined in 13.0 mm diameter to make it resonant in TM_{01} cut off mode at about 17.7 GHz. The unloaded and loaded quality factors of the microwave resonator were 12,000 and 9,000, respectively.

The problem of such an optics is the strong influence of atomic velocity, that is, the chromatic aberration. The only solution to overcome this problem is to use monochromatic atoms, that is, to observe only images made by a group of atoms having velocities in rather narrow range, at the present stage of matter. Such selection can be made by opening the gate of the detector only in a given time interval after the atoms are released by the 599 nm laser pulse.

Numerical calculation and also equation (6) gave a focal length of about 20 cm for an input microwave power of 3 W. Experimental demonstration worked reasonably well as shown in Fig. 3. The parameters have not been optimized yet, but we see clearly that the atoms are detected in a broader range without applying the field, as shown in (a), and larger number in the focal range than the surrounding are when the microwave field was supplied as shown in (b). Although the intensity at the focus point is saturated in reproduced picture, total number of the detected atom is the same in (a) and (b), and this understanding may help in distinguishing the difference .

4. Discussion

Demonstration described above is a test of focusing only, but the present system is very much similar to the optical lens in principle, and there should be no problem in using such a system to make images. The most important problem left for further consideration is the aberration corresponding to the spherical aberration and the chromatic aberration in the spherical lens in optics. It may be possible to remove the aberration by arranging the positive and negative lenses of various geometry and field intensity. A negative lens having axial symmetry may be provided in a dc multi-pole cylindrical electrodes. Detailed discussions are left for further investigation.

References

1. O. Carnal and J. Mlynek, Phys. Rev. Lett. **66**, 2689 (1991).
2. D.W. Keith, C.R. Ekstrom, Q.A. Turchtte, and D.E. Pritchard, Phys. Rev. Lett. **66**, 2693 (1991).
3. M. Kassevich and S. Chu, Phys. Rev. Lett. **67**, 181 (1991).
4. F. Shimizu, K. Shimizu, and H. Takuma, Phys. Rev. **A 46**, R17 (1992).
5. H. Takuma, K. Shimizu, and F. Shimizu, *Quantum Control and Measurement* , *North-Holland Delta Series*, H. Ezawa and Y. Murayama, ed., p 289 (North-Holland, Amsterdam, London, New York, Tokyo, 1993).
6. M. Kassevich and S. Chu, Appl. Phys. **B 54**, 321 (1992).
7. F. Riehle, Th. Kisters, A. Witte, and J. Helmucke, Phys. Rev. Lett. **67**, 177 (1991).
8. O. Carnal, M. Siegel, T. Slator, H. Takuma, and J. Mlynek, Phys. Rev. Lett. **67**, 3231 (1991).
9. F. Shimizu, *Atomic Physics 13*, *AIP Conf. Proc.* **275**, H. Walther, T.W. Hänsch, and B. Neizert, ed., p. 157 (American Institute of Physics, New York, 1992).
10. F. Shimizu, K. Shimizu, and H. Takuma, Jpn. J. Appl. Phys. 31, L 436 (1992).

Matter-Wave Optics and Interferometry with Laser-Cooled Atoms

M. Christ, J.H. Müller, D. Bettermann, R. Deutschmann, V. Rieger, M. Schiffer, A. Scholz, K. Sengstock, U. Sterr, and W. Ertmer

Institut für Angewandte Physik der Universität Bonn,
Wegelerstr. 8, D-53115 Bonn, Germany

1. Introduction

Laser manipulation of atoms has proven to be a valuable tool for preparing atomic samples for atom interferometry. Spontaneous light forces may be used to slow, deflect, compress, cool and trap atoms; thus creating either nearly monochromatic atomic beams or atom clouds almost at rest, both of high densities and with de Broglie wavelengths up to several hundred nanometers. These ensembles show sufficient spatial and temporal coherence to exhibit matter wave interference and diffraction.

While dissipative processes provide mechanisms for the manipulation of the atomic velocity distribution, coherent atom-light interaction are the key to create atom optical elements, e.g. beam splitter and interferometers. In this article we describe two separate experiments where the techniques of laser cooling have been successfully integrated with methods from atom optics to observe atomic diffraction and interference, respectively.

The first section describes reflection and diffraction of cold atoms from evanescent light-fields. A well-collimated beam of slow atoms with a low velocity width was coherently split through the mechanical action of a standing evanescent light wave [1].

In the second section, we present the measurement of topological Aharonov-Bohm phases on laser trapped atoms using four pulse Ramsey interferometry. Cooling and trapping techniques are used both for the preparation of the atomic sample and the efficient detection of the interference [2].

2. Mirror and reflection grating for metastable neon atoms

The evanescent wave provides particularly interesting features if applied to coherent reflection and diffraction of atomic de Broglie waves. A versatile model system for reflective diffraction consists of a two-level atom, coherently interacting with a standing evanescent laser wave. If the frequency of this laser is sufficiently detuned to the blue side of the atomic transition, $\delta=\nu_{laser}-\nu_{atom}>0$, reflection is possible. Whereas the periodic optical potential of the standing wave induces multiple coherent exchanges of photon momenta $n\hbar\boldsymbol{Q}$ parallel to the wavevector $\boldsymbol{Q}$, the exponential envelope of the evanescent wave introduces a *smooth* variation of the atom-field coupling *perpendicular* to the surface. Applying the model given in [3], each diffraction order corresponds to one adiabatic potential. In the regime of

Springer Proceedings in Physics, Vol. 77 Quantum Optics VI
Editors: D.F. Walls · J.D. Harvey

grazing incidence and far from the surface the asymptotic values of these potentials are separated by approximately the Doppler energy $h\delta_{Doppler}=\hbar Q v_{||}$ ($v_{||}$: velocity component parallel to $\boldsymbol{Q}$). Closer to the surface, the increasing atom-field coupling strongly shifts the potential values. Thus a light shift considerably larger than $h\delta_{Doppler}$ is necessary to provide both classical turning points for reflection and avoided crossings between the potentials, responsible for motion induced nonadiabatic transitions between the diffraction orders.

As the interaction time may be comparable with the lifetime of the upper atomic level and as most of the potentials are non-reflective [3], there is an appreciable probability for incoherent scattering by both spontaneous photon emissions as well as surface collisions. In the experiment described below these incoherent events are discriminated using an open-cycle transition.

Labeling the diffraction orders by the net number n of photon momenta transferred to an incoming atom in the lower state (for *reflection* n is limited to zero or one). The outgoing momentum components expressed by the incoming momentum components $p_{||}$ and $p_{\perp}$ in the diffraction order n are given by

$$p_{||,n}^{out} = p_{||}^{in} + n\,\hbar\,Q$$
$$p_{\perp,n}^{out} = \sqrt{(p_{\perp}^{in})^2 - 2mh\,(n\,\delta_{Doppler} + n^2\,\delta_{recoil} - \epsilon\,\delta)} \qquad (1)$$

with $h\delta_{recoil}=(\hbar Q)^2/2m$ and a Kronecker symbol $\varepsilon=0$ for n even, and $\varepsilon=1$ for n odd. The diffracted parts of the atom leave in the lower state with n even and in the excited state with n odd.

2.1 Experimental setup

An overview of the whole experimental setup for observation of diffraction is given in Fig. 1. The metastable atoms are produced in a DC-discharge. They are decelerated in a Zeeman-slower. To reduce the divergence and beam diameter as well as to increase the density of the beam, the slowed atoms pass a strongly tilted two dimensional magneto-optical funnel (2D-MOF) [4]. Its axis of symmetry, defining the direction of the free atomic motion, is tilted by 68° versus the original atomic beam axis. Only atoms in the 3s[3/2] 2 state are deflected and compressed down to about 150 µm. The compressed atomic beam used for the reflection and diffraction experiments reported below has a flux of 10^8 atoms/s, a longitudinal velocity v_{long}=25 m/s, a longitudinal velocity spread Δv_{long}^{FWHM}=12 m/s, and a transverse velocity spread of v_{trans}^{FWHM}=0.75 m/s (meanwhile an additional running molasses decreases the longitudinal velocity spread to Δv_{long}^{FWHM}=3 m/s).

The evanescent wave for the reflection and diffraction experiment is prepared by total internal reflection in a quartz prism (n=1.46). The prism is mounted with its base surface horizontally, which yields an atomic deflection in vertical direction.

The 2D-beam-profile detector consists of a two-stage MCP with successive phosphoric screen imaged on an intensified cooled slow-scan CCD camera.

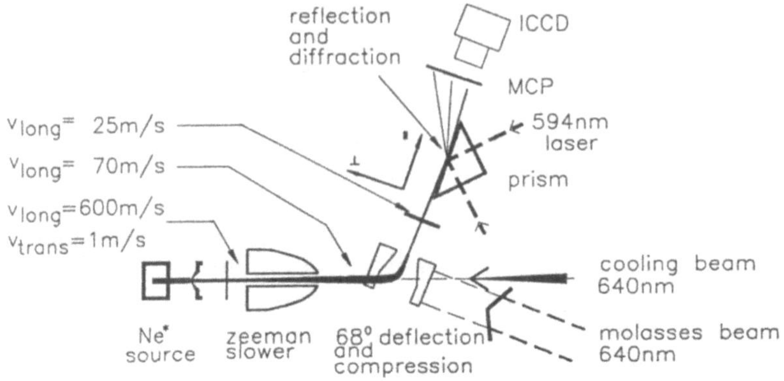

Fig 1: Experimental setup for the observation of coherent diffraction of a metastable neon beam

2.2 Experiment: Reflection and Diffraction

We observe reflection with a steep increase in efficiency for a laser detuning near $\delta \approx +500$ MHz, an optimum at $\delta = +900$ MHz and a smooth decrease for larger values up to $\delta = +2$ GHz. For a provided laser power of 120mW reflection can be achieved for grazing incidence angles α up to $\alpha = 74$ mrad which corresponds to $v_\perp = 1.9$m/s. In the following the outgoing angles are given relative to the prism base surface. Fig. 2 shows a vertical cut for different reflection angles at $\delta = +900$ MHz and p-

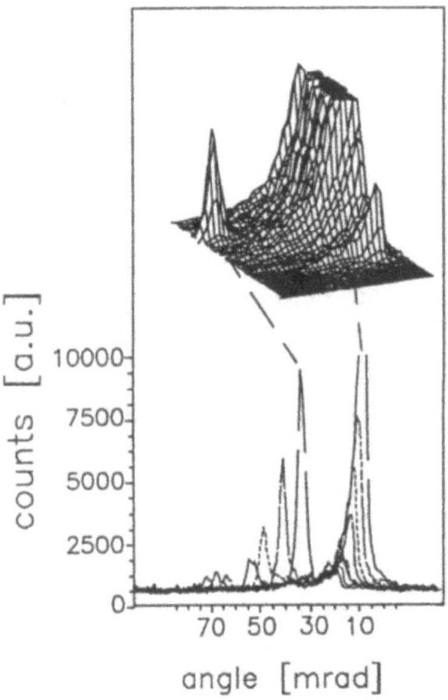

Fig.2: Experimental results for atomic reflection; vertical cuts from CCD-camera images taken for different angles of grazing incidence α, show two peaks. One peak is due to reflected atoms (between 30 mrad to 80 mrad) and one is due to atoms missing the prism (at 10 mrad). Reflection was evident up to an angle α of 74 mrad ($v_\perp = 1.9$m/sec perpendicular to the prism base).

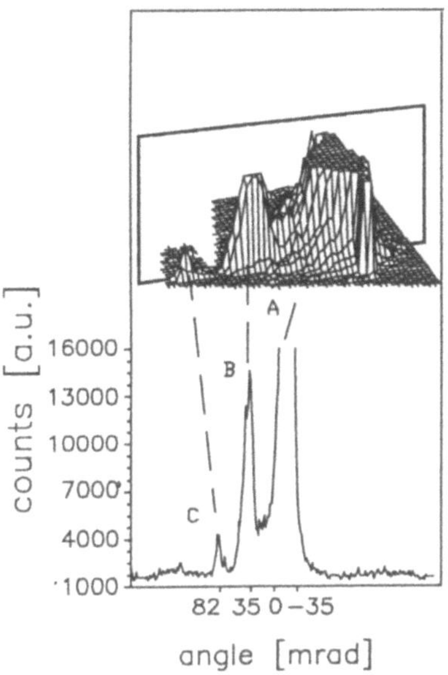

Fig.3: Experimental results for atomic diffraction obtained for two counter-propagating laser waves with a ratio of intensities of 1.64:1; analogous to Fig. 2 the diffraction pattern shows three peaks: (A) transmitted atoms, (B) reflected atoms at 36 mrad, and (C) diffracted lower-state atoms in the n=-2 order at 84 mrad.

polarization. The group of peaks in Fig. 2 around 10 mrad originates from atoms missing the prism. With increasing angle α, the downstream edge of the prism cuts off more and more from those atoms. The peaks between 30 and 80 mrad correspond to specular reflection for various incident angles α. Two effects arise from the gaussian laser beam envelope. First, for increasing incidence angle α the vertical velocity $v_\perp$ increases and reduces the effective mirror surface of sufficient potential height to invert the vertical motion. Thus the count rates decrease for increasing angles α. Second, the slow atoms with v_{long}=25m/s sense the light gradient locally contrary to thermal atomic beams. Therefore the atoms exhibit the gaussian shaped curvature of the mirror surface. A computer simulation of classical atomic trajectories based on this argument is in quantitative agreement with the experimental data.

In order to form a reflective diffraction grating, the laser beam is retroreflected after total internal reflection and adjusted to an intensity ratio of 1.64 between both running waves at the prism. It turned out, that meeting this optimum condition as well as the alignment were very critical to generate a diffraction pattern. The incoming 594 nm laser beam was set to a power level of 150 mW at the prism surface and a detuning of δ=+900 MHz. Fig. 3 shows the original image data for an incident angle of α=36 mrad and a vertical cut of this image. It reveals one peak due to atoms missing the prism (A), the specular reflected atoms at 36 mrad (B), and the diffracted part (C) at 84 mrad corresponding to the diffraction order n=-2. Normalized to the flux of atoms matching the mirror surface (approximately 450·300μm²), up to 70% of these atoms are detected in the spot of specular reflection and 1.5-3% form the diffraction pattern (C). Fig. 4 shows the measured angular position of the n=0 and n=-2 diffraction orders versus different angles α in comparison with the theoretically expected locations. The diffraction pattern showed a similar behaviour in respect to the detuning as the reflection.

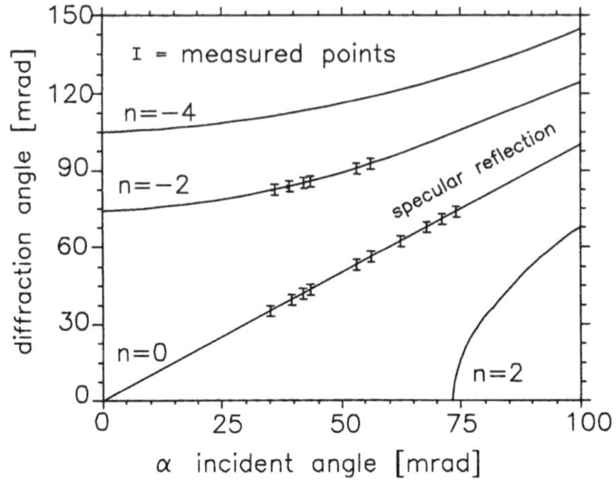

Fig. 4: Measured angular positions of the n=0 and n=-2 diffraction orders versus different angles α compared with the theoretical locations calculated for the parameters v_l=25m/sec and $\delta_{Doppler}$=45MHz (see Eq. 1). The measured points (error bars of ±1.4 mrad) are in good agreement with the theoretical curves.

2.3 Two Port Beam splitter based on an Evanescent Wave

While the *standing* wave grating generally couples many outgoing channels by successive photon-momentum transfers, we propose a two port beam splitter based on the interaction between an evanescent *running* wave of arbitrary polarization and a near resonant atomic $J_g \rightarrow J_e$-transition, the Zeeman degeneracy of which is lifted by an additional magnetic field [5].

Close to the surface the atomic sublevels are shifted by the AC-Stark-effect. Therefore the potentials are sensitive to Clebsch-Gordon-coefficients as well as relative amplitudes of the π- and $\sigma^{\pm}$-polarization components present in the evanes cent wave. The latter include two independent parameters to prepare one single avoided crossing with a transition probability in the range from 0 to 1 (Fig. 5).

An incoming atomic de Broglie wave in the m_g^{in}=0 substate splits coherently at the crossing between the adiabatic m_g=0 and m_g=1 channel, After reflection at classical turning points the partial waves interfere at the crossing thus leading to Stueckelberg oscillation between the populations in the outgoing m_g=0 and m_g=1 channel. The resulting inelastic energy transfer (m_g^{in} - m_g^{out}) $\hbar\omega_{Lamor}$ leads via energy-momentum-conservation to an entanglement between the internal m_g-substate and the external atomic momentum eigenstate. For incoming ground-state atoms of mass M and sublevels m_g^{in}=0 (the parallel momentum $p_{||}$ remains unchanged during the interaction with a running wave)

$$p_{\perp}^{out} = \sqrt{(p_{\perp}^{in})^2 - 2M\, m_g^{out}\, \hbar\, \omega_{Lamor}} \quad . \tag{2}$$

One essential advantage of this beam splitter scheme is its population transfer exclusively between reflective ground-state Zeeman-sublevels, thus minimizing tunneling and spontaneous losses. Additionally it renders possible to adjust the splitting ratio and splitting angle by controlling the light polarization and the magnetic field strength.

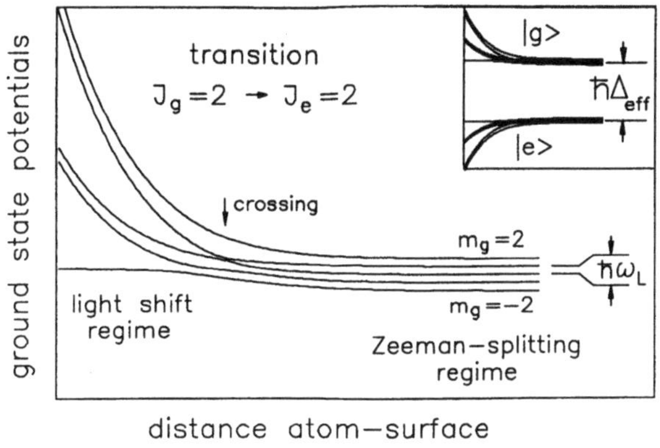

Fig. 5: Adiabatic potentials for a specially choosen polarization of the evanescent wave to produce only one single crossing between the m_g=0 and the m_g=1 channel.

Experimental study of this scheme is in progress using the decelerated and compressed metastable neon beam. For a longitudinal atomic velocity of 25 m/s we expect large splitting angles in the order of milliradian per gauss.

3. Atom interferometry with magnesium atoms

Perhaps the simplest coherent beam splitter consists just of a laser pulse resonantly interacting with a two-level atom. The atomic wave function of an atom in the ground level after the interaction with a laser pulse is divided in two different parts, one part propagating undisturbed in the ground state and the other part shifted in momentum space by one photon recoil propagating in the excited state. Unlike to the behaviour of optical beam splitters the phase difference transferred to the splitted wave function is variable and is equal to the total phase of the laser field at the space-time point of the interaction.

To construct a matter wave interferometer, a sequence of two counterpropagating pairs of collinear laser pulses splits, redirects and subsequently recombines the atomic wave function [6]. A recoil diagram (Fig. 6) of this four-pulse Ramsey interaction scheme shows two closed paths corresponding to the two recoil components of classical saturation spectroscopy. Both closed paths have two exit ports differing in momentum and more important in the internal state of the atom. The intensity at the two exit ports and hence the probability P for the atom to leave the interferometer in the excited state varies periodically as a function of the phase difference between the two arms

$$P \propto \cos[2T(\Delta \pm \delta) + \Phi] \quad . \qquad (3)$$

Here Δ denotes the detuning between the laser frequency and the undisturbed atomic transition, δ denotes the recoil correction ($\delta = \hbar k^2/2m$) and T is the time delay between two copropagating pulses. The additional phase Φ contains the sum of the phase differences between the collinear laser pulses. The dependence of the interferometric signal on the laser frequency provides simple means to scan the

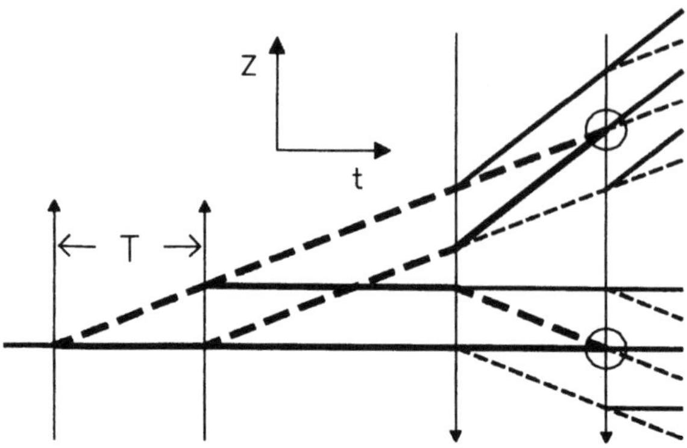

Fig. 6: Recoil diagram for the four pulse Ramsey excitation. The recoil induced separation of the atomic wave packets is shown vs. time. Dashed and solid lines represent atoms travelling in the excited and ground state, resp. The top (bottom) closed trapezoid formed by the thick lines contributes to the low (high) frequency recoil component of the Ramsey fringes.

phase of the output signal for measurement purposes and allows the use of the Ramsey setup as a highly sensitive frequency discriminator. The frequency resolution is just governed by the time T which can be chosen by the experimenter.

The phase difference between the two arms of the interferometer can be influenced by external potentials. In general the phase evolution during the time between two laser pulses can be described by the action S along the classical path Γ of the particle

$$\Phi_\Gamma = S/\hbar = 1/\hbar \int_\Gamma L(r,p,t)\,dt = 1/\hbar \int_\Gamma (p\,dr - E\,dt)\ . \qquad (4)$$

Here L denotes the Lagrangian with p the momentum and E the energy of the atom. For the special case of a time dependend scalar potential V(t) with no spatial variation the momentum is conserved, i.e. no force is acting on the particle, and the application of Eq. 4 shows for the additional phase $\delta\Phi$:

$$\partial\Phi = -1/\hbar \int V(t)\,dt\ . \qquad (5)$$

For a potential that acts selectively on one arm of the interferometer a net phase difference and thus a shift of the interference pattern results. One realizes a situation exactly in the sense of the original proposal of Aharonov and Bohm [7] to demonstrate the significance of potentials for measurable quantities in quantum mechanics.

A convenient way to act selectively on one arm of the interferometer is to use a state-selective potential, e.g. a light shift potential, as the atom propagates in different internal states along the arms of the interferometer (see Fig. 6). This way the need for a spatial separation of the interferometer arms, which is possible in principle, can be circumvented.

3.1 Experimental setup

Only atoms with a long living excited state provide long coherent interaction times and thus high interferometric sensitvity and frequency resolution. The alkaline earth elements with their two-electron level scheme offer both narrow intercombination transitions and for the case of Mg also a closed fast transition in the singulett system for laser cooling and trapping. The $^3P_1$-$^1S_0$ transition of ^{24}Mg with an upper level lifetime of 5.1 ms at a wavelength of 457 nm is an ideal candidate. Cold atoms prepared in a magneto-optical trap provide an ensemble that stays inside a volume of a few mm^3 for interaction times in the range of 10^{-3} s.

Mg atoms are captured in a magneto-optical trap from a thermal atomic beam and cooled down to temperatures below 3 mK (Fig. 7). The trapped cloud contains some 10^3 atoms and has a Gaussian shape with a width of 200 μm. The uv-laser light at 285 nm for trapping is generated by frequency doubling the output of a ring dye laser at 570 nm. At the start of a measurement cycle the magnetic field and the trapping laser light are switched off and the four beam splitting pulses irradiate the freely expanding cloud. The blue laser light for interferometry is provided by a high precision dye laser. The four light pulses are cut from two counterpropagating laser beams with the aid of acousto-optic modulators. This way the influence of the laser phase is canceled for each pulse pair separately [2].

As stated above, the output signal of the interferometer can be read out by monitoring the population of the internal atomic state. Instead of watching the weak fluorescence from the $^3P_1$ state, monitoring the decrease of the $^1S_0$ state population via the decrease of the strong trap flourescence when the unaffected atoms are recaptured gives a quantum amplified measure of the weak absorption of the blue laser light. Repeating this measurement cycles many times during a mean lifetime of the trap, in our case at a rate of 94 Hz with a trap lifetime of 0.3 s, results in a further signal amplification as the ground state atoms are 'recycled' many times.

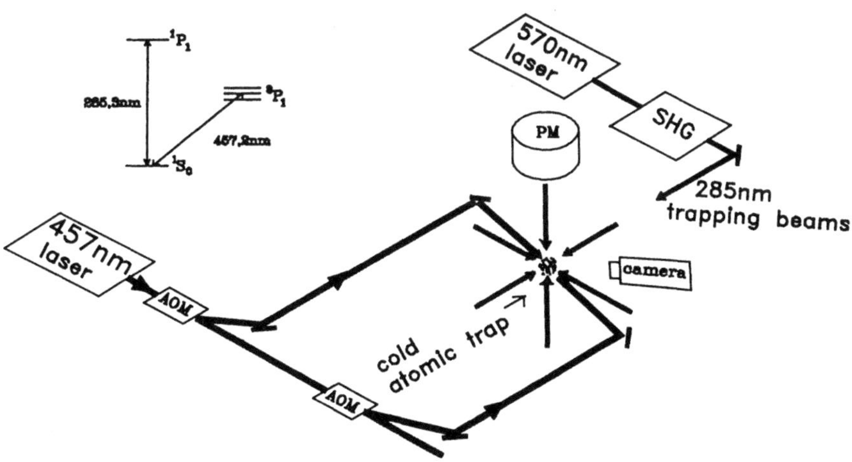

Fig 7: Experimental setup for interferometry with laser trapped magnesium atoms.

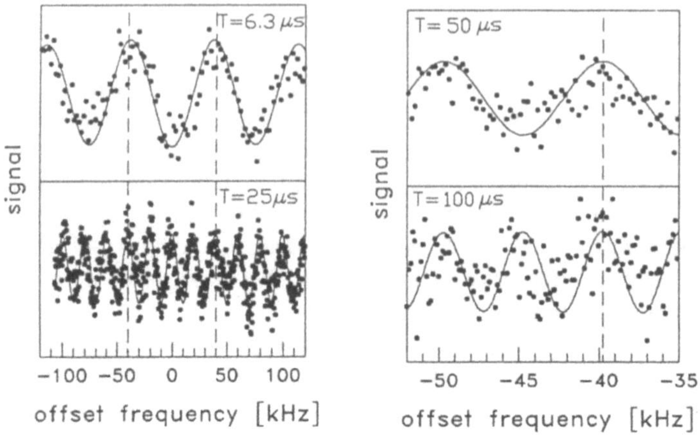

Fig 8: Examples of Ramsey fringes of different resolutions. The pulse separation time T varies in T=6.3μs, 25μs, 50μs, 100μs, corresponding to a minimal periodicity of 5 kHz.

Interferometric signals for different delays T between the pulses of each pair are shown in Fig. 8. A broad background consisting of the Doppler profile and a Lamb-dip has been subtracted from the data. The minimum fringe periodicity in frequency space achieved so far is 5 kHz [2].

A state selective potential in the interferometer is realized by pulses from a laser detuned several linewidths from the trapping transition. The uv pulse is applied between the first and the second beam splitter pulse. The spatial profile of the laser is widened, so that the relative intensity variation over the trap volume is less than $2 \cdot 10^{-2}$. The interaction of the atoms with the laser field shifts the ground state by

$$\Delta E = \hbar/2\,[(\Delta^2 + \Omega^2)^{1/2} - \Delta]\,. \tag{6}$$

This energy shift is a potential for the external degree of freedom of the atom. Its spatial derivative, which is negligible here, would be the wellknown dipole force.

Expanding Eq. 6 for detunings Δ large compared to the Rabi frequency Ω and using Eq. 5 yields for the detectable phase difference the proportionality

$$\varphi \propto I/\Delta \cdot \tau \tag{7}$$

with I the intensity and τ the width of the uv-pulse. Fig. 9a shows two interference patterns recorded simultaneously, one with and one without the additonal uv-laser pulses, where the phase shift is directly visible. The proportionality stated in Eq. 7 has been tested for several different values of the relevant parameters as is shown in Fig. 9b.

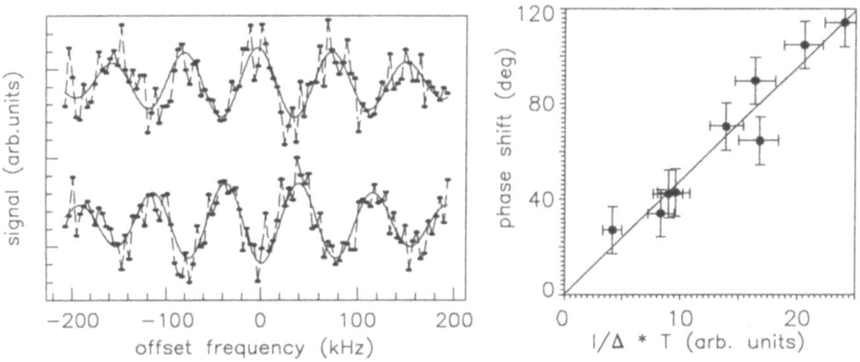

Fig 9: **(a, left)** Ramsey fringe signal with and without an additional scalar uv-laser potential (see text). **(b, right)** Measured phase shift vs. normalized potential parameter.

References

[1] M. Christ, A. Scholz, M. Schiffer, R. Deutschmann, and W. Ertmer submitted to Optics Comm..

[2] K. Sengstock, U. Sterr, G. Hennig, D. Bettermann, J. H. Müller, and W. Ertmer, Optics Comm. **103**, 73 (1993).

[3] R. Deutschmann, W. Ertmer, and H. Wallis, Phys.Rev.A, **47**, 2169 (1993).

[4] A. Scholz, M. Christ, D. Doll, J. Ludwig, and W. Ertmer, submitted to Optics Comm.

[5] R. Deutschmann, W. Ertmer, and H. Wallis, Phys. Rev. A **48**, 4023 (1993)

[6] Ch.J. Bordé, Phys.Lett. A **140**, 10 (1989)

[7] Y. Aharonov, D. Bohm, Phys.Rev. **115**, 485 (1959)

Towards a Cold-Atom Matter-Wave Interferometer

P. Hannaford[1], *R.J. McLean*[1], *G.I. Opat*[2], *W.J. Rowlands*[2], *and A. Sidorov*[2]

[1]CSIRO Division of Materials Science & Technology, Clayton, Australia 3169
[2]School of Physics, The University of Melbourne, Parkville, Australia 3052

Abstract. Matter-wave interferometers based on beams of ultra-slow laser-cooled atoms are potentially highly sensitive inertial sensors of accelerations, rotations and fields, including gravitational fields. We report on progress towards the development of a cold atom matter-wave interferometer with spatially separated beams for conducting fundamental physics experiments and for detecting gravity fields.

1. Introduction

Recent experiments in Paris, Stanford, Boulder and Gaithersburg [1-4] have demonstrated that near-resonant laser light can be used to cool a cloud of atoms down to temperatures of microkelvins, corresponding to atomic velocities (a few cm s^{-1}) of just a few times the single-photon recoil velocity. At such low velocities, the de Broglie wavelength of the atoms approaches the wavelength of the light used to cool the atoms, and this opens up the possibility of performing matter-wave interferometry with beams of ultra-cold laser-cooled atoms using mirrors and beamsplitters with flatness and rigidity requirements comparable with optical interferometry.

Matter-wave interferometers based on beams of ultra-slow atoms are potentially highly sensitive inertial sensors of accelerations, rotations and fields, including gravitational and magnetic fields. Possible applications include the development of a sensitive accelerometer or gravimeter; the development of an atom gyroscope for measurements of rotations; and various fundamental physics experiments including tests of the principle of equivalence and general relativity, precision measurements of the Aharanov-Casher phase shift, measurements of Berry's topological phase, and measurements of Newton's gravitational constant.

In this paper we briefly review the new field of atom interferometry and report progress on the University of Melbourne-CSIRO project to develop a cold atom matter-wave interferometer for conducting fundamental physics experiments and for detecting gravity fields. The use of atom interferometry for detecting gravity fields has previously been proposed by Opat [5] and Clauser [6].

Springer Proceedings in Physics, Vol. 77 Quantum Optics VI
Editors: D.F. Walls · J.D. Harvey

2. Phase Shifts and Sensitivity of a Cold Atom Interferometer

We consider a simple idealised Mach-Zehnder interferometer (Fig. 1) in which the de Broglie waves of a beam of monoenergetic atoms are coherently split between two spatially separated paths, ABC and ADC, and then recombined at C prior to detection. The atoms are assumed to follow straight line trajectories.

When the interferometer experiences an acceleration **g** in the -y direction (Fig. 1), atoms moving in the top arm (ABC) of the interferometer lose kinetic energy between A and B; so that compared with atoms moving along AD the wavenumber for atoms moving along BC is reduced by

$$\Delta k = (mv/\hbar)[1-(1-2gH/v^2)^{1/2}],$$

where m is the mass of the atoms, v is their initial velocity, and H is the height of the interferometer. For $mgH \ll mv^2/2$, which is a condition required to have straight line trajectories, the above expression reduces to

$$\Delta k \approx (m/\hbar v)(gH).$$

The difference in phase between the two arms of the interferometer with enclosed area A=HW is then

$$\begin{aligned}\Delta\Phi_{accel} &= \Delta k\ W \\ &\approx (m/\hbar v)(gA) = 2\pi gA/\lambda_{deB}v^2.\end{aligned}$$

Thus, for a caesium atom interferometer in which $v = 10$ cm s^{-1} and $A = 1$ cm^2, an acceleration of only $5\mathrm{x}10^{-10}$ cm s^{-2}, or $5\mathrm{x}10^{-13}$ of the Earth's gravitational acceleration, is required to yield a phase shift of 0.01 mrad.

When the interferometer is rotated with angular frequency Ω normal to the xy plane (Fig. 1), the atoms in each of the two arms of the interferometer experience a path difference

$$\Delta s = WH\Omega/v.$$

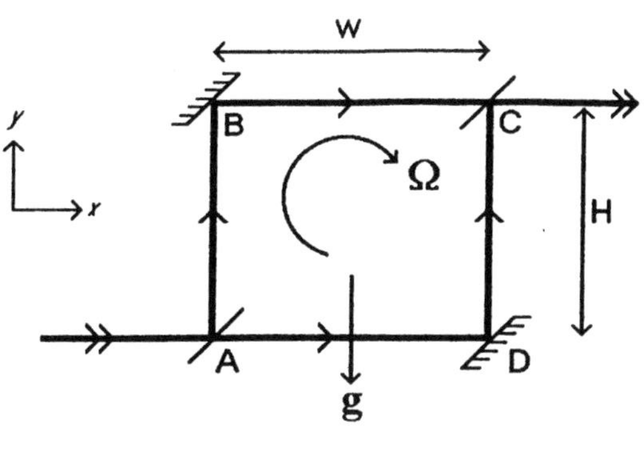

Fig. 1. Idealised Mach-Zehnder interferometer

The difference in phase between the two arms of the interferometer (Sagnac effect) is then

$$\Delta\Phi_{rot} = k\, 2\Delta s$$
$$= (2m/\hbar)\Omega A = 4\pi\Omega A/\lambda_{deB} v.$$

Thus, for a caesium atom interferometer in which $v = 10$ cm s^{-1} and $A = 1$ cm^2, a rotation of only 3×10^{-11} rad s^{-1} is required to yield a phase shift of 0.01 mrad.

To date, densities of laser-cooled atoms of up to $N \approx 8 \times 10^{11}$ cm^{-3} have been generated in atom traps [7]. This would appear to be close to the maximum density usable in atom interferometers, even after beam collimation, because of limits imposed by atom-atom interactions. The beam flux for 10^{12} atoms cm^{-3} moving at 10 cm s^{-1} in a cross sectional area of 1 mm^2 is 10^{11} s^{-1}, which should be sufficient to permit the measurement of phase shifts of about 0.01 mrad during an integration time of 1 s.

3. Comparison with Neutron and Optical Interferometers

For measurement of accelerations, the phase shift of a matter wave interferometer scales as m/v, and so the phase shift for a caesium atom interferometer in which $v = 10$ cm s^{-1} is potentially about 10^6 times larger than that of a neutron interferometer with thermal neutrons moving at about 10^5 cm s^{-1}. For measurement of rotations, the phase shift scales as m, and so the phase shift for a caesium atom interferometer is potentially about 10^2 times larger than that of a neutron interferometer. Furthermore, the particle fluxes from laser-cooled atom sources (up to about 10^{11} s^{-1}) are 10^8 to 10^{11} times higher than those used in neutron interferometers, and this offers the prospect of a further 10^4-10^5 gain in sensitivity for a given integration time. Also, the magnetic moment of an atom is about $\mu_B/\mu_N = 1836$ times that of a neutron, which means that phase shifts of milliradians observed, for example, in an Aharanov-Casher experiment by neutron interferometry [8] translate into phase shifts of radians in atom interferometry. Finally, cold atom sources are very much more compact and more readily available than neutron sources, which involve a nuclear reactor.

A comparison of the relative sensitivities of atom and optical interferometers depends to a large extent on the particular measurement. Scully and Dowling [9] have considered the relative sensitivities for the measurement of rotations. The ratio of the mass of a caesium atom to the effective mass of a photon ($m_\gamma = \hbar\omega/c^2$) is 6×10^{10}, which suggests that the phase shift for a caesium atom interferometer may be potentially about 10^{11} times larger than for an optical interferometer. However, in a laser ring gyroscope the photons make about 10^4 trips around the ring, compared with only one round trip for atoms in an atom interferometer, and this increases the phase shift by 10^4. Also, the photon flux in a laser gyroscope, about 10^{16} s^{-1} (for 3 mW beams), is about 10^5 times higher than the maximum atom flux in an atom interferometer, and this increases the sensitivity by a further factor of 300. On this basis, an atom interferometer would appear to be potentially about 10^5 more sensitive to rotations than a laser ring gyroscope.

For the measurement of accelerations, comparison of the relative sensitivities of atom and optical interferometers is even less straight forward. For identical configurations, the phase shift scales as m/v, which suggests that the phase shift for a caesium atom interferometer may be potentially about 10^{20} larger than for an optical interferometer, or 10^{14} larger if we allow for the higher fluxes and the larger number of round trips in optical interferometers. However, measurement of the frequency shift of light is a rather insensitive method of measuring accelerations. Current methods based on a falling corner cube in one of the arms of a Michelson interferometer [10] can measure gravitational accelerations to an accuracy approaching a few parts in 10^9. Our estimates in Sect. 2 suggest that an atom interferometer may be about 10^4 times more sensitive to accelerations than this.

4. Existing Atom Interferometers

In 1991 Carnal and Mlynek [11] in Konstanz constructed a simple atom interferometer based on a Young double-slit configuration. Using a a thermal beam of metastable helium atoms having very short de Broglie wavelengths (≤ 0.1 nm) and a double slit with 8 μm slit separation, fringe visibilities of up to 30% were observed. More recently, Shimizu *et al* [12] have developed a similar system based on laser-cooled metastable neon atoms. The neon atoms were cooled to about 2.5 mK in a magneto-optical trap, then dropped 7.6 cm onto a double slit with 6 μm slit separation, and finally detected 1.1 cm below. Using a time-of-flight technique, interference fringes were observed at several atomic velocities in the range 30-200 cm s^{-1}.

Keith *et al* [13] at MIT reported the operation of a Mach-Zehnder atom interferometer based on three microfabricated 400 nm-period transmission diffraction gratings as beam splitter, mirror, and beam recombiner. Using a highly collimated supersonic beam of sodium atoms ($\lambda_{deB} \sim 0.016$ nm) and a 1.2 m long interferometer, spatial separations of 27 μm, about equal to the beam width, were achieved. To obtain reasonable fringe contrast, the relative rms motion of the gratings was reduced to 40 nm. In a more recent experiment, in which 200 nm period transmission gratings were used, complete spatial separation of the two arms was achieved, permitting the two interferometer beams to pass on opposite sides of a stretched metal foil. By applying a uniform electric field to one of the arms, precision measurements were made of the polarisability of the ground state of sodium [14].

Riehle *et al* [15] at Braunschweig have constructed an atom interferometer based on the interaction of an atomic beam with four transverse travelling laser fields in a Ramsey separated optical fields configuration. The four interacting laser fields and the associated atomic recoil play the role of coherent beam splitter, mirrors, and beam recombiner. For a thermal beam of calcium atoms, the atomic deflection after a photon absorption or stimulated emission process leads to diffraction angles of about 20 μrad. A feature of this interferometer, and also of the stimulated two-photon Raman-pulse interferometer described below, is that the atoms in the two arms are labelled by different internal quantum numbers and therefore the two paths of the interferometer do not have to be spatially resolved. The observed optical Ramsey fringes have been used to measure shifts caused by the rotation of the interferometer (Sagnac effect). Sengstock et al [16] have recently used a similar interferometer, based on beams of laser-cooled Mg atoms, to measure the scalar Aharonov-Bohm phase shift [17].

Robert *et al* [18] at Paris Nord have constructed an atom interferometer based on the longitudinal Stern-Gerlach effect. A thermal beam of metastable 2s hydrogen atoms is spin-polarised by a transverse magnetic field B_P, then passed through a nonuniform longitudinal magnetic field $B_\parallel(z)$ having two zones of opposite field gradient, and finally through a transverse quenching field B_Q which selects one of the spin states. The rapid passage from the transverse B_P fringe field to the longitudinal $B_\parallel(z)$ fringe field induces transitions between the two spin states and prepares the atoms in linear superposition states. In the first field-gradient zone of $B_\parallel(z)$ the wave packets associated with the two spin states are accelerated or retarded depending on the orientation of their magnetic moments, leading to a spatial separation, and in the second (opposite) field-gradient zone the velocity change is reversed. Due to the different phase shifts of the two wave packets, their recombination during passage from $B_\parallel(z)$ to B_Q leads to phase-dependent populations of the two spin states.

Kasevich and Chu [19,20] at Stanford have developed a light-pulse atom interferometer based on stimulated two-photon Raman transitions between two ground-state hyperfine levels, $|g_1>$ and $|g_2>$, in laser-cooled sodium or caesium atoms. An initial $\pi/2$-pulse prepares the atom in a coherent superposition of $|g_1>$ and $|g_2>$ states, and the momentum recoil $2\hbar k$ associated with the two-photon Raman pulse coherently splits the wave packet. A π-pulse then acts as a mirror to redirect the trajectory of each wave packet. A final $\pi/2$-pulse recombines the two wavepackets, putting the atom into either the $|g_1>$ or $|g_2>$ state depending on the relative phases of the wavepackets. Interference fringes have been observed for wave packets separated by as much as 2.4 mm. This interferometer has recently been used to make precision measurements of the gravitational acceleration of a laser-cooled sodium atom in an atomic fountain with a sensitivity of 3×10^{-8} [20] and of the photon recoil of a caesium atom [21].

5. The Present Project

The present goal is to develop a sensitive atom interferometer having spatially separated beams of laser-cooled caesium atoms, using optical elements (mirrors, and gratings as beam splitters) based on spatially varying periodic magnetic fields [22].

5.1 The Cold Atom Source

An ultra-high vacuum system, consisting of a multi-port stainless steel chamber pumped by a 450 ℓs^{-1} turbomolecular pump for rapid pump-down and an 80 ℓs^{-1} ion pump and liquid-nitrogen cooled titanium sublimation pump for final pump-down, provides a relatively collision-free environment ($3x10^{-10}$ Torr) for trapping and cooling caesium atoms. The caesium atoms are trapped directly from a vapour in a magneto-optical trap [23]. The trap is formed by the intersection of three orthogonal pairs of counterpropagating laser beams having opposite circular polarisations, together with a pair of anti-Helmholtz coils which provide a spherical quadrupole magnetic field with a gradient of 5-20 G cm^{-1} passing through zero at the point of intersection of the laser-cooling beams. The three pairs of laser-cooling beams are tuned about 10 MHz to the low frequency side of the $F_{lower}=4$ to $F_{upper}=5$ hyperfine transition. A second laser, tuned to the $F_{lower}=3$ to $F_{upper}=4$ transition, repumps caesium atoms that accumulate in the $F_{lower}=3$ level back into the $F_{lower}=4$ to $F_{upper}=5$ system. The quadrupole magnetic field provides a shallow confining potential ($\sim$1K deep), allowing the caesium atoms to be trapped directly from the low-velocity tail of the Maxwellian distribution [3].

The tunable laser light at 852 nm is provided by an actively stabilised titanium:sapphire laser and a number of extended-cavity diode laser systems [24] which are locked to the sides of narrow saturated absorption lines.

In our preliminary experiments, the density of trapped laser-cooled caesium atoms, determined from the absorption of a probe laser, was $3x10^8$ cm^{-3}, using laser beams of 7 mm diameter, and the diameter of the cloud of trapped caesium atoms was about 0.4 mm. The temperature of the laser-cooled atoms was determined by driving the atoms in the trap with an oscillatory magnetic field and using the observed frequency dependence of the phase shift of the oscillating fluorescence to determine the spring constant of the trap [25]. The temperature of the caesium atoms confined in the trap was 70 μK, corresponding to an rms velocity of 11 cm s^{-1}.

The next stage is to launch the atoms from the trap, using a moving molasses scheme [1], to generate a beam of ultra-slow caesium atoms for atom optics experiments.

5.2 Atom Optics

Atomic mirrors and beamsplitters required for the atom interferometer are being developed using schemes involving spatially varying periodic magnetic fields laid down on ferromagnetic substrates. This type of optical element, which is applicable to slow atoms, avoids the complication of spontaneous emission, which is inherent in optical elements based on evanescent light fields [26]. The theory, as outlined briefly below, has been described in detail by Opat *et al* [22].

We consider the spatial dependence of the magnetic field density arising from a potential $\phi(x,y)$ which is periodic in the x direction on the xz plane. For the case of a surface potential with a single harmonic, $\phi(x,0)=V_1\cos(2\pi x/a)$, the energy density at distances y much larger than the spatial period a takes the form

$$\mathcal{E} = (\pi V_1/a)^2 \exp(-4\pi y/a).$$

Thus, at distances y much greater than a, the energy density falls off exponentially with distance y from the surface and is not periodic in x. For a constant value of y, the fields rotate with constant magnitude as x changes.

We now consider a beam of ultra-slow laser-cooled atoms moving towards this surface, and assume the atoms are moving adiabatically, i.e., sufficiently slowly in relation to the magnetic field gradients that they always remain in the same Zeeman substate. Atoms which are prepared in a positive Zeeman substate, for example by optical pumping, will then be repelled by the increasing magnetic energy density, and thus the surface behaves as a magnetostatic mirror.

When a uniform magnetic field B_0 is superimposed in the y direction (normal to the surface), the energy density becomes

$$\mathcal{E} = B_0^2/2 + 2\pi B_0(V_1/a)\cos(2\pi x/a)\exp(-2\pi y/a) + 2(\pi V_1/a)^2\exp(-4\pi y/a).$$

The first term is a constant and exerts no force. The second term is periodic in x and dies away slowly with y compared with the (non-periodic) third term. The surface can now behave as a diffraction grating.

A number of methods for constructing magnetic mirrors and gratings are currently under investigation, including the recording of periodic magnetic fields on ferromagnetic substrates as in sound recording. Magnetic fields of the order of tens of gauss are required to reflect laser-cooled atoms. The magnetic materials must be capable of being used in ultra-high vacuum.

An electrostatic mirror based on periodically varying electric fields has already been constructed and used to reflect neutral polar molecules of chloromethane [27].

5.3 Beam Curvature

The trajectories of laser-cooled atoms in general will not be straight, especially in the presence of the Earth's gravitational field. For example, in order to satisfy the condition, $mgH \ll mv^2/2$ (Sect. 2) using $v=10$ cm s^{-1} and $H=1$ cm, background accelerations are required to be much less than 5% of the Earth's gravitational acceleration. Thus in order for the interferometer to accommodate background gravitational (or magnetic) fields and also to have a reasonable dynamic range as an inertial sensor, it is important to be able to correct for beam curvature. This can be accomplished by applying appropriate magnetic field gradients such that the compensating force on the atoms cancels the deflecting inertial force in the interferometer [6].

6. Concluding Remarks

The development of a Mach-Zehnder atom interferometer based on laser-cooled atoms, static atom optics, and spatially separated atom beams offers the prospect of a highly sensitive inertial sensor for measurement of accelerations, rotations and fields and for conducting various fundamental physics experiments. The major challenge is going to be to fabricate efficient atomic mirrors and efficient wide-angle beamsplitters such as diffraction gratings. Mirrors and gratings based on spatially varying periodic magnetic fields may provide a possible solution.

Acknowledgements. The authors would like to thank J. Kimble and P. Grangier for helpful comments on the comparison of the sensitivities of atom and optical interferometers, and the Australian Research Council for their support.

References

1. C. Salomon, J. Dalibard, W.D. Phillips, A. Clairon, S. Guellati, Europhys. Lett. **12**, 683 (1990)
2. M. Kasevich and S. Chu, Phys. Rev. Lett. **69**, 1741 (1992)
3. C. Monroe, W. Swann, H. Robinson and C. Wieman, Phys. Rev. Lett. **65**, 1571 (1990)
4. P.D. Lett, W.D. Phillips, S.L. Rolston, C.E. Tanner, R.N. Watts and C.I. Westbrook, J. Opt. Soc. Am. B **6**, 2084 (1989)
5. G.I. Opat, Proc. Third Marcel Grossmann Meeting on General Relativity, Ed. Hu Ning (Science Press and North Holland Publ., 1983), p 1491
6. J.F. Clauser, Physica B **151**, 262 (1988)
7. W. Ketterle, K.B. Davis, M.A. Joffe, A. Martin and D.E. Pritchard, Phys. Rev. Lett. **70**, 2253 (1993)
8. A. Cimmino, G.I. Opat, A.G. Klein, H. Kaiser, S.A. Werner, M. Arif and R. Clothier, Phys. Rev. Lett. **63**, 380 (1989)
9. M.O. Scully and J.P. Dowling, Phys. Rev. A **48**, 3186 (1993); see also discussion Phys. Reports **219**, pp 166-173 (1992)
10. J.E. Faller and I. Marson, Metrologia **25**, 49 (1988)
11. O. Carnal and J. Mlynek, Phys. Rev. Lett. **66**, 2689 (1991)
12. F. Shimizu, K. Shimizu and H. Takuma, Phys. Rev. A **46**, R17 (1992)
13. D.W. Keith, C.R. Ekstrom, Qu.A. Turchette and D.E. Pritchard, Phys. Rev. Lett. **66**, 2693 (1991)
14. D.E. Pritchard, C.R. Ekstrom, J. Schmiedmayer, M.S. Chapman and T.D. Hammond, In *Laser Spectroscopy*, Eds. L. Bloomfield, T. Gallagher and D. Larson (AIP Press, NY, 1993) p 61
15. F. Riehle, Th. Kisters, A. Witte, J. Helmcke and Ch.J. Bordé, Phys. Rev. Lett. **67**, 177 (1991)

16. K. Sengstock, U. Sterr, G. Hennig, D. Bettermann, J.H. Müller and W. Ertmer, Opt. Commun. **103**, 73 (1993)
17. W. Ertmer, Proceedings of this symposium.
18. J. Robert, Ch. Miniatura, S. Le Boiteux, J. Reinhardt, V. Bocvarski and J. Baudon, Europhys. Lett. **16**, 29 (1991)
19. M. Kasevich and S. Chu, Phys. Rev. Lett. **67**, 181 (1991)
20. M. Kasevich and S. Chu, Appl. Phys. B **54**, 321 (1992)
21. D.S. Weiss, B.C. Young and S. Chu, Phys. Rev. Lett. **70**, 2706 (1993)
22. G.I. Opat, S.J. Wark and A. Cimmino, Appl. Phys. B **54**, 396 (1992)
23. E.L. Raab, M. Prentiss, A. Cable, S. Chu and D.E. Pritchard, Phys. Rev. Lett. **59**, 2631 (1987)
24. R.J. McLean, P. Hannaford, P.L. Larkins and W.J. Rowlands, Opt. Commun. **102**, 43 (1993)
25. P. Kohns, P. Buch, W. Süptitz, C. Csambal and W. Ertmer, Europhys. Lett. **22**, 517 (1993)
26. J.V. Hajnal and G.I. Opat, Opt. Commun. **71**, 119 (1989)
27. S.J. Wark and G.I. Opat, J. Phys. B **25**, 4229 (1992)

Uncertainty-Principle-Limited Position Measurement of Moving Atoms

J.E. Thomas, J.R. Gardner, G.R. Welch, M.L. Marable, T.A. Savard, and L.J. Wang

Physics Department, Duke University, Durham, NC 27708-0305, USA

Abstract. We have developed all-optical techniques for position measurement and localization of moving atoms with suboptical wavelength precision. Based on resonance imaging in the potential of an intense spatially varying light-shifting field, the methods have achieved spatial resolution of 200 nanometers, with linearity over several microns. Centroids of localized distributions, which are created and measured in the experiments, are determined with an accuracy of $\pm$ 20 nanometers. Using larger light-shift gradients and single atom detection, these imaging methods can scale to yield nanometer spatial resolution limited by the uncertainty principle. Applications include the characterization of atom-optical elements, atomic interferometry, and the direct imaging of spatial distributions for ultracold trapped atoms, confined in the wells of a polarization gradient or in a dipole force microtrap. This method of ultra-high resolution resonance imaging also can be applied to measure two-point spatial correlation functions, which may prove useful for studies of the quantum statistical features of ultracold atoms.

1. Introduction

Recently, there have been many new developments in the study of the de Broglie wave optics of neutral atoms, which is emerging as one of the most exciting new fields in atomic and optical physics [1]. Atom-optical elements including beam splitters and gratings for atoms, as well as complete interferometers have been demonstrated. With the advent of laser cooling to very low temperatures, quantum mechanical features of the atomic center of mass motion are becoming evident in the Raman spectra of atoms trapped in periodic potentials due to optical fields (optical lattices) [2,3,4,5]. A gravito-optical cavity for cold atoms has been demonstrated which uses an evanescent wave as an atomic mirror at the bottom of the trap, and the gravitational potential to provide an upper bound for the atoms [6,7]. The low temperatures and high densities which have been achieved in current experiments suggest that the exploration of many body effects will be possible in the near future and theoretical treatments of nonlinear atom-optical effects have already begun [8,9].

In many of the atom optics experiments which have been performed to date, atomic spatial distributions are created which vary over suboptical wavelength scales. This has provided motivation for developing all-optical, ultra-high reso-

Springer Proceedings in Physics, Vol. 77 Quantum Optics VI
Editors: D.F. Walls · J.D. Harvey

lution methods for directly measuring moving atom spatial distributions. While mechanical methods (i.e. hot wires, pinholes, slits, etc.) have been used in the past, in general, such methods will perturb the system prior to measurement, and therefore are not satisfactory in all cases. All-optical methods for measuring moving atom position distributions can be adapted to a variety of applications in both atomic beams and in traps, and permit rapid time dependent measurement when needed, readily variable spatial resolution and state dependent measurement. In the following sections, the basic physics of moving atom position measurement by resonance imaging in high field gradients is reviewed. Then, a number of applications will be described, including precision atomic localization and position measurement, measurement of atomic interference patterns, and correlated atomic position measurements.

2. Precision Atomic Position Measurement by Resonance Imaging

The essential features of precision atomic position measurement by resonance imaging can be understood using heuristic arguments [10,11]. The basic resonance imaging method, Fig. 1, employs a Raman or rf transition between two long lived states $|i\rangle$ and $|f\rangle$. Atoms in the initial state $|i\rangle$ are assumed to be in some interesting spatial pattern which is to be measured. The final state $|f\rangle$ is pre-emptied by optical pumping. A spatially varying potential shifts the state $|f\rangle$, causing the resonance frequency for the transition to vary as a function of atomic position. After interaction with the rf (or Raman) field, subsequent detection of an atom in state $|f\rangle$ indicates that an atom must have been present in the initial state near the point in space where the transition is tuned into resonance with the applied rf field.

Over short distances, the potential energy of state $|f\rangle$ is taken to vary linearly as

$$V_f(x) = -(x - x_o)F, \tag{1}$$

where F is the force exerted on the final state $|f\rangle$, here taken to be constant in time.

The spatial resolution of the resonance imaging method is determined by the spectral resolution, $\Delta\omega$. The smallest resolvable position change, Δx, is such that the corresponding change in the $i \to f$ transition frequency due to the potential is equal to the spectral resolution, i.e. $F\Delta x/\hbar = \Delta\omega$. In the experiments, the optical fields which induce Raman transitions copropagate, and are generated by acousto-optic modulation from a common laser source. In this case, the laser jitter cancels, and due to the low frequency of the Raman (or rf) transition, the difference in the Doppler shifts is also negligible. Hence, the spectral resolution for the method is limited only by the interaction time, T, between the transition inducing fields and the atom, so that $\Delta\omega = 1/T$. The maximum spatial resolution is therefore

$$\Delta x = \frac{\hbar}{FT}. \tag{2}$$

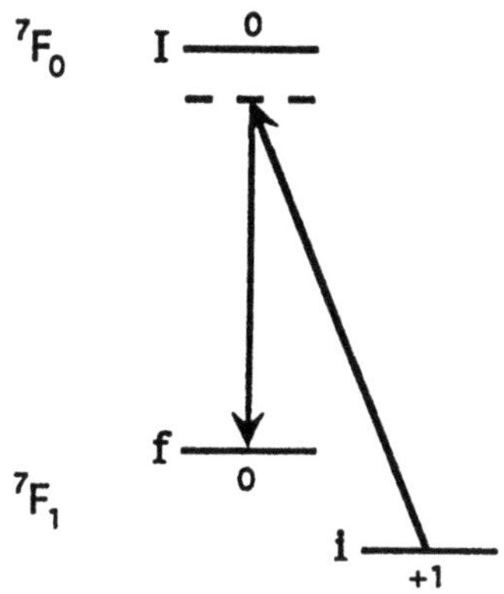

Figure 1: Atomic position measurement by resonance imaging. An atom initially in state $|i\rangle$ makes a transition to a pre-emptied state $|f\rangle$ for which the energy is dependent on the atomic position x. Atoms near the point in space where the potential tunes the atomic transition into resonance with the transition inducing field arrive in state $|f\rangle$ where they are detected by any convenient method. Measuring the total number of atoms in state $|f\rangle$ as a function of the frequency of the transition inducing field determines the spatial distribution of atoms in state $|i\rangle$. (Level structure shown for ^{152}Sm.)

According to Eq. 2, for a given force, F, the spatial resolution improves as the interaction time T increases. However, the interaction time T cannot be arbitrarily chosen because the atoms move along the measurement, x, axis during the measurement. If the statistical velocity spread of the atomic distribution is dominant, then the maximum T should be chosen so that $v_xT = \Delta x$, where v_x is the atomic velocity along the x axis. Using $T = \Delta x/v_x$ in Eq. 2 shows that the velocity limited resolution is of order $\sqrt{\hbar v_x/F}$. A more careful calculation [11] for a gaussian pulse (instead of a square pulse of duration T) shows that the optimum velocity spread limited spatial resolution in this case is given by

$$\Delta x_{VEL} = \sqrt{\frac{2\hbar v_x}{F}}, \tag{3}$$

where Δx_{VEL} is the half width at $1/e$ of a gaussian shaped region in which transitions occur.

For a sufficiently small statistical velocity spread, the interaction time T is limited by diffraction in the initial state or acceleration in the final state. In the latter case, one can ask what is the width, Δx, of the smallest final state wave packet in which an atom can emerge after a transition from state $|i\rangle$. For an atom of mass M the maximum interaction time must be chosen so that $FT^2/2M = \Delta x$ in order that the final state not be spread by the force over distances larger than Δx. Solving for T in terms of Δx and using Eq. 2 yields the optimum acceleration limited spatial resolution as

$$\Delta x_{ACCEL} = \left(\frac{\hbar^2}{2MF}\right)^{\frac{1}{3}}. \tag{4}$$

In the diffraction limit, the velocity spread due to localization effectively limits the interaction time. For an atom of mass M, one can ask what is the smallest width Δx into which an atom in the initial state can be localized prior to a measurement of the position distribution, and still obtain for a measurement of the width by resonance imaging a result comparable to Δx. Due to the localization, the initial state packet spreads with a velocity $\hbar/(M\Delta x)$. Hence, the maximum measurement time, T, should be chosen so that $(\hbar/M\Delta x)T \simeq 2\Delta x$, or $T = 2M\Delta x^2/\hbar$. With Eq. 2, this yields the diffraction limited resolution

$$\Delta x_{DIFF} = \left(\frac{\hbar^2}{2MF}\right)^{\frac{1}{3}}, \tag{5}$$

which is identical to the acceleration limited resolution. Using in Eq. 3 for Δx_{VEL} a value of v_x approximately equal to the diffraction velocity, and solving for Δx yields the same result, which shows that the diffraction limit is a just a special case of velocity limited resolution. It is not surprising that the same optimum spatial resolution is obtained in the acceleration and diffraction limits, since there is only one natural length scale for a particle of mass M in a linearly varying potential. For a spatially varying potential such that $F/h = 10^{12}$ Hz/cm, as can be obtained using a light-shift gradient, Eq. 4 yields a spatial resolution of 7 nanometers for ^{154}Sm atoms.

In each of the above cases, the optimum interaction time is determined such that the atom moves a distance which is approximately equal to the optimum spatial resolution during the measurement time. If the resonance imaging method has high spatial resolution compared the relevant length scales, the atoms will therefore move a negligible distance during the measurement. Hence, measurements can be made in traps with the trapping fields turned off during the interaction time T in situations where the position measurement would be perturbed by the trapping potential. In a far off resonance dipole microtrap [12], for which all ground state sublevels are shifted approximately equally by the trapping potential, the trapping fields can remain on.

An interesting feature of the resonance imaging method in the limit of high collimation or low temperature, where the acceleration limit is attained, is that minimum uncertainty final state wave packets can be created. An atom which is initially in a broad wave packet in state $|i\rangle$ will make a transition to the final state and emerge in a wavepacket localized in a region of width $\Delta x = \hbar/(FT)$ according to Eq. 2. The corresponding momentum spread, Δp, is due to the spread in time, T, during which a transition can take place. Since the force F acts only on the final state, if a transition occurs near the beginning of the interaction period, a maximum momentum FT is imparted to the final state. However, if the transition occurs near the end of the interaction period, almost no momentum is imparted. Hence, the momentum spread is approximately FT, so that $\Delta x \Delta p = \hbar$. Under appropriate circumstances, with gaussian pulses, a minimum uncertainty product $\hbar/2$ can be obtained [11].

2.1 The Spatial Resolution Function

The spatial resolution of the resonance imaging method depends generally on the statistical velocity spread as well as the velocity spread due to localization (diffraction). Hence, the Wigner phase space distribution for the initial state, $\rho_i(x,p)$ must be specified in order to determine N_f, the total number of atoms which make a transition to the final state. N_f can be written in the simple form

$$N_f = \int_{-\infty}^{\infty} dp \int_{-\infty}^{\infty} dx\, R(x - \bar{x}_o, p)\, \rho_i(x,p). \tag{6}$$

The spatial resolution function $R(x - \bar{x}_o, p)$ determines the momentum (p) dependent shape of the spatial region which is sampled. It is sharply peaked as a function of position, x, at the point $\bar{x}_o$, which is adjustable by changing the frequency of the transition inducing field. The shape of the resolution function is derived for a constant force F in Ref.[11], where the case of a gaussian pulse is discussed in detail. In Ref.[13] a concise derivation of the general case is given.

3. Applications

In the following sections, we discuss some applications of precision atomic position measurement by resonance imaging which are currently being explored in our laboratory.

3.1 Resonance Imaging in a Magnetic field Gradient: Atomic Interferometry by Spatial Rephasing

In the first experiments to demonstrate high resolution resonance imaging of atoms in a samarium atomic beam, a pair of copper tubes (see region (3), Fig. 2), spaced by 6 mm with equal currents of 100 A was used to generate a magnetic field gradient of 500 G/cm along an axis perpendicular to the atomic beam. For the ${}^7F_1 \rightarrow {}^7F_0$ transition of ${}^{152}Sm$ the ground state g-factor is 1.5, yielding a tuning rate of 10^9 Hz/cm. A Raman transition was induced by two optical fields and achieved transit time limited resolution. For the optimum diameter of the Raman region, a spatial resolution of 1.7 micrometers was achieved in the experiments, in agreement with the expected ideal velocity limited resolution for the experimental conditions [14].

Currently, this imaging method is being used to characterize an atomic interferometer based on spatial rephasing, as proposed by Chebotayev [15]. In

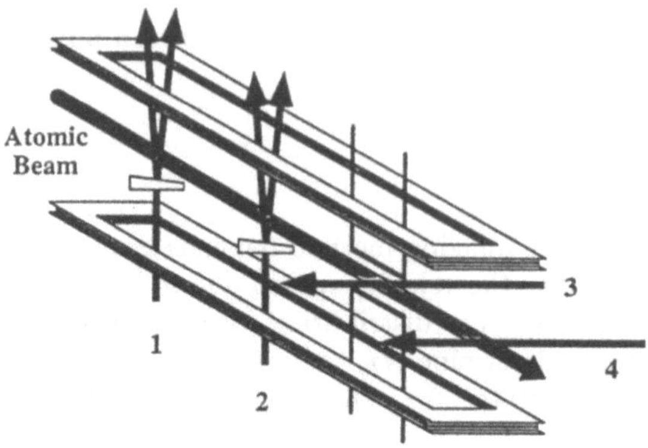

Figure 2: Atomic Interferometry by Spatial Rephasing. Samarium atoms cross diffraction gratings comprised of crossed optical fields from birefringent wedges (1,2). The σ_+ component of the optical fields is resonant with an optical transition and coherently depopulates the ground state in a spatially periodic manner. After traversing both gratings, an atomic interference pattern rephases and is observed by resonance imaging (3). The final state is detected by resonance fluorescence (4). (Optical pumping regions before and after the grating regions are omitted for clarity.)

the experiments, Fig. 2 atoms traverse two atomic diffraction gratings which are generated using birefringent wedges. The σ_+ components of the $\hat{x}$ and $\hat{y}$ polarized fields which cross near the surface of the wedges interfere to form a σ_+ polarized standing wave which varies in intensity along an axis x perpendicular to the atomic beam. A uniform magnetic field splits the ground state Zeeman sublevels so that only the σ_+ fields are resonant with the $^7F_1 \rightarrow ^7F_0$ optical transition in ^{152}Sm. This field coherently excites the atoms in a spatially periodic manner, thereby periodically depleting the ground state population. After traversing both gratings, a spatial interference pattern which is independent of the transverse velocity, v_x, rephases donwstream. Including optical pumping effects, the lowest order interference pattern for the optimum intensites in the grating regions is given by

$$\langle N_+(y,x)\rangle_{v_x} = \frac{N}{3}\left\{0.79 - 0.049\left\langle \cos\left[2q\left(x - \frac{v_x}{v_y}(y - 2y_2 + y_1)\right)\right]\right\rangle\right\}, \quad (7)$$

where the first grating is located at position y_1 along the atomic beam axis and the second at y_2. The pattern rephases at the point $y = 2y_2 - y_1$ where a resonance imaging region is located. In the experiments, $2q \equiv 2\pi/\lambda$, and λ is 20 micrometers. With spatial resolution of 2 micrometers, both the lowest order interference pattern and higher order patterns, which are obtained at higher intensities in the grating regions, should be observable. An interesting feature of the experiments is that the magnetic field gradient can be reduced so that the entire atomic beam profile can be imaged. This is helpful in aligning the gratings and in determining the optimum laser intensities in the grating regions. Once the interference pattern is characterized, it is possible to employ two separated Raman beams to generate a detection pattern in the form of a broad, spatial Ramsey grating with a period equal to that of the interference pattern. This method can be used to dramatically increase the signal.

3.2 Resonance Imaging in a Light-Shift Gradient

Suboptical wavelength spatial resolution is readily obtained by resonance imaging in the potential of a spatially varying light shift [16]. In the first demonstration of this method, Fig. 3, two sets of Raman fields are employed in region (2). The first set creates a spatial hole in the initial state. The second set, which is downstream from the first probes the spatial distribution set up by the first. When the second region is tuned to sample the hole due to the first, a decrease in the final state fluorecence signal occurs. This allows a determination of the spatial resolution of the method. In the experiments, a light-shift gradient of $F/h = 1.16$ MHz/μm is obtained, as determined from the light-shifting beam parameters and the matrix element of the light-shifting transition. In addition, a mechanical calibration was employed to confirm this result [16]. From the data, the spatial resolution (half width at 1/e) of the gaussian distributions

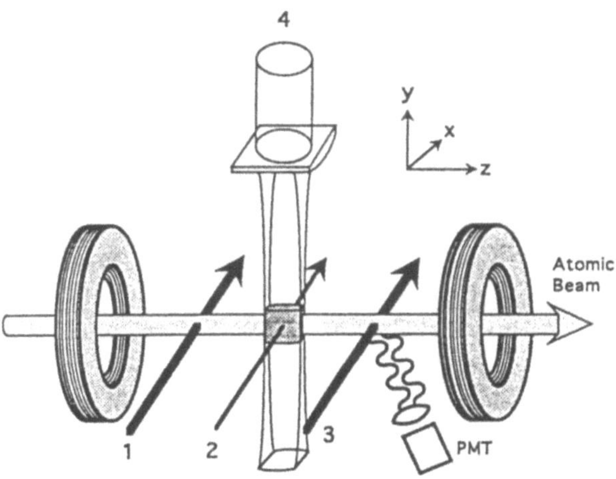

Figure 3: Resonance imaging in a Light-shift gradient. An intense, off-resonant light field (4) is focussed with a cylindrical lens to yield a spatial variation of the energy of the final state, with a constant gradient over a few microns. Raman fields (2) induce transitions between the initial and final state in a localized region along the $\hat{x}$ axis. The number of atoms emerging in the final state, whcih is initially emptied by optical pumping (1), is measured by resonance fluorescence (3) as a function of the Raman difference frequency to determine the spatial distribution in the initial state.

for the creation and measurement regions (which are convoluted according to Eq. 6) is estimated to be 200 nanometers, in excellent agreement with the predicted optimum spatial resolution for the experimental conditions [16]. The spatial resolution in these experiments was limited by the signal to background ratio obtainable by the resonance fluorescence detection method. Substantial improvement can be obtained by exploiting single atom detection techniques based on photon burst for alkali atoms or by photonionization.

Applications include imaging of atoms in dipole force microtraps [12] and in one dimensional light crystals.

3.3 Correlated Atomic Position Measurements by Resonance Imaging

The method of resonance imaging in a light-shift gradient is readily generalized to measure jointly at two spatially separated points by using more than one frequency for the transition inducing fields. Ideally, transitions at each point can be induced to different final states allowing the final state counts to be state selectively measured and correlated. A theoretical treatment of this method is given in Ref. [13] where it is shown that the joint probability, $P^{(2)}$ of detecting one atom at each of two points x_1 and x_2 is given by

$$P^{(2)} = \int dx_1\, dp_1 \int dx_2\, dp_2\, R_1(x_1, p_1)\, R_2(x_2, p_2)\, W^{(2)}(x_1, p_1; x_2, p_2), \qquad (8)$$

where $W^{(2)}$ is the two point analogue of the Wigner distribution. The resolution function, R_i is peaked at the point $\bar{x}_{oi}$, $i = 1, 2$. Eq. 8 is readily generalized to three dimensions.

For ultracold lithium atoms in a far off resonance dipole microtrap [12], it is possible to perform spatial correlation measurements with high spatial resolution to measure directly the effects of quantum statistics on the spatial distributions. As shown in Ref. [13], the relevant length scale at low density is $\delta x = \hbar/\sqrt{M k_B T}$, which is 300 nm for Li at 1 μK.

References

[1] For a recent review, see the special issue on Optics and Interferometry with Atoms, edited by J. Mlynek, V. Balykin, and P. Meystre, [Appl. Phys. B **54** (1992)].

[2] D. Grison, B. Lounis, C. Salomon, J.-Y. Courtois, and G. Grynberg, Europhys. Lett. **15**, 149 (1991).

[3] P. Verkerk, B. Lounis, C. Salomon, C. Cohen-Tannoudji, J.-Y. Courtois, and G. Grynberg, Phys. Rev. Lett. **68**, 3864 (1992).

[4] P. S. Jessen, C. Gerz, P. D. Lett, W. D. Phillips, S. L. Rolston, R. J. C. Spreeuw, and C. I. Westbrook, Phys. Rev. Lett. **69**, 49 (1992).

[5] A. Hemmerich and T. W. Hansch, Phys. Rev. Lett. **70**, 410 (1993).

[6] K. Helmerson, S. L. Rolston, L. Goldner, and W. D. Phillips, in Optics and Interferometry with Atoms, Book of Abstracts, WE-Heraeus-Seminar, Konstanz, 1992 (unpublished).

[7] C. G. Aminoff, A. M. Steane, P. Bouyer, P. Desboilles, J. Dalibard, and C. Cohen-Tannoudji, Phys. Rev. Lett. **71**, 3083 (1993).

[8] G. Lenz, P. Meystre, and E. M. Wright, Phys. Rev. Lett. **71**, 3271 (1993).

[9] W. Zhang, D. F. Walls, and B. C. Sanders, Phys. Rev. Lett. **72**, 60 (1994).

[10] J. E. Thomas, Opt. Lett. **14**, 1186 (1989).

[11] J. E. Thomas, Phys. Rev. A **42**, 5652 (1990).

[12] J. D. Miller, R. A. Cline, and D. J. Heinzen, Phys. Rev. A **47**, R4567 (1993).

[13] J. E. Thomas and L. J. Wang, Phys. Rev. A **49**, 558 (1994).

[14] K. D. Stokes, C. Schnurr, J. R. Gardner, M. Marable, G. R. Welch, and J. E. Thomas, Phys. Rev. Lett. **67**, 1997 (1991).

[15] V. P. Chebotayev, B. Ya. Dubetsky, A. P. Kasantsev, and V. P. Yakovlev, J. Opt. Soc. Am. **B 2**, 1791 (1985).

[16] J. R. Gardner, M. L. Marable, G. R. Welch, and J. E. Thomas, Phys. Rev. Lett. **70**, 3404 (1993).

Atoms Optics with Standing Waves of Light

H.-A. Bachor, Q. Li, B.W. Stenlake, I.C.M. Littler, K.G.H. Baldwin, and D.E. McClelland

Laser Physics Centre and Physics Department, Faculty of Science, The Australian National University, Canberra, Australia ACT 0200

Abstract. Near resonant standing waves can be used to manipulate the spatial distribution of atomic beams. We report two distinct effects: channelling of atoms in a perpendicular standing wave and the reflection of atoms from a grazing incidence mirror formed by a moving evanescent standing wave. The first arrangement allows us to focus an atomic beam over a large distance. The latter arrangement has the potential to demonstrate diffraction of fast atoms and could lead to an application as a beamsplitter.

1. Introduction

One aim of atom optics is to develop the equivalent of optical components for atoms, such as lenses, mirrors, beamsplitters and gratings. Many such components have already been demonstrated [1]. The fundamental idea is to use the forces which are exerted on atoms by the spatial gradients in the amplitude of laser beams tuned close to an atomic resonance. In the work presented here we concentrate on the forces produced inside a standing wave of light. The standing wave can be viewed as a periodic potential. For large detunings the atoms interact with the light through stimulated processes only, resulting in a conservative gradient force. For small detunings spontaneous emission has to be taken into account, which adds dissipative effects such as cooling and heating [2,3]. These forces are strongly dispersive, that means they depend on the velocity of the atom and thus the deBroglie wavelength. All our work uses fast, thermal atomic beams. Velocity selective detection allows us to observe the effects on a group of atoms with fixed, well defined deBroglie wavelength. We investigate two cases:

The first case uses atoms propagating orthogonally to the optical axis of a standing wave formed by two counter propagating beams. We demonstrate the well established effect of channelling [4] and show how this can lead to a focussing effect [5]. We show that the addition of spontaneous forces adds the additional option of changing the effective focal length of this system.

In the second case we show how a standing wave evanescent light field can be used as a mirror and a grating for atoms. The conditions for diffraction are discussed and we demonstrate Dopplerons, an effect due to the interaction with both light fields. Unlike experiments with slow atoms [6] we cannot find diffraction within the experimental errors. However, we derive the condition for which diffraction should be visible, even for thermal beams using the ideas of moving light gratings.

Springer Proceedings in Physics, Vol. 77 Quantum Optics VI
Editors: D.F. Walls · J.D. Harvey

2. Focussing and collimation in a transverse standing wave field

Consider a standing wave formed by two counter-propagating laser beams arranged perpendicular to a well collimated atomic beam. The standing wave produces a stationary periodic potential U(y). The atoms experience a transverse force, along the optical axis, which depends on the derivative of the potential at any given location. For red detunings the force will point towards the antinodes of the standing wave. For large detunings the force can be described by a simple,velocity independent gradient force [7]

$$F(y) = - \delta U(y) / \delta y . \qquad (1)$$

Close to an antinode the atoms will experience a parabolic potential well which collimates the atoms towards the antinode [8]. This focusses the atoms directly inside the standing wave. This effect is the basis of experiments in photolithography [9] for the deposition of very fine atomic structures on surfaces. However, we are interested in the effects over a large distance. We are use a beam with a diameter many times the period (300 nm) of the wave. Here focussing of the beam can be observed, both for large and small detunings. The schematic experimental arrangement is shown in figure 1. The source is a Na oven operating at about 500 C, the atoms escape through a nozzle of 0.5 mm diameter. Apertures collimate the beam, to a size at the standing wave of about 1mm. The atoms are detected 600mm downstream from the interaction region [10]. A second laser beam intersects the atomic beam at an angle of 60 degrees. Atoms with the correct Dopplershift, and thus velocity, will fluoresce. In this way the narrow linewidth of the laser (2MHz) is translated into a velocity selection Δv_z of < 5 m/s. The spatial distribution of the beam is measured with a resolution of 0.1 mm by scanning the slit and the detection optics perpendicular to the atomic beam (y' axis).

A typical result for a spatial distribution is shown in figure2. The beam has a narrow distribution with a width of about 0.6 mm. For comparison the beamshape without the interaction (laser off) is shown. The atomic profile is narrowed by a factor of 3.

This effect can be interpreted as a consequence of channelling. Atoms that enter the standing wave experience a gradient force towards the potential minima. If their transverse velocity is less than the escape velocity v_{esc} (transverse kinetic energy = $1/2\, mv_{esc}^2$ = height of the potential). The gradient force acts like a restoring force hence the atoms

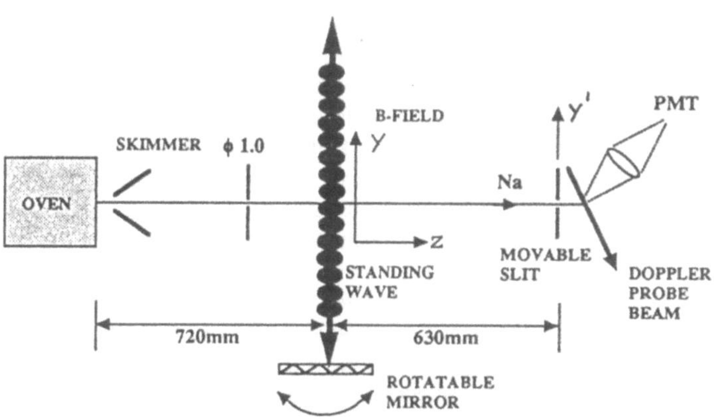

Figure 1: Schematic experimental arrangement

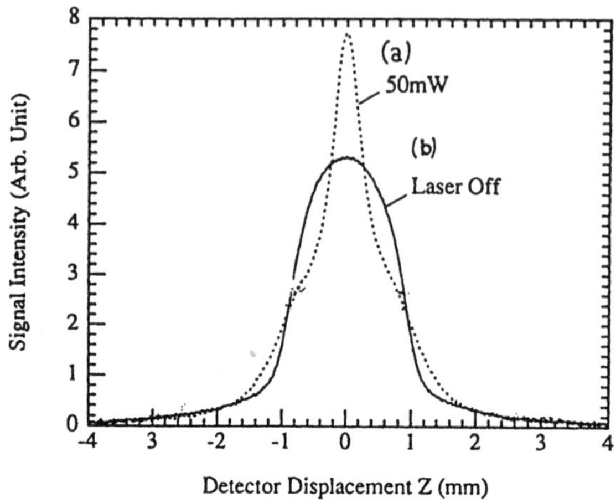

Figure 2:
Experimental result.
Spatial profile,
(a) with standing wave
(b) without

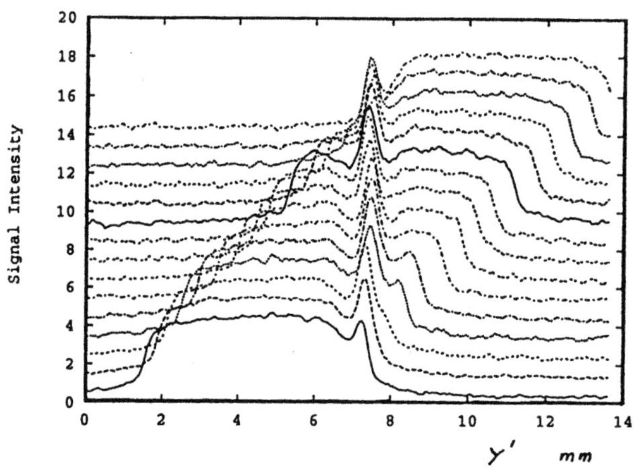

Figure 3:
Experimental results. Spatial profiles for various tilted atomic beams.

will oscillate in the transverse direction, around the minimum [4]. The effect of channelling is clearly demonstrated by the series of experimental traces shown in figure3. Here the atomic beam has a wide range of transverse velocities, 6 times wider than v_{esc}. The atoms within transverse velocities < v_{esc} are concentrated in one peak, the other atoms move over the potential hills and continue on their original trajectory. For the various traces in figure 3 the collimation slit in front of the standing wave is moved, thereby generating beams tilted in respect to the standing wave. Clearly, atoms within the same range of transverse velocities are selected by channelling, independent of the tilt of the beam.

The effect of the standing wave can be simulated by calculating the classical trajectories of atoms from the source through the interaction region to the detector. For large detunings the force is described by the gradient force (equ.1). The starting parameters (position and transverse velocity v_{in}) are selected and the trajectory of the atom is calculated in many small steps. At the end of the interaction region the output velocity v_{out} is calculated. This uniquely determines the position y' of the atom in the detection

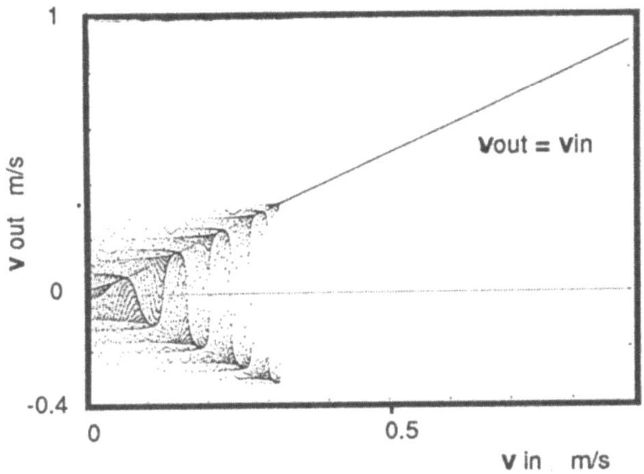

Figure 4:
Numerical simulation. Output velocity v_{out} versus input velocity v_{in} for one position at the source.

plane. The details of this calculation are described by Li et.al.[10]. A typical result is given in figure 4. Each point on the diagram represent v_{out} and v_{in} for one trajectory. All trajectories are for a constant position in the source. Without the interaction, the result would be $v_{out} = v_{in}$ for all atoms. The simulation shows that the magnitude of the velocity remains largely unchanged, a consequence of the conservative force. The points are clustered together near the two lines $v_{out} = v_{in}$ and $v_{out} = - v_{in}$. On their path through the standing wave some atoms describe a full spatial oscillation, the others have an additional half oscillation. Thus about half the atoms leave the standing wave without a change in their direction of propagation, the others are redirected towards the optical axis, keeping the magnitude of their transverse velocity approximately constant. These atoms will converge to one point on the optical axis located at a distance from the standing wave equal to the distance from the source. The standing wave acts like a 1:1 imaging systems for these atoms. For a different position in the source the features of figure 4 remain the same, only the velocities at which the switch in the sign of v_{out} occurs is shifted. For the full simulation many such traces are averaged, and the above conclusions remain correct.

How would this result change when the detuning is reduced such that spontaneous emission cannot be neglected? This simulation is far more involved. The force can no longer be described by the gradient force (equ. 1). It can be evaluated using the continued fraction method [11].

$$F = F_o + \Sigma_{n=1} \left(F_n^c \cos(2\,nky) + F_n^s \sin(2\,nky) \right) . \qquad (2)$$

Here F_o is a constant term while F_n^c and F_n^s describe the higher, spatially varying components of the force. This force depends strongly on the transverse velocity and is no longer conservative. The spontaneous forces introduce dissipation. Standing waves with small blue detunings have been used by others [2,3] for the transverse cooling of the beam, while red detunings would result in transverse heating. Both effects appear to be counterproductive for imaging. However, a simulation for a small red detuning (figure 5) shows remarkable results: the output velocity is certainly increased (heating), but again, v_{out} oscillates between positive and negative values. The results are now clustered fairly close to a line with $v_{out} = 2.4\ v_{in}$. Half the atoms experience heating and are accelerated

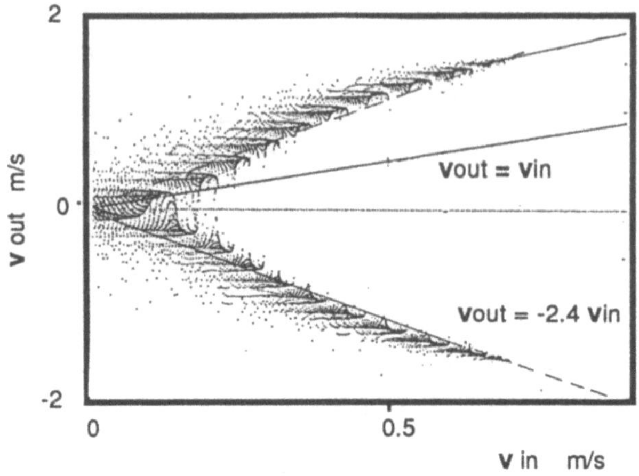

Figure 5:
Numerical simulation.
Similar to figure 4,
but for small detunings .

away from the optical axis. The other half are accelerated towards the atomic beam axis. Since v_{out} / v_{in} is roughly constant, trajectories of these atoms converge at one point on the beam axis, close to the standing wave. The atoms are again focussed, but this time with an image ratio less than 1:1. One could describe this as a lens with variable focal length, determined by the detuning. Our experiments show a good qualitative agreement with these simulations. The imaging properties of this system are not as good as of the system without heating. But the ability to change the focal length with detuning could have useful consequences.

3. Atom reflection and diffraction from an evanescent light field

An additional effect of a standing wave of light on atoms is diffraction, the initial atomic beam is split into a series of diffraction orders propagating in different directions. This can be interpreted in two complementary ways: In a photon picture the atom can interact with both the counter-propagating beams. In particular it can absorb m photons from one beam and followed by the stimulated emission of m photons into the other beam. This leads to a momentum transfer of 2mhk parallel to the light beams, with the atom propagating into a new direction, called the mth diffraction order.

Alternatively, one can consider the wavelike nature of the atom, interpreting a well collimated beam of atoms as a plane deBroglie wave. This wave interacts with the periodic potential of the standing light wave. The change in the direction of propagation into the mth diffraction order is described by the diffraction angle Φ_m which is linked to the periodicity $d = 1/2\, \lambda_{opt}$ of the grating, the wavelength of the deBroglie wave λ_{dB} and the incident angle α. Φ_m can be determined from the classical grating equation

$$\cos(\Phi_m) - \cos(\alpha) = m\, \lambda_{dB} / d\,. \qquad (3)$$

For thermal atoms (typically 1000 m/s , $\lambda_{dB} = h / (mv) = 1.7\ 10^{-11}$ m for Na) each momentum transfer hk corresponds to a velocity change of 3 10^{-3} m/s. For a transverse standing wave at right angles (α =90 degrees) the diffraction angles are very small , of the order of micro radians. This effect was first demonstrated in beautiful, high resolution experiments by the group at MIT [12], confirming the wavelike nature of the atoms. More recently a trick (the magneto-optic effect) has been found to increase the number of momenta transferred by exposing the atoms to a magnetic field parallel to the standing wave. A transfer of +/- 42 hk has been observed [13].

We followed a different idea to increase the diffraction angle. It is well known from classical optics that grazing incidence (very small α) results in larger diffraction angles, in particular for $d >> \lambda_{dB}$. A mirror for atoms with a periodic structure is required. Hajnal and Opat [14] proposed to use an evanescent light field of counter-propagating waves for this purpose.

An evanescent field of light is formed by a beam of light which is totally internally reflected at a surface. The field has an amplitude decaying exponentially over about one optical wavelength. This field forms a repulsive potential barrier for atoms in the groundstate, provided the beam is detuned to the blue (higher frequency) side of an atomic resonance. This has been demonstrated by a number of groups [15,16]. It should be noted that the same field provides an attractive potential for an atom in an excited state. The evanescent wave reflects all those atoms with a kinetic energy, $1/2\ mv_y^2$, less than the height of the potential. The potential increases with the intensity of the light and decreases with the detuning $\Delta\nu = \nu_1 - \nu_{atom}$. However, spontaneous emission has to be avoided otherwise the direction of the atoms will be randomised. Thus an optimum detuning exists, typically $\Delta\nu$ = 2-3 GHz. For the available power of 500 mW and a spot size of 0.3 mm diameter the potential corresponds to a maximum incoming velocity v_y of 2.5 m/s, corresponding to a maximum input angle α_{max} of 2.5 mrad.

Diffraction of atoms into the mth order will only occur when the diffraction angle Φ_m is less than α_{max}. The grating equation can be used to predict the range of deBroglie waves, or input angles, which can lead to diffraction. For very small input angles (< 2 mrad) and for thermal atoms there is no solution. However, it should be noted that equation (3) has to be applied in the frame of reference of the grating. The relative velocity v_{rel} between the grating and the atoms then determines the appropriate deBroglie wavelength and $v_{rel}^2 = v_y^2 + v_z^2$. For a grating stationary in the laboratory $v_{rel} = v_z$. The grating can be moved, relative to the laboratory, by detuning one of the counter propagating waves with respect to the other. In the extreme case ($\nu_2 - \nu_1 = 2\nu_{Doppler}$) the grating can be made to travel with the same velocity as the atoms and $v_{rel} = v_y$. The detuning $\nu_2 - \nu_1$ provides a mechanism for controlling both the input angle and the deBroglie wavelength. In figure 6 the results for small relative velocities are shown. A relative velocity of < 40 m/s is required to allow diffraction for our case with α_{max}= 2.5 mrad.

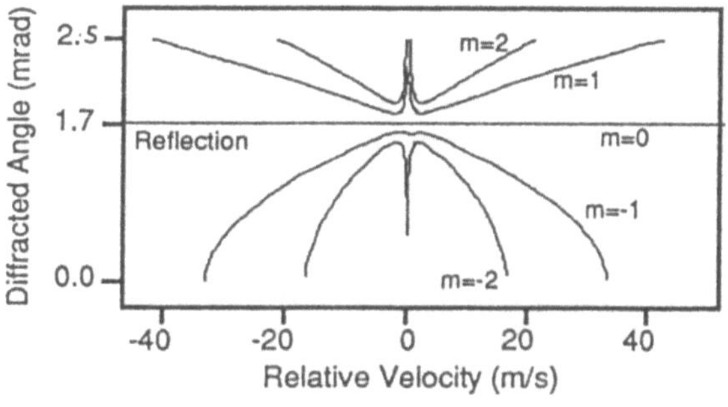

Figure 6:. The output angle as a function of the relative velocity.

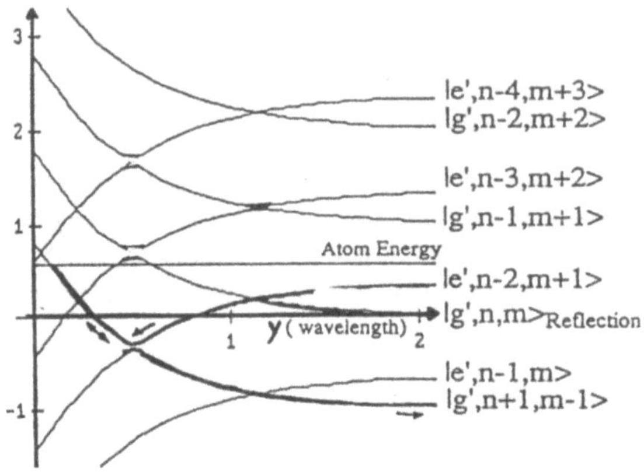

Figure 7: Momentum eigenvalues of the atom approaching the surface.

The above models cannot determine the intensity of the diffraction peak. For this purpose the interaction of the atom with the light along the trajectory through the evanescent field has to be evaluated. This requires a quantum mechanical calculation. One approach is to directly solve the Schrödinger equation. The collimated atom beam is represented by a plane deBroglie wave and the interaction Hamiltonian contains both the free propagation as well as the dipole interaction with the light. Spontaneous emission is neglected. Such calculations were performed by Savage [17] showing the spreading of the function in both momentum and position space and the emergence of several components. Some of these move towards the surface; they correspond to excited states of the atom attracted to the surface. Others split off and represent diffracted orders. These calculations require considerable computing time.

In order to gain more physical insight an alternative model has been developed by the group in Bonn [18]. They describe the system in terms of combined momentum eigenstates for the atoms and the light. Similar to dressed states these states can be classified by the excitation of the atom far from the field, and the number of photons, n and m, in the two fields. An exchange of photons between the fields, corresponds to the transition from one eigenstate |g',n,m> to the other |g',n-1,m+1>, which have different momenta. These momentum eigenvalues can be evaluated for an atom approaching the surface. An example is shown in figure 7. An atom which is initially in the groundstate will have its momentum reduced in the field. It eventually comes to rest and reverses. In figure 7 this appears like a particle moving up the potential hill formed by the evanescent wave. Atoms initially in the excited state are accelerated, crashing into the surface. The interesting point is the crossing of such states. The eigenstates approach each other and form avoided crossings.

The process of diffraction corresponds to the following path through this system of states; the atom enters in the state |g',n,m> and is slowed down. At the first crossing three photons are exchanged |e',n-2,m+1> and the atom is accelerated towards the surface. At the next crossing stimulated emission occurs |g',n-1,m+1>. The atom is slowed down and negotiates another crossing, remaining in this eigenstate. It is reflected and moves away from the surface with different velocity, into the first diffraction order. This process requires the exchange of 2 photons, thus is called even diffraction. If an odd number of photons is exchanged the atom will finally undergo spontaneous emission, which randomises the direction under which it emerges.

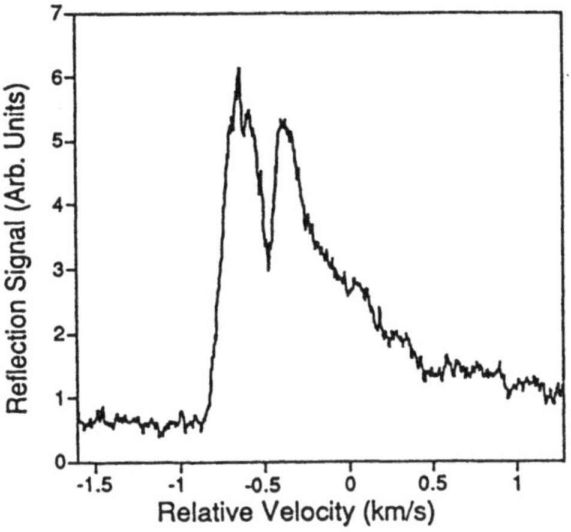

Figure 8: Detected intensity of the specularly reflected atoms as a function of grating velocity v_{rel}

This model can be used to evaluate the probability of diffraction. For an accurate calculation the total probability of a transition between eigenstates integrated along the particle trajectory has to be determined. This will depend on the length of time an atom spends in the condition of an avoided crossing. Such calculations have been carried out by the groups in Bonn and Melbourne [18,19]. However, simple estimates already show that for even diffraction orders five avoided crossings have to be passed, each with a probability of about 0.5 for the required transition. The total probability is thus about $0.5^5 = 0.03$. Detailed calculations show that the diffraction peak should contain about 2 - 5% of the reflected atoms [6,18].

We carried out experiments with a moving evanescent grating of variable speed [20]. The experimental arrangement is shown in Fig.1, with the only change that the transverse standing wave is replaced by an evanescent light field. The counter-propagating laser beam was simultaneously used as the velocity selective fluorescence detector and was kept at a fixed frequency ν_1, detecting atoms at 1100 m/s. This corresponds to a detuning of - 1.5 GHz. The co-propagating beam has a variable frequency ν_2. For this beam the Dopplershift is positive; it has a positive detuning (+2.5 GHz) when the beams have the same frequency ($\nu_2 = \nu_1$). For these detunings only the counter-propagating beam reflects the atoms. Figure 8 shows the intensity of the reflected beam versus the relative velocity (scanning ν_2). At $v_{rel} < 700$ m/s both lasers have a negative detuning, the reflection sets in. The reflection gets weaker for increasing detuning. A striking feature is the sharp drop of reflectivity near 500m/s. The Dopplershifts just compensate the detuning, a condition known as a Doppleron [21]. The atom is resonant for a three photon transition into the excited state involving two photons from the counter-propagating and one photon from the co-propagating wave. As a consequence, the atoms find themselves in an attractive state close to the surface and disappear from the reflected beam. The Doppleron is experimental evidence that the atom interacts with both laser beams, a necessary condition for diffraction.

Finally we searched the parameter space for diffraction signals. A number of scans with fixed detector position y' and variable laser frequency ν_2 were combined with a number of scans with fixed ν_2 and variable y'. The result is the map shown in figure 9. The trace shown in figure 8 is a cross section through this map for specular reflection ($\Phi = \alpha = 1.75$ mrad). The Doppleron is clearly visible as the valley in this map. Does this

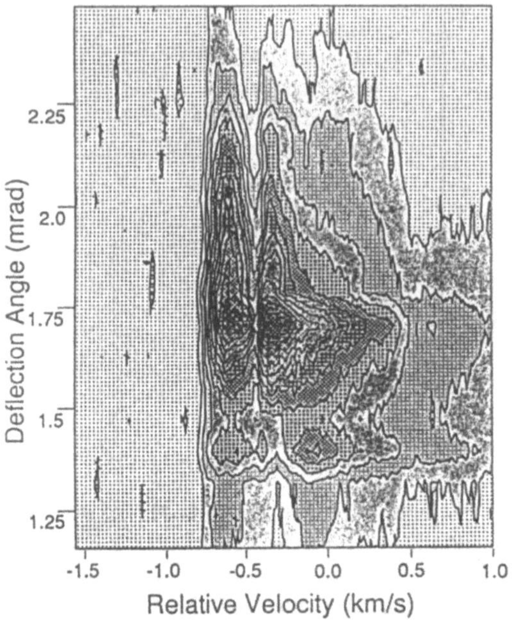

Figure 9:
Detected atomic intensit as a function of detector position Φ and grating velocity.

map show diffraction? The grating equation predicts diffraction in a narrow velocity regime of +/-40 m/s. The position of the diffracted peak should vary rapidly with the velocity (see figure 6). We have no evidence for diffraction. The drifts in laser frequency were too large to resolve such small features. They would have been washed out in the time it took us to record the map. There are interesting features at $\Phi = 1.4$ mrad. The signal is reproducibly increased , forming islands on the map. We interpret these islands as evidence for atoms which exchanged an odd number of photons. These atoms emerge from the evanescent wve after one spontaneous emission, their trajectory is less well defined due to the time spent in the excited state.

The experiments at Bonn [6], using slow atomic beams, have already shown clear evidence for diffraction. The position and strength of the diffracted peak is in good agreement with the models presented here.

4. Summary

Thermal beams in combination with velocity selective detection can be used to demonstrate atom optics with standing waves. For transverse waves both cooling as well as focussing is possible. We point out that the combination of gradient and dissipative forces can yield new interesting results, in particular the imaging by a lens with variable focal length.

The evanescent standing wave opens the opportunity to demonstrate diffraction with enlarged (mRad) diffraction angles. In principle the moving grating can achieve the same diffraction angles as a slow beam. However, the exact synchronisation of the motion of atoms and grating requires much better frequency control than has been used in the experiments presented here.

The future of these these experiments will rest on new ideas for influencing the internal momemtum states of the atoms while it traverses the evanescent field. A good example is the proposal to use Zeeman shifting and optical pumping [22] to generate two output beams with equal intensity.

Acknowledgment

We like to thank C.M.Savage and P.T.H.Fisk for many stimulating discussions and B.Brown for his expert technical assistance during the course of this work. This project has been supported by the Australian Research Council.

References

[1] Review articles such as
C.S.Adams, M.Sigel, J.Mlynek "Atom optics" Phys.Rep. in print (1994)
[2] A.Aspect, J.Dalibard, A.Heidmann, C.Salomon, C.Cohen-Tannoudji Phys.Rev.Lett.**57**, 1688 (1986)
[3] C.E.Tanner, B.P.Masterson, C.E.Wieman, Opt.Lett.**13**, 357 (1988)
[4] M.G.Prentiss, S.Ezekiel, Phys.Rev.Lett.**56**, 46 (1986)
V.I.Balykin, V.S.Letokhov, Yu.B.Ovchinnkov, A.I.Sidorov,S.V.Shul'ga, Opt.Lett.**13**, 958 (1988)
[5] T.Esslinger, A.Hemmerich, T.W.Hänsch Opt.Comm.**93**, 49 (1992)
[6] M.Christ, A.Scholz, M.Schiffer, R.Deutschmann, W.Ertmer submitted to Opt.Comm.Nov 93
[7] J.Dalibard, C. Cohen-Tannoudji J.O.S.A.**B2**, 1707 (1985)
[8] T.Sleator, T.Pfau, V.Balykin, J.Mlynek Appl.Phys.**B54**, 375 (1992)
[9] G.Timp, R.E.Behringer, D.M.Tennant, J.E.Cunningham Phys.Rev.Lett.**69**, 1636 (1992)
[10] Qiming Li, B.W.Stenlake, I.C.M.Littler, H-A.Bachor, K.G.H.Baldwin, D.E.McClelland to appear in Laser Physics (1994)
[11] V.G.Minogin, O.T.Serimaa Opt.Comm.**30**, 373 (1979)
[12] P.L.Gould, G.A.Ruff, D.E.Pritchard Phys.Rev.Lett.**56**, 827 (1986)
P.J.Martin, B.G.Oldaker, A.H.Miklich, D.E.Pritchard Phys.Rev.Lett.**60**, 515 (1988)
[13] T.Pfau, C.Kurtsiefer, C.S.Adams, M.Sigel, J.Mlynek Phys.Rev.Lett. (1994)
[14] J. Hajnal, G.Opat Opt.Comm.**71**, 119 (1989)
[15] R.J.Cook, R.K.Hill Opt.Comm.**43**, 258 (1982)
V.I.Balykin, V.S.Letokhov, Yu.B.Ovchinnikov, A.Sidorov Phys.Rev.Lett.**60**, 2137 (1988)
[16] S.Feron, J.Reinhardt, S.Leboiteux, O.Gorceix, J.Baudon, M.Ducloy, J.Robert, C.Miniatura, S.N.Chormaic, H.Haberland, V.Lorent Opt.Comm.**102**, 83 (1993)
[17] C.M.Savage private communication, unpublished (1993)
[18] R.Deutschmann, W.Ertmer, H.Wallis Phys.Rev.A **47**,2169 (1993)

[19] J.Murphy private communication (1993)
[20] B.W.Stenlake, I.C.M.Littler, H-A.Bachor, K.G.H.Baldwin, P.T.H.Fisk Phys.Rev.A **49**, R16 (1994)
[21] E.Kyrölä, S.Stenholm Opt.Comm.**22**, 123 (1977)
[22] R.Deutschmann, H.Wallis, M.Christ, A.Scholz, W.Ertmer EQEC 93 Poster QOFrP39 Tenchnical digest p.721

Part II

Atomic Optics Theory

Quantum Effects in Atomic Diffraction and Reflection

S.M. Tan and D.F. Walls

Physics Department, The University of Auckland,
Private Bag 92019, Auckland, New Zealand

In this paper we consider the effects of spontaneous emission on two simple problems in atomic optics, namely the diffraction of an atomic wave which impinges at normal incidence on a classical optical standing wave and the reflection of an atomic wavepacket from an evanescent travelling light field. Spontaneous emission is modelled using a fully quantum mechanical model based on the Monte Carlo technique and includes the effects of atomic recoil. The spontaneous emission leads to a broadening of the outgoing wave packet and mechanisms for this broadening are discussed.

I. INTRODUCTION

The recent experimental realizations of light-induced diffraction gratings [1–3] and evanescent wave mirrors [4,5] for atoms have led to considerable theoretical interest in understanding how these devices may possibly be employed as optical elements for atomic interferometry. To date, most analyses have been concerned with preserving atomic coherence throughout the interaction and have thus sought conditions under which spontaneous emission may be avoided [6–8]. This usually requires a large detuning between the optical and atomic frequencies, reducing the strength of the interaction. As the detuning is decreased, spontaneous emission becomes important, and its effects need to be included. In this paper, we use the method of Monte Carlo wave functions [10] in order to investigate the effects of spontaneous emission on the distribution of outgoing momentum for two simple configurations in atomic optics in the situation where both internal and external degrees of freedom of the atom are treated quantum mechanically. The analysis treats the time-dependent evolution of atomic wavepackets as they interact with the field.

II. PROBLEM FORMULATION WITHOUT SPONTANEOUS EMISSION

We wish to consider the effects of spontaneous emission on the problems of atomic diffraction from a standing wave light field (see Figure 1(a)) and of atomic reflection from an evanescent travelling light wave (see Figure 1(b)). In both of these cases, the optical field will be treated classically. The centre of mass motion of the atoms will be treated quantum mechanically so that the analysis will apply to cold atoms.

Springer Proceedings in Physics, Vol. 77 Quantum Optics VI
Editors: D.F. Walls · J.D. Harvey

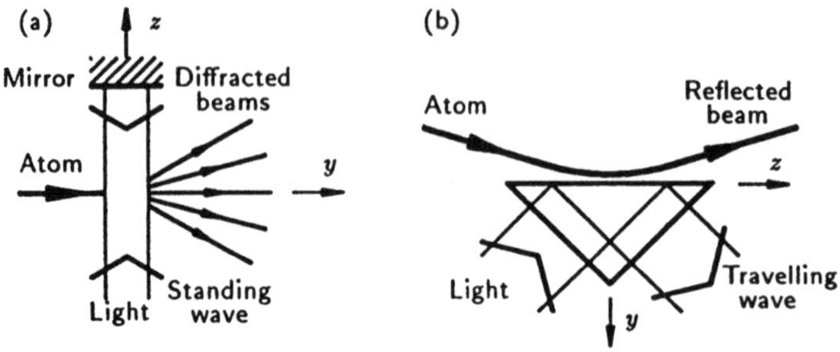

FIG. 1. (a) Atomic diffraction and (b) Atomic reflection configurations

In the absence of spontaneous emission, the Hamiltonian in the rotating wave and electric dipole approximations has the form

$$H = \frac{p^2}{2m} + \hbar\epsilon_1 |1\rangle\langle 1| + \hbar\epsilon_2 |2\rangle\langle 2| - \left[\int \mathbf{D}^{(-)}(\mathbf{r}) \cdot \mathbf{E}_L^{(+)}(\mathbf{r},t)\, d^3\mathbf{r} + \text{c.c.}\right], \quad (1)$$

where p is the centre of mass momentum of the two-level atoms with mass m, and $|1\rangle$ and $|2\rangle$ are the lower and upper atomic levels with energies $\hbar\epsilon_1$ and $\hbar\epsilon_2$ respectively. $\mathbf{D}^{(-)}(\mathbf{r})$ is the dipole moment operator of the atom and is given by

$$\mathbf{D}^{(-)}(\mathbf{r}) = \boldsymbol{\mu}\, |2\rangle\langle 1| \otimes |\mathbf{r}\rangle\langle\mathbf{r}|\, \delta(\mathbf{r} - \mathbf{r}_a) \quad , \quad (2)$$

where $\boldsymbol{\mu}$ is the dipole moment vector and $\mathbf{r}_a$ is the position of the atom. The electric field of the light is $\mathbf{E}_L^{(+)}(\mathbf{r},t)$ where

$$\mathbf{E}_L^{(+)}(\mathbf{r},t) = \begin{cases} \mathbf{E}_0 F(y) \exp[i(k_L z - \omega_L t)] \\ \mathbf{E}_0 F(y)[\exp(ik_L z) + \exp(-ik_L z)]\exp(-i\omega_L t) \end{cases} \quad (3)$$

for a travelling or standing wave respectively. $F(y)$ specifies the transverse field profile, ω_L is the angular frequency and k_L is the wavenumber of the light.

Working in an interaction picture with free Hamiltonian $\hbar(\epsilon_1 + \omega_L)|2\rangle\langle 2| + \hbar\epsilon_1 |1\rangle\langle 1|$ and defining the wave functions ϕ_1 and ϕ_2 via

$$|\psi\rangle = \int d^3\mathbf{r}\, [\phi_1(\mathbf{r},t)\, |\mathbf{r}\rangle \otimes |1\rangle + \phi_2(\mathbf{r},t)\, |\mathbf{r}\rangle \otimes |2\rangle] \quad (4)$$

we obtain the Schrödinger equations

$$i\hbar\frac{\partial \boldsymbol{\phi}}{\partial t} = -\frac{\hbar^2}{2m}\left[\frac{\partial^2 \boldsymbol{\phi}}{\partial y^2} + \frac{\partial^2 \boldsymbol{\phi}}{\partial z^2}\right] + \begin{bmatrix} 0 & -[\boldsymbol{\mu}\cdot\mathbf{E}_L(\mathbf{r},0)]^* \\ -[\boldsymbol{\mu}\cdot\mathbf{E}_L(\mathbf{r},0)] & -\hbar\Delta \end{bmatrix} \boldsymbol{\phi}, \quad (5)$$

where $\Delta = \omega_L - (\epsilon_2 - \epsilon_1)$ is the detuning and $\boldsymbol{\phi}$ is a vector whose components are ϕ_1 and ϕ_2.

III. SPONTANEOUS EMISSION VIA THE QUANTUM MONTE CARLO TECHNIQUE

We follow the quantum Monte Carlo wave function approach for spontaneous emission introduced by Dum, Zoller and Ritsch [9] and by Molmer, Castin and Dalibard [10]. Although spontaneous emission leaves the system in a mixed state, the quantum Monte Carlo technique follows the evolution of the system conditioned on a particular photodetection history which allows us to follow the dynamics using a wave function rather than a density matrix. When we average over different photodetection histories, the expectation values of all quantities tend to those calculated via density matrices and the master equation.

The quantum Monte Carlo technique allows us to include Liouvillians of the form [10]

$$\mathcal{L}\rho = \sum_m C_m \rho C_m^\dagger - \frac{1}{2} C_m^\dagger C_m \rho - \frac{1}{2} \rho C_m^\dagger C_m \quad , \tag{6}$$

where C_m denotes a "collapse" operator which, when applied to the system wave function performs an elementary quantum jump. For the situation of spontaneous emission, the summation over m corresponds to an integral over the permitted directions of emission $\mathbf{k}$ and a summation over outgoing polarizations $\mathbf{e}$. The explicit form for $C_{(\mathbf{k},\mathbf{e})}$ is

$$C_{(\mathbf{k},\mathbf{e})} = \sqrt{3\Gamma/(8\pi)}(\mathbf{e}^* \cdot \boldsymbol{\mu}) \exp(-i\mathbf{k} \cdot \mathbf{r}) \left|1\right\rangle \left\langle 2\right| \quad , \tag{7}$$

where the factors on the right represent a rate factor which specifies how efficiently an atom with dipole moment $\boldsymbol{\mu}$ radiates spontaneously with polarization $\mathbf{e}$, the momentum kick of $-\mathbf{k}$ imparted to the atom due to the recoil and $\left|1\right\rangle \left\langle 2\right|$ lowers the atom into the ground state.

The quantum Monte Carlo method proceeds from one quantum jump to the next using the following procedure. A random number r uniformly distributed in the interval $(0, 1)$ is generated. The wave function $\left|\psi\right\rangle$ is evolved using a Schrödinger equation constructed from the non-Hermitian Hamiltonian

$$H' = H - i\hbar/2 \sum_m C_m^\dagger C_m = H - i\hbar\Gamma/2 \left|2\right\rangle \left\langle 2\right| \quad , \tag{8}$$

where H is the system Hamiltonian in the absence of the spontaneous emission and the last equality follows by carrying out the integration and summation over the operators in Eq. (7). As this evolution proceeds, the norm of ψ decreases. When $\left|\psi\right\rangle \left\langle \psi\right| = r$, a spontaneous emission is deemed to have occured. To perform the quantum jump, one of the collapse operators C_m is chosen according to probabilities proportional to $\left\langle \psi\right| C_m^\dagger C_m \left|\psi\right\rangle$. This collapse operator is applied to $\left|\psi\right\rangle$ and the result is renormalized. i.e., we set

$$\left|\psi\right\rangle \rightarrow C_m \left|\psi\right\rangle / \sqrt{\left\langle \psi\right| C_m^\dagger C_m \left|\psi\right\rangle} \quad . \tag{9}$$

At this stage a new random number is generated to find the time of the next quantum jump.

It is possible to obtain explicit formulae for the relative probabilities with which spontaneous emission will occur into different directions. We assume that the dipole moment $\boldsymbol{\mu}$ of the atom is aligned with that of the classical light field and choose a direction $\mathbf{k}$ for the outgoing photon. The probability of emission into unit solid angle $d\Omega$ is proportional to the sum $|\boldsymbol{\mu}\cdot\mathbf{e}_1|^2+|\boldsymbol{\mu}\cdot\mathbf{e}_2|^2$ where $\mathbf{e}_1$ and $\mathbf{e}_2$ are two polarization vectors orthogonal to $\mathbf{k}$. The probability density is normalized to give an integral of unity when taken over all outgoing directions.

For example, with a circularly polarized classical light field, the probability for spontaneous emission into solid angle $d\Omega$ along a direction which makes an angle θ with the incident radiation is

$$\Pr[\text{spontaneous emission into } d\Omega] = 3/(16\pi)(1+\cos^2\theta)\,d\Omega\ . \tag{10}$$

For a linearly polarized classical light field, the probability depends on the angle ψ between the polarization direction of the incident field and the wave vector $\mathbf{k}$ of the emitted light. We find that

$$\Pr[\text{spontaneous emission into } d\Omega] = 3/(8\pi)\sin^2\psi\,d\Omega\ . \tag{11}$$

It is straightforward to use a random number generator to choose a direction of spontaneous emission to conform to these angular distributions.

IV. THE ATOMIC DIFFRACTION PROBLEM

This situation is depicted in Figure 1(a). We shall consider a Gaussian atomic wavepacket initially in the ground state travelling along the y direction and normally incident on the standing wave. The interchange of momentum with the field occurs along the z direction and so the y motion may be treated classically. We suppose that the atomic beam enters with a well-defined longitudinal velocity v_y. If we define new wave functions $\tilde{\phi}_j$ which vary slowly with y by removing the systematic motion in the y direction, i.e., by setting $\phi_j(y,z,t) = \tilde{\phi}_j(y,z,t)\exp\left[(i/\hbar)(mv_y y - mv_y^2 t/2)\right]$, these may be substituted into the Schrödinger equation and the second derivatives of $\tilde{\phi}_j$ with respect to y ignored. This yields

$$i\hbar\left(\frac{\partial}{\partial t}+v_y\frac{\partial}{\partial y}\right)\tilde{\phi} = -\frac{\hbar^2}{2m}\frac{\partial^2}{\partial z^2}\tilde{\phi} + \begin{bmatrix} 0 & -\hbar\Omega F\cos k_L z \\ -\hbar\Omega F\cos k_L z & -\hbar\Delta - i\hbar(\Gamma/2)\end{bmatrix}\tilde{\phi}. \tag{12}$$

The quantity Ω is a Rabi frequency which satisfies $\Omega = \boldsymbol{\mu}\cdot\mathbf{E}_0/\hbar$ and the non-Hermitian matrix element $-i\hbar(\Gamma/2)$ has been included to model the spontaneous emission. The convective derivative term in parentheses on the left hand side may be replaced by a time derivative $\partial/\partial t'$ in a frame moving with the atoms. The transverse intensity profile on the optical field translates into a time dependent field strength $F(t')$ seen by the atoms in the moving frame. This allows the dependence on y to be dropped and we may regard $\tilde{\phi}_j$ as being functions of z and t' alone.

It is usual to treat this problem in two stages. Firstly, there is a relatively short time during which the atom interacts with the field (i.e., while the atom crosses the light field) followed by a longer period during which the atom flies freely to the detector screen without interacting with the light. If the interaction time with the field is short enough, the effect of the interaction with the light is simply to transfer momentum to the atom without appreciably changing its position. This defines the Raman-Nath regime and it allows us to drop the kinetic energy term on the right hand side. If the profile $F(t')$ is rectangular, there is a straightforward analytic solution which can be found by using Laplace transforms.

In the region of free propagation following the light field, we set $F(y) = 0$. Once again, these equations may be solved analytically using a combination of Fourier and Laplace transforms.

Modelling a spontaneous emission event involves choosing a momentum kick $\hbar k'$ drawn from the projection of the angular distribution of spontaneous emission onto the z axis and setting the new value of $\tilde{\phi}_1$ to a normalized version of $\tilde{\phi}_2 \exp(-ik'z)$ and the new value of $\tilde{\phi}_2$ to zero.

V. COMPARISON WITH EXPERIMENTAL RESULTS

Experimental results for the atomic diffraction problem have been obtained by Gould *et al.* [3] using a monoenergetic, highly collimated atomic sodium beam. By varying the detuning, they were able to investigate the transition between the diffractive regime and the diffusive regime in which spontaneous emission is important. These results have been analysed theoretically using a variety of methods based on the density matrix [11,12]. In Figure 2, we show the experimental data together with results of the quantum Monte Carlo

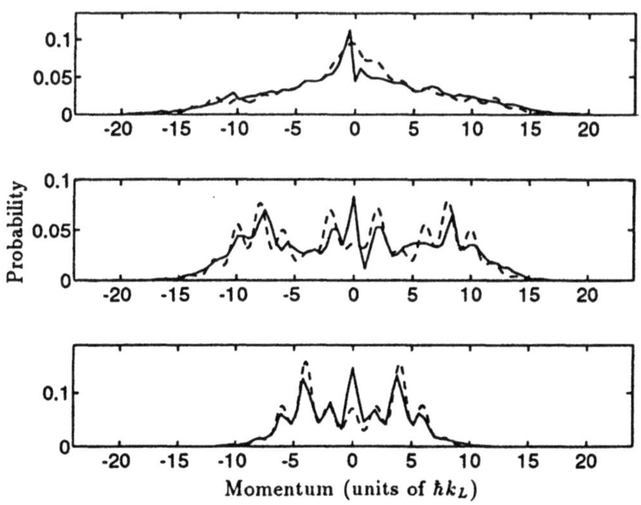

FIG. 2. Experimental results of Gould *et al.* (solid line) together with results of Monte Carlo simulation (dashed line). The average number of spontaneous emissions during the interaction time is 4.5 (top graph), 1.2 (middle graph), 0.4 (bottom graph).

simulation in the Raman-Nath regime and using a rectangular approximation to the laser profile. There is good agreement between the experimental results and this simulation and also between the simulation and those performed using density matrix methods. We note that the broadening of the peaks introduced by the spontaneous emission is on the same order as $\hbar k_L$, the amount of the photon recoil.

VI. THE EVANESCENT WAVE ATOMIC MIRROR

The situation under consideration is shown in Figure 1(b). For a positive detuning Δ, ground state atoms are repelled by a gradient force produced by the intensity variation in the evanescent light field. A quantum mechanical analysis can be carried out by working in momentum space in the z direction and in position space in the y direction. If the z component of the momentum $\hbar k_z = mv_z$ is well-defined for the incoming ground-state atoms, the effect of the travelling light-field is to couple atoms in states $|1\rangle \otimes |k_z\rangle$ to states $|2\rangle \otimes |k_z + k_L\rangle$. The general description of the atomic state may be written as

$$|\psi\rangle = \int dy\, [\phi_1(y,t)\, |1\rangle \otimes |k_z\rangle \otimes |y\rangle + \phi_2(y,t)\, |2\rangle \otimes |k_z + k_L\rangle \otimes |y\rangle] \,. \tag{13}$$

Schrödinger's equation in the absence of spontaneous emission (and neglecting unimportant phase factors) may now be written as

$$i\hbar \frac{\partial \phi}{\partial t} = -\frac{\hbar^2}{2m}\frac{\partial^2 \phi}{\partial y^2} + \begin{bmatrix} 0 & -\hbar\Omega F(y)/2 \\ -\hbar\Omega F(y)/2 & -\hbar\Delta_{\text{eff}} \end{bmatrix} \phi \,, \tag{14}$$

where $\hbar\Delta_{\text{eff}} = \hbar(\Delta - k_L v_z) - \hbar^2 k_L^2/(2m)$. In this problem, the exponential profile shape $F(y)$ is important as its variation leads to the force and the kinetic energy term must be retained since there is significant atomic motion during the interaction with the field. The equations may be solved numerically using the split operator technique [13] in which calculations are performed alternately in position space and momentum space.

If the evolution of a Gaussian wave packet of atoms initially in the ground state is plotted as it interacts with the evanescent field, the packet is seen to approach the barrier and then bounce off, reversing its direction of travel. At intermediate times, there is interference between waves propagating in the forward and reverse directions.

It is convenient to use the concept of dressed states to understand the action of the atomic mirror. If we diagonalize the light-induced portion of the Hamiltonian which is the matrix in Eq. (14), the eigenvectors are called dressed states and the eigenvalues are called quasipotentials. If the intensity $F(y)$ changes slowly as the atom traverses the field, the adiabatic theorem [14] states that an atom placed within one of these dressed states will remain in that dressed state as the atom enters and leaves the light field. As the light amplitude $F(y)$ changes, the dressed states correspond to different linear superpositions of the states $|1\rangle$ and $|2\rangle$. Figure 3(a) shows the variation of the quasipotentials with y. At distances far from the mirror, the dressed states $|-\rangle$ and $|+\rangle$ coincide

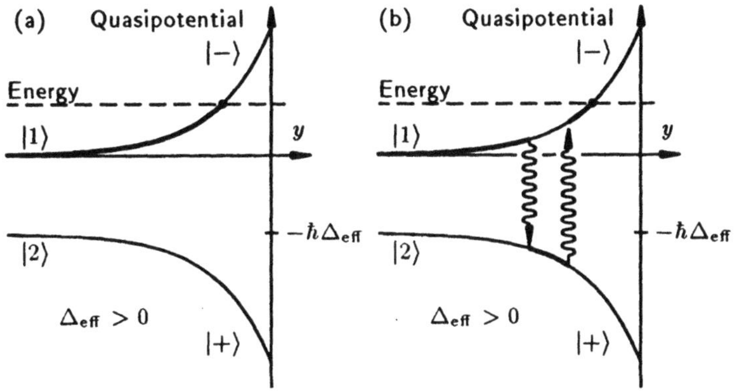

FIG. 3. (a) Quasipotential curves and "path" of atom incident on mirror in the ground state, (b) Transitions between dressed states caused by spontaneous emissions

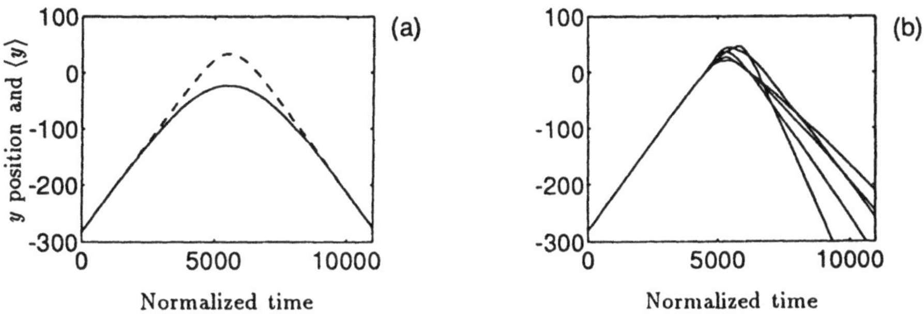

FIG. 4. (a) Classical trajectory (dashed line) and the mean position of a quantum mechanical Gaussian wavepacket $\langle y \rangle$ (ground state only, solid line) as it encounters the mirror. (b) Several "classical Monte Carlo" trajectories showing the fluctuation in force due to transitions between dressed states

with the ground and excited states $|1\rangle$ and $|2\rangle$ respectively. The gradient force arises from the negative derivative of the quasipotentials. For positive Δ_{eff}, this force is repulsive for state $|-\rangle$ and attractive for state $|+\rangle$.

Given the gradient force as a function of position in the evanescent wave, we can calculate the classical dependence of y and p_y on time by integrating the force. Quantum mechanically, these variables cannot be measured simultaneously but we can calculate how their expectation values vary with time. Superimposing the classical and quantum calculations (Figure 4(a)) shows good agreement away from the mirror, but the mean of the wavepacket does not penetrate the field as deeply as the classical turning point.

Experimentally [15] it has been found that spontaneous emission leads to greater spreading in the reflected atomic beam than can be accounted for by the recoil associated with emission. In reference [15], a semiclassical analysis is carried out based on the quasipotentials. The atom is considered as a point

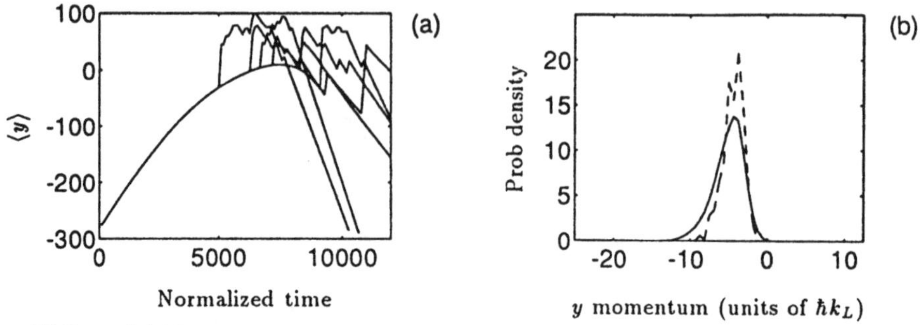

FIG. 5. (a) Mean position of wavepacket $\langle y \rangle$ for several quantum mechanical trajectories, showing discontinuities at the times of spontaneous emission. (b) Probability density of y momentum of outgoing atoms using the quantum (solid line) and classical (dashed line) Monte Carlo simulation.

particle with a classical trajectory. As the atom enters the light field, it is assumed to adiabatically follow the $|-\rangle$ dressed state and to experience the appropriate gradient force. Once it is within the light, the $|-\rangle$ state contains some excited state component and so there is some probability of spontaneous emission from this state. As a result of this spontaneous emission, the atom is given a momentum kick and is projected into the ground state which is a linear combination of the two dressed states. Instead of following this linear superposition, they probabilistically decide on whether state $|+\rangle$ or state $|-\rangle$ is to be used. If the atom makes a transition to the $|+\rangle$ state, it experiences an attractive gradient force, which strongly deflects it towards the mirror. Unless a further spontaneous emission occurs, such an atom will be lost to the mirror surface. Figure 3(b) shows a possible semiclassical "path" for the atomic state, where the vertical transitions indicate spontaneous emissions. Atoms which are reflected can be deflected through large angles due to such large fluctuations in the force. They call this method a "classical Monte Carlo" calculation since the times and consequences of each spontaneous emissions are determined probabilistically and it is necessary to average over many realizations. Figure 4(b) illustrates five typical realizations of classical Monte Carlo trajectories. They find good agreement between their experimental results and the simulations.

The classical Monte Carlo trajectories are necessarily continuous since only discontinuous momentum kicks are imparted to the atom. This is appropriate when the atom can be treated as a point particle rather than an extended wave packet. When this is not the case, it is necessary to perform the full quantum calculation. It is easy to include the effects of spontaneous emission into the quantum Monte Carlo simulation for this situation by modifying Eq. 14 in an analogous way to that described in connection with the atomic diffraction problem. Figure 5(a) shows the variation of the expectation values of $\langle y \rangle$ with time for five typical realizations of such a simulation. The striking difference between these quantum stochastic trajectories and the semiclassical ones in Figure 4(b) is that $\langle y \rangle$ has discontinuities at the positions of the quantum jumps. If the light field amplitude varies on a spatial scale comparable to these jumps, this leads to an additional broadening mechanism for the outgoing momenta which is not present in the classical Monte Carlo analysis.

It is not difficult to understand the reason for these jumps if one examines the composition of the wavepacket as it encounters the light field. If the ground state and the excited state populations within the wavepacket is plotted as the packet encounters the light field, it is found that the excited state population peaks closer to the mirror than for the ground state population. This can be understood from adiabatic following and the dressed state picture, since the composition of the $|-\rangle$ dressed state contains more of the excited state as $F(y)$ becomes larger. Since the quantum Monte Carlo algorithm requires that the new ground state wave function be set to a normalized version of the old excited state wave function (together with the effects of a momentum kick) when a spontaneous emission occurs, this causes the jumps in $\langle y \rangle$.

As a result of averaging over many quantum and classical Monte Carlo trajectories, one obtains the y momentum distribution of the outgoing atoms. This is shown in Figure 5(b). The classical result is obtained by averaging over runs in which the incoming momentum is varied over the range of momenta for the quantum mechanical wavepacket. It is apparent that the quantum result is broader with a tail of higher momentum atoms.

VII. SUMMARY

In summary, the Quantum Monte Carlo technique is useful for treating spontaneous emission as it straightforwardly models both the effects of momentum recoil on the atom as well the stochastic nature of the emission process without the use of large density matrices. Good agreement with experiment is found in the situation of atomic diffraction from a standing wave, and in the case of the evanescent wave atomic mirror, the time- dependent solutions give insight into the mechanisms causing the broadening of the outgoing momentum distribution.

ACKNOWLEDGMENTS

We thank the New Zealand Foundation of Research, Science and Technology and the University of Auckland Research Committee for financial support.

References:

[1] P.E. Moskowitz, P.L. Gould, S.R. Atlas and D.E. Pritchard, Phys. Rev. Lett. **51**, 370 (1983).

[2] P.L. Gould, G.A. Ruff and D.E. Pritchard, Phys. Rev. Lett. **56**, 827 (1986).

[3] P.L. Gould, P.J. Martin, G.A. Ruff, R.E. Stoner, J.L. Picque and D.E. Pritchard, Phys. Rev. A **43**, 585 (1991).

[4] V.I. Balykin, V.S. Letokhov, Yu. B. Ovchinnikov and A.I. Sidorov, Phys. Rev. Lett. **60**, 2137 (1988).

[5] M.A. Kasevich, D.S. Weiss and S. Chu, Opt. Lett. **15**, 607 (1990).

[6] R.J. Cook and R.K. Hill, Opt. Comm. **43**, 258 (1982).
[7] J.V. Hajnal and G.I. Opat, Opt. Comm. **71**, 119 (1989).
[8] R. Deutschmann, W. Ertmer and H. Wallis, Phys. Rev. **A47**, 2169 (1993).
[9] R. Dum, P. Zoller and H. Ritsch, Phys. Rev. A **45**, 4879 (1992).
[10] K. Molmer, Y. Castin and J. Dalibard, J. Opt. Soc. Am. B **10**, 524 (1993).
[11] S.M. Tan and D.F. Walls, Phys. Rev. A **44**, 2779 (1991).
[12] M. Wilkens, E. Schumacher and P. Meystre, Opt. Commun., **86**, 34 (1991).
[13] R. Kosloff, J. Phys. Chem., **92**, 2087 (1988).
[14] A. Messiah, *Quantum Mechanics*, Chap. 17 (North-Holland, Amsterdam, 1961)
[15] W. Seifert, C.S. Adams, V.I. Balykin, C. Heine, Yu. Ovchinnikov, and J. Mlynek, submitted to Phys. Rev. A

Theory of Interactions Between Laser-Cooled Atoms

M.J. Holland, K.A. Suominen, and K. Burnett

Clarendon Laboratory, University of Oxford, Parks Rd.,
Oxford OX1 3PU, England

In this paper, we calculate the resulting momentum distribution for the relative motion of two atoms from a cold atomic collision in the presence of a driving field. We consider temperatures well below the Doppler limit of laser cooling. The collisional process acts as a heating mechanism which competes with the frictional forces in the trap. This limits the possible temperatures and densities and has important implications for the formation of a Bose-Einstein condensate in such a trap.

I. OPTICALLY INDUCED COLLISIONS

Collisions between atoms undergoing laser cooling or trapping limit the densities and temperatures that can be achieved. Establishing the precise limits is an extremely complex matter as one has to treat cold collisions in the presence of a laser driving field. At the very cold kinetic temperatures that have been reached the de Broglie wavelengths are large compared to atomic dimensions. The interactions between the atoms also occur on very large characteristic lengths since atoms with very little kinetic energy feel their mutual influence at large separations. This means that the quantal motion of the atoms has to be followed on a scale far larger than that associated with conventional thermal energy collisions [1].

The problem is difficult to solve in this parameter region because the time scale of each collision is comparable to the time scale for the atom to spontaneously decay so that one needs to combine excitation and decay processes with the quantum motion of the colliding atoms. There are, however, certain regimes in which semiclassical or perturbative approaches can be used. The first approach to the problem [2,3] employed a combination of a local excitation model with classical motion on the excited surface. This model, further developed by Julienne and Vigué [4] and by Smith, Burnett, and Julienne [5], was able to give a fairly accurate description of collisions at temperatures $T \gtrsim 1$ mK as long as saturation effects could be ignored. The optical Bloch equation approach of Band and Julienne [6] remedied the defect in the semiclassical approach for what is now considered a moderately high temperature. At the very low temperatures now available both these approaches are known to fail even in the weak field regime [7,8]. For very weak fields a probe field approach based on optical potentials can be used rather successfully [7].

In the regime we shall address in this paper, namely ultracold temperatures with laser saturation effects, a full treatment of the density matrix is required and this is what we can now provide [9]. The simulation of the measurement of spontaneously emitted photons is necessary to reduce the complexity of the dynamics to a computable level. Our study produces a statistical approximation to the solution of the full density matrix by combining wave packet dynamics with the Monte Carlo state vector method. We also present a quantitative semiclassical treatment based on the Landau-Zener model, and show that it agrees with the quantal results in certain parameter regions. However, for low temperatures and strong laser fields the full quantum treatment is needed. The results of the simulations show that low energy collisions produce appreciable translational heating of the atoms involved without necessarily contributing significantly to trap loss processes. We examine the dependence of the induced heating on the strength and frequency of the driving field, and on the initial momentum of the colliding atoms.

II. THE BINARY COLLISION MODEL

At the low temperatures achievable by laser cooling, collisions, even though relatively long ranged, can still be regarded as binary events since the time between collisions is several times longer than both the spontaneous decay time scale $1/\Gamma$ and the duration of a single collision [10]. We have chosen the Cesium atom as the basis of our study. The slowly colliding atoms can be regarded as a quasi-molecule system, in which the motion takes place on the corresponding molecular potential surfaces along the axis connecting the two atoms. We can assume for red detunings that our excitation from the 0_g^+ molecular ground state ($2S_{1/2}$+$2S_{1/2}$) occurs dominantly to the attractive 0_u^+ molecular state ($2S_{1/2}$+$2P_{3/2}$) of the Cs_2 quasi-molecule. For simplicity we assume that the ground state potential is flat with an infinite repulsive core at $R = 0$, where R is the interatomic distance coordinate. The excited state potential is assumed to consist solely of the dipole-dipole interaction, which dominates the motion on the excited state surface in the regions important for laser-induced interactions. The attractive potential has the form $\hbar\omega_0 - C_3/R^3$, where $\hbar\omega_0$ is the asymptotic energy difference between the ground state and the excited state.

The wave packet motion is found by solving the time-dependent Schrödinger equation for the ground state ψ_g and excited state ψ_e

$$i\hbar\frac{\partial}{\partial t}\begin{pmatrix}\psi_g\\ \psi_e\end{pmatrix} = H\begin{pmatrix}\psi_g\\ \psi_e\end{pmatrix} \tag{1}$$

where H is the matrix representation of the full system Hamiltonian. There are several algorithms for evolving these equations, usually based on the split-operator technique [11], where the Hamiltonian is divided into parts. During one step of time each part operates on the state vector on its turn. One such division is the separation into the kinetic and potential energy operator parts, as was done in Refs. [12,13]. Another way of splitting the Hamiltonian is based on separating wave packet motion (wpm) and field-induced interaction (int). We have adopted this approach and divide our Hamiltonian in the following way

$$H = H_{\text{wpm}} + H_{\text{int}} \tag{2}$$

where

$$H_{\text{wpm}} = \begin{pmatrix} T(R) + U_g(R) & 0 \\ 0 & T(R) + U_e(R) \end{pmatrix} \tag{3}$$

$$H_{\text{int}} = \begin{pmatrix} -\hbar\Delta & \hbar\Omega \\ \hbar\Omega & -i\hbar\frac{\Gamma}{2} \end{pmatrix} \tag{4}$$

and the kinetic and potential energy terms are given by

$$\begin{aligned} T(R) &= -\frac{\hbar^2}{2\mu}\frac{\partial^2}{\partial R^2} \\ U_g(R) &= 0 \\ U_e(R) &= -\frac{C_3}{R^3}. \end{aligned} \tag{5}$$

Here Ω is the Rabi frequency for the driving field, Δ is the detuning of the laser field from the atomic transition frequency, and μ is the reduced mass of the Cs_2 system. The non-Hermition term in H_{int} containing the spontaneous emission rate Γ is part of the Monte Carlo method and corresponds to the null measurement result from the simulated measurements of the reservoir excitations. The casual evolution described by the effective Hamiltonian is interrupted by a sequence of quantum jumps which projects the system from the excited state into the ground state when a spontaneous photon is emitted. A random number generator is used to determine the times at which these decay processes occur. This depends only on Γ and the total probability that the system is in the excited state. We scale our spatial variables with the transition wave length λ divided by $2\pi \simeq 136$ nm. This distance unit is equal to $2560a_0$, where a_0 is the Bohr radius. Momentum is measured in units of $\hbar k$ where the wave number is $k = 2\pi/\lambda$. This corresponds to the recoil from the emission of a single resonant photon.

III. LANDAU ZENER THEORY

The attractive potential corresponding to the excited state of the system compensates exactly for the detuning of the driving field at some value of R and at this point the system is resonantly driven. This value of the interatomic distance is equivalent to a crossing point for the potential surfaces associated with the ground and excited state. Regarding the wave packet moving on the

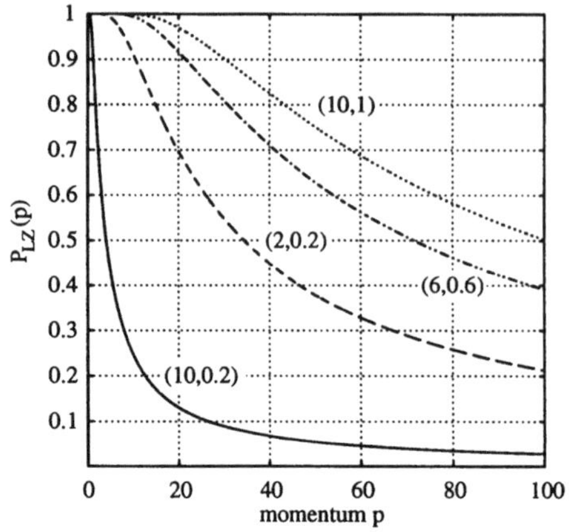

FIG. 1. The momentum dependence of the Landau-Zener transition probabilities. The crossing process is adiabatic when $P_{\rm LZ}$ is close to unity. In the (2,1) case $P_{\rm LZ}$ is practically equal to unity for $p < 100$, and is therefore omitted. The lines correspond to the different (detuning,coupling) combinations as marked in the graph.

potential surface as a classical object allows us to apply the Landau-Zener curve crossing probability to calculate the output momentum distribution providing that we can neglect spontaneous emission in the crossing region. If the classical object is moving with momentum p, the probability it will switch from the flat ground state to the excited surface is

$$P_{\rm LZ}(p) = 1 - \exp\left(-\frac{\pi\Omega^2}{|p|\alpha}\right) \tag{6}$$

where the slope α is determined by the dipole-dipole coupling and detuning

$$\alpha = 3\left(\frac{\Delta^4}{C_3}\right)^{1/3} \tag{7}$$

In Fig. 1 we plot Eq. (6) as a function of momentum for selected parameters. We use the notation (m, n) for different parameter combinations: $\Delta = m\Gamma$ and $\Omega = n\Gamma$.

Clearly, with values of p close to 10 units the wave packet evolves and moves onto the excited state adiabatically as it passes through the interaction region. However, if it is strongly accelerated by the radiative heating mechanism, it has a clear non-zero probability to remain in the ground state when it goes through the crossing for a second time after a reflection at the core.

If the wave packet does switch at a crossing point, the excited components then decay back to the ground state with a characteristic rate Γ. This results in the following probability distribution over momentum states

$$P(p') = \frac{\Gamma}{\alpha}\exp\left(-\frac{\Gamma}{\alpha}(p' - p)\right) dp'. \tag{8}$$

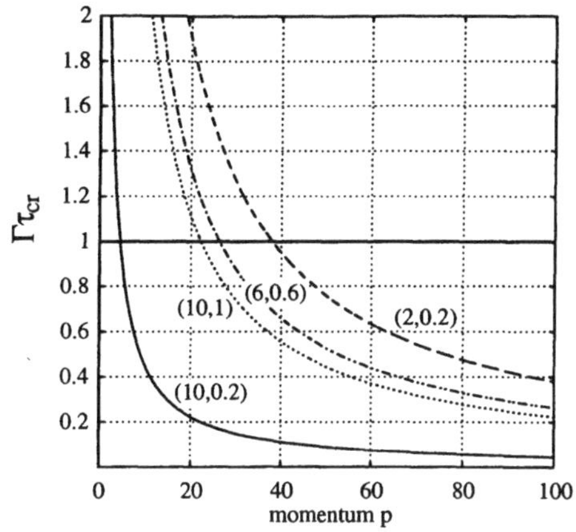

FIG. 2. The ratio of the crossing time τ_{cr} to the averaged decay time $1/\Gamma$ for different parameter combinations. In the (2,1) case $\Gamma\tau_{cr} \gg 1$ for $p < 100$ and is therefore beyond the range used in the graph. The line styles are as in Fig. 1.

This distribution can be integrated over the initial momentum distribution and combined with the possible trajectories for the two sequential crossings as the atoms come together and then move apart. This provides a simple method for estimating the heating induced by the collision. If the initial kinetic energy is too low, the probability that a spontaneous photon is emitted in the vicinity of the crossing region becomes appreciable and the Landau-Zener theory will fail. It is then necessary to track the full quantum motion using the Monte Carlo method. In Fig. 2 we show the ratio of the time it takes to traverse the crossing (p unchanged during the process) to the mean decay time $1/\Gamma$. If this ratio is larger than one, we can expect recycling of population between the ground state and the excited state during the crossing process.

IV. RESULTS

In this section we present our main results, obtained from the fully quantal approach and from the Landau-Zener approach. In Fig. 3 we give two examples of how the wave packet evolves for the (6,0.6) case. In (a), the initial wave packet spreads while moving towards the crossing, goes through the heating process and returns with a spread that extends well beyond the free one. A part of the wave packet remains trapped in the upper adiabatic state potential well. The trapped population appears as a tall but narrow peak: its weight is very small if compared with the large main part of the wave packet. In (b) the wave packet moves fast and spreads very little before entering the crossing. Since the crossing is not fully adiabatic at the initial mean momentum $p_0 = 50$, the initial wave packet forms a visible part of the final wave packet. The total diffusion of the wave packet, however, still clearly exceeds the free spreading.

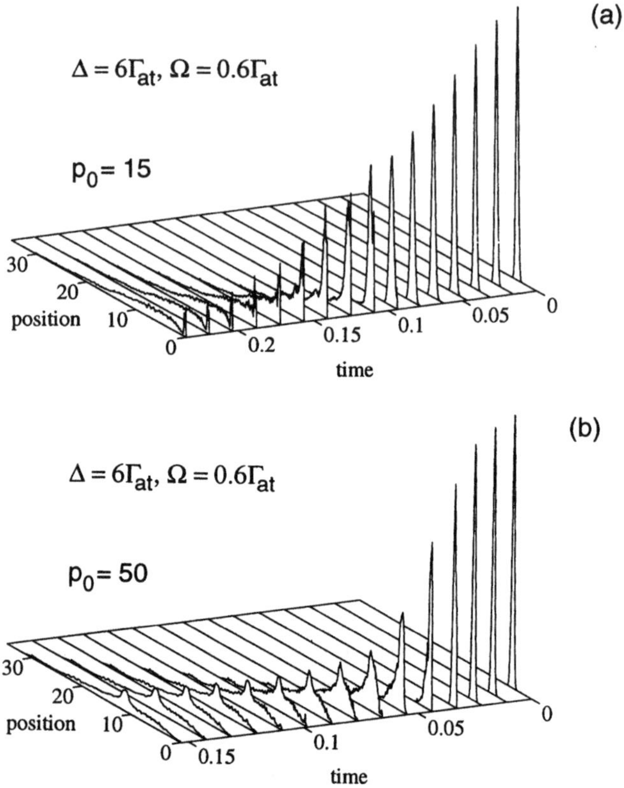

FIG. 3. Examples of wave packet evolution on the ground state in the position representation. The parameters are as indicated in the figure. The initial Gaussian wave packet is located at $R = 4$ and moves towards $R = 0$ with initial momentum p_0. It enters the main collision region ($R < 1$) and then re-emerges, except in (a) for the part that is trapped at $R < R_{cr}$ (which eventually escapes later, but in (a) the time evolution is plotted only up to $\tau = 0.24$ in order to show the details of the dominating collision process).

Examples of the wave packet evolution in the momentum representation are shown in Fig. 4 for two $\Delta = 2\Gamma$ cases with $p_0 = 10$. In both figures we see how the sharp initial distribution is depleted as its components experience excitation and core reflection. In (a) the heating clearly dominates as the excitation process is practically adiabatic, and we see peaks appearing in the momentum distribution well separated from the initial value. The core-reflected components reappear at $p > 0$ and eventually move towards smaller values of $|p|$. This is because of deceleration at the second crossing, which is still adiabatic. The net effect is clear heating combined with strong diffusion: we have a final mean momentum $p \simeq 30$ compared to the initial mean momentum of $p_0 = 10$. In (b) we see only slight heating and mainly diffusion for the (2,0.2) case. At the first crossing the initial wave packet obtains only a small

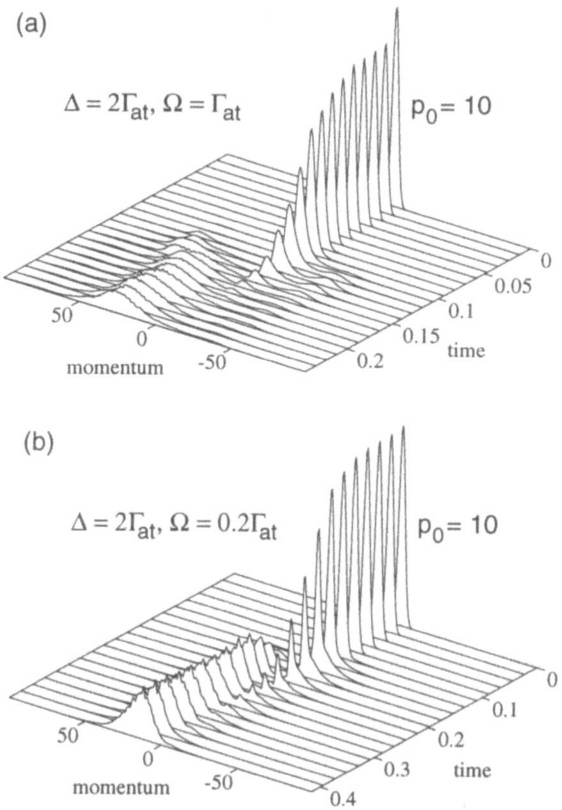

FIG. 4. Examples of wave packet evolution on the ground state in the momentum representation. The parameters are as marked in the graph. The initial Gaussian wave packet is located at $\langle p \rangle = -p_0$, but the crossing process severely alters its shape and position. In the absence of the coupling to the excited state the initial distribution would merely shift smoothly to $\langle p \rangle = p_0$ and regain its shape, as the wave packet in the position representation is reflected by the repulsive core at $R = 0$.

tail towards higher $|p|$, signifying very modest heating. Because of the lesser heating, the collision lasts longer than in the corresponding (2,1) case of (a).

In Fig. 5 we illustrate output momentum distributions for various values of p_0 for the case (2,0.2). The excitation probability is large enough to observe modest heating for small p_0. The Landau-Zener prediction is not too far off, especially for $p_0 > 20$.

Fig. 6 shows a clear breakdown of the Landau-Zener argument for all values of momentum in the (2,1) case. Because of the strong adiabaticity the recycling effect is present at both crossings. The main failure of the Landau-Zener approach at large p_0 is the prediction for spreading. The linearization argument is far from valid for both the excitation process and the semiclassical trajectory since the levels refuse to decouple at large separation.

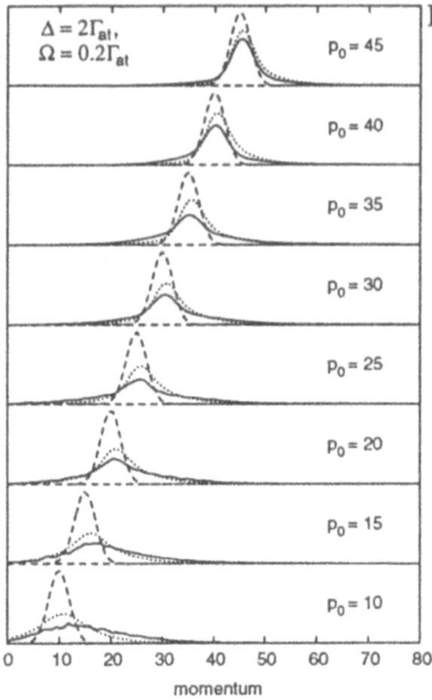

Fig. 5

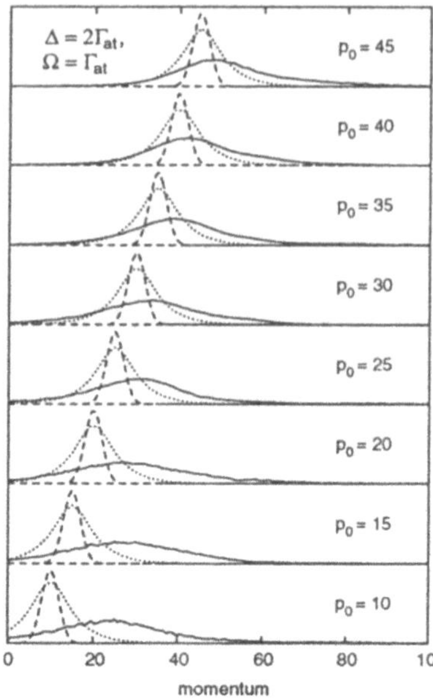

Fig. 6

FIG. 5. The final momentum distributions for the (2,0.2) case. The momentum values are as indicated on the graph. The solid lines show the simulation results, the dashed lines the initial distributions, and the dotted ones the corresponding Landau-Zener predictions. The distributions are normalized to have the same area.

FIG. 6. The final momentum distributions for the (2,1) case. The layout is as in Fig. 5.

V. CONCLUSION

Cold collisions in the presence of a driving field is a significant heating mechanism which limits the phase space density achievable in experiments on laser cooling and trapping. We have shown how it is now possible to solve the binary collision problem with a full quantum wave packet and spontaneous decay by applying a measurement simulation approach. The computation, although very considerable, is still well within the range of the capabilities of today's workstations. In the ultracold temperature regime we have seen that the momentum spread can become very large compared to the initial distribution. This goes hand in hand with significant reexcitation (population recycling) and precludes any simple quantitative approach in this region.

References:

[1] P. S. Julienne and F. H. Mies, J. Opt. Soc. Am. B **6**, 2257 (1989).

[2] A. Gallagher and D. E. Pritchard, Phys. Rev. Lett. **63**, 957 (1989).

[3] P. S. Julienne, Phys. Rev. Lett. **61**, 698 (1988).

[4] P. S. Julienne and J. Vigue, Phys. Rev. A **44**, 4464 (1991).

[5] A. M. Smith, K. Burnett, and P. S. Julienne, Phys. Rev. A **46**, 4091 (1992).

[6] Y. B. Band and P. S. Julienne, Phys. Rev. A **46**, 330 (1992).

[7] H. M. J. M. Boesten, B. J. Verhaar, and E. Tiesinga, Phys. Rev. A **48**, 1428 (1993).

[8] K.-A. Suominen, M. J. Holland, K. Burnett, and P. S. Julienne, to appear in Phys. Rev. A.

[9] M. J. Holland, K.-A. Suominen, and K. Burnett, (unpublished).

[10] M. Trippenbach, Bo Gao, J. Cooper, and K. Burnett, Phys. Rev. A **45**, 6555 (1992).

[11] W. H. Press, B. P. Flannery, S. A. Teukolsky, and W. T. Vetterling, *Numerical Recipes in C* (Cambridge University Press, Cambridge, 1988).

[12] W. K. Lai, K.-A. Suominen, B. M. Garraway, and S. Stenholm, Phys. Rev. A **47**, 4779 (1993).

[13] K.-A. Suominen, B. M. Garraway, and S. Stenholm, Phys. Rev. A **45**, 374 (1992).

Many-Body Theory of Atom Optics

G. Lenz and P. Meystre

Optical Sciences Center, University of Arizona, Tucson, AZ 85721, USA

We present a many-body theory for a system of N identical two-level atoms interacting with classical and quantum mechanical electromagnetic fields. We derive a master equation for the density operator. From that master equation we derive a BBGKY-hierarchy for the reduced density matrices, and show that the reduced one-particle density matrix contains a new term in the form of a complex potential.

1 Introduction

The study of the physics of ultra-cold atoms opens up a broad new field of fundamental and applied research. It plays an essential role in understanding the fundamental limitations of atomic cooling, and hence of devices such as atomic clocks. The many-body effects that are predicted to occur when the atomic temperature is near the recoil limit are bound to lead to shift and broadening mechanisms, and possibly nonlinear heating, thereby influencing the achievable performance of these devices. Many-body effects are expected to be important in lithographic applications as well. In addition to these immediate practical concerns, the availability of ultra-cold atoms is of considerable interest in atom interferometry, since it naturally leads to the possibility of coherently splitting atomic wave functions with a large separation between the arms of the interferometer. Cold atoms also have profound implications in the physics of collisions, since they lead us to a regime where standard approximations cease to be valid [1].

The net effect of any two-body interaction $V_{12}(r)$ is given by its overlap with the atomic wave functions, properly symmetrized in the case of indistinguishable particles. For short range interactions, a sizeable overlap between the atomic wave functions, and hence a high atomic density, is necessary to achieve noticeable effects, but in the case of the long range dipole-dipole interaction between coherently driven atoms, many-body effects can be expected already at relatively modest densities. As we shall see, manybody effects can lead to an effective nonlinear behaviour of the single atoms. This regime, that we call "nonlinear atom optics" [2], is to de Broglie optics what nonlinear optics is to conventional optics. It suggests that novel effects such as the generation of atomic solitons [2, 3] and solitary waves, atomic wave mixing, and atomic phase conjugation could be possible. In addition, sufficiently high atomic densities in

Springer Proceedings in Physics, Vol. 77 Quantum Optics VI
Editors: D.F. Walls · J.D. Harvey

optical lattices could lead to spontaneous pattern formation and the creation of other stable long-range structures, such as possibly super-lattices. Even more intriguing perhaps, the combination of nonlinearities with dissipation, in the form of spontaneous emission, and of a "pump" mechanism, suggests the possibility of a "coherent atomic beam generator", which could losely be considered as the atom optics equivalent of a laser.

In this paper, we consider the many-body description of an ensemble of N atoms interacting with an electromagnetic field, e.g. a single standing-wave laser mode acting as a diffraction grating for atoms. We derive a master equation for the N-particle system. We then construct a BBGKY-hierarchy for the reduced density matrices. We break the hierarchy after the equation for the one-particle reduced density matrix. The resulting equations are time-dependent Hartree-Fock equations modified due by the presence of a damping term of the Lindblad form.

2 Model

The starting point of our model is the multipolar Hamiltonian of an atom in an electromagnetic field, which is obtained by performing a Power-Zienau transformation on the minimal-coupling Hamiltonian [4, 5]

$$\hat{H} = \sum_{i,\alpha_i} \frac{1}{2m_{\alpha_i}} \left(\hat{p}_{\alpha_i} - \frac{q_{\alpha_i}}{c}\hat{\mathcal{A}}(\hat{r}_{\alpha_i})\right)^2 + \frac{1}{8\pi}\int dr(\hat{\mathcal{B}}^2(r) + \hat{\mathcal{E}}_\perp^2(r))$$

$$+ \frac{1}{2}\int dr d\acute{r} \frac{\hat{q}(r)\hat{q}(\acute{r})}{|r-\acute{r}|}, \tag{1}$$

where the index i runs over the atoms and α_i over its constituents. We assume charge neutrality, $\sum_{\alpha_i} q_{\alpha_i} = 0$. The charge density is $\hat{q}(r) = \sum_{i,\alpha_i} q_{\alpha_i}\delta(r - \hat{r}_{\alpha_i})$, and we assume that it makes sense to talk about individual atoms. Effects of electron exchange can not be handled with the formalism developed in the following, because these processes are excluded in the setup. The Power-Zienau transformation is defined by the unitary operator

$$\begin{aligned}
\hat{U} &:= \exp\left[-\frac{i}{\hbar c}\int dr \hat{P}(r)\cdot\hat{\mathcal{A}}(r)\right], \\
-\nabla\cdot\hat{P}(r) &:= \hat{q}(r), \\
\hat{P}(r) &= \sum_{i,\alpha_i} q_{\alpha_i}(\hat{r}_{\alpha_i} - R_i)\int_0^1 d\lambda\delta(r - R_i - \lambda(\hat{r}_{\alpha_i} - R_i)), \\
&=: \sum_i \hat{P}_i(r).
\end{aligned} \tag{2}$$

The new Hamiltonian is given by $\hat{\acute{H}} = \hat{U}\hat{H}\hat{U}^{-1}$, since $\hat{U}$ is not explicitly time-dependent. This means that we fix the reference configuration $\{R_i\}$ of the atoms. If we don't do this all kinds of currents occur [6], but for the systems that we consider, currents play no role at all. It is easy to prove the following relations :

$$\hat{U}\left(\frac{\hat{p}_{\alpha_i}}{m_{\alpha_i}} - \frac{q_{\alpha_i}}{m_{\alpha_i}c}\hat{A}(\hat{r}_{\alpha_i})\right)\hat{U}^{-1} = \frac{\hat{p}_{\alpha_i}}{m_{\alpha_i}}$$

$$+\frac{q_{\alpha_i}}{m_{\alpha_i}c}\int_0^1 d\lambda\lambda(\hat{r}_{\alpha_i} - R_i) \times \hat{\mathcal{B}}(R_i + \lambda(\hat{r}_{\alpha_i} - R_i)),$$

$$\hat{U}\hat{\mathcal{E}}_\perp(r)\hat{U}^{-1} = \hat{\mathcal{E}}_\perp(r) - 4\pi\hat{P}_\perp(r). \tag{3}$$

In these two equations, the right-hand sides are the representations of the velocity of particle α_i and of the transverse electric field, respectively. These physical quantities are represented in the old representation by the operators between the transformation matrices $\hat{U}$,$\hat{U}^{-1}$ on the left-hand side. Hence, it is clear that by writing $\hat{p}_{\alpha_i}$ and $\hat{\mathcal{E}}_\perp$ on both sides of Eq. (3) we mean the same mathematical operator, but that this operator does not describe the same physical quantity. Consequently, in the new representation we will continue to write $\hat{p}_{\alpha_i}/m_{\alpha_i}$ and $\hat{\mathcal{E}}_\perp(r)$, with the understanding that these operators belong to the conjugate momentum of x_{α_i} and $A(r)$, respectively (and *not* to the velocity v_{α_i} and the transverse part of the electric field $E_\perp(r)$).

The Power-Zienau transformation yields the Hamilton operator :

$$\hat{H} = \sum_i(\hat{H}_i + \hat{V}_i) + \frac{1}{2}\sum_{i\neq j}\hat{V}_{ij} + \hat{\mathcal{H}}_f, \tag{4}$$

where

$$\begin{aligned}
\hat{H}_i &= \sum_{\alpha_i}\frac{1}{2m_{\alpha_i}}\left(\hat{p}_{\alpha_i} + \frac{q_{\alpha_i}}{c}\int_0^1 d\lambda\lambda(\hat{r}_{\alpha_i} - R_i) \times \hat{\mathcal{B}}(R_i + \lambda(\hat{r}_{\alpha_i} - R_i))\right)^2 \\
&\quad +2\pi\int dr\hat{P}^2_{\perp,i}(r) + \frac{1}{2}\int drd\acute{r}\frac{\hat{q}_i(r)\hat{q}_i(\acute{r})}{|r-\acute{r}|}, &(5)\\
\hat{V}_i &= -\int dr\hat{P}_i(r)\cdot\hat{\mathcal{E}}_\perp(r), &(6)\\
\hat{V}_{ij} &= 4\pi\int dr\hat{P}_{\perp,i}(r)\cdot\hat{P}_{\perp,j}(r) + \int drd\acute{r}\frac{\hat{q}_i(r)\hat{q}_j(\acute{r})}{|r-\acute{r}|}, \\
&= 4\pi\int dr\hat{P}_i(r)\cdot\hat{P}_j(r), &(7)\\
\hat{\mathcal{H}}_f &= \frac{1}{8\pi}\int dr(\hat{\mathcal{B}}^2(r) + \hat{\mathcal{E}}^2_\perp(r)). &(8)
\end{aligned}$$

In Eq. (7) we have used the fact that $-\nabla\cdot\hat{P}_i(r) = \hat{q}_i(r)$ and

$$\frac{\nabla_{i,r}\nabla_{j,\acute{r}}}{4\pi}\frac{1}{|r-\acute{r}|} = \delta^{\|}_{ij}(r-\acute{r}).$$

After dropping all magnetic multipole terms, we replace Eq. (5) by a two-level system. In addition, Eq. (6) is split into a part containing the classical electromagnetic fields and one containing the quantum mechanical fields. The first part, together with Eq. (7), is retained in $\hat{\mathcal{H}}_s$. The polarization of atom i is then replaced by $\hat{P}_i(r) = (\mathbf{d}_{12}\sigma^+ + \mathbf{d}_{12}^*\sigma^-)\delta(r - \hat{r}_i)$, so that the contact interaction reads, for our two-level systems :

$$\hat{V}_{12} = 3\pi\hbar\gamma_0(\hat{\sigma}^+ \otimes \hat{\sigma}^- + \hat{\sigma}^- \otimes \hat{\sigma}^+)\delta(|k_0|(\hat{r}_1 - \hat{r}_2)). \tag{9}$$

Note that in this equation we have dropped nonresonant terms to be consistent with the Rotating Wave Approximation that will later on be performed. We also introduced the spontaneous decay rate $\gamma_0 = 4|d_{12}|^2k_0^3/3$, and took into account the fact that the delta function is a homogeneous function, $\delta(r) = c^{-dim}\delta(cr)$. In this manner, we obtained a form of the contact interaction that depends on the same dimensionless parameter k_0r, and has the same scale $\hbar\gamma_0$, as interactions that will appear in the following.

No instantaneous interaction appears in that approach. Rather, all interactions are mediated by retarded fields and the contact term. Hence, we deal from the beginning with an interacting system. However, the following calculations are not harmed by this or any other type of interaction, provided that the basic electronic transition is not significantly changed by it.

We conclude this section by introducing a few useful definitions and notations. First, we note that in the following, *single*- or *two*-particle operators in the Hilbert space of fixed particle number are indicated by a hat, e.g. $\hat{a}, \hat{b}, \ldots, \hat{A}, \hat{B}, \ldots$ Calligraphic letters $\hat{\mathcal{A}}, \hat{\mathcal{B}}, \ldots$ are reserved for operators in Fock space. Also, we introduce the positive frequency component of the "electromagnetic field" :

$$\hat{\mathcal{E}}_\perp^+(r) = \sum_{\{\kappa\}} i\epsilon(\kappa)\hat{a}_\kappa e^{ik\cdot r}, \quad \epsilon(\kappa) = \sqrt{\frac{\hbar\omega_k 2\pi}{V}}\epsilon_i(k), i = 1, 2, \tag{10}$$

κ being a compound index containing both the polarization i and wave vector k of a field mode, and $\epsilon_i(k)$ being the usual polarization vectors. In second quantization, the operator for the dipole moment of the system is :

$$\hat{\mathcal{P}}(r) = \int d\{i\}\langle 1|\hat{d}(r)|2\rangle\hat{\Psi}^\dagger(1)\hat{\Psi}(2), \quad \hat{d}(r) = (\vec{d}_{12}\hat{\sigma}^+ + \vec{d}_{12}^*\hat{\sigma}^-)\delta(r - \hat{r}). \tag{11}$$

When "integrating over numbers" such as $d1, d2$, we mean a summation over a complete set of quantum numbers for the single particle system under consideration. In our case, this could be e.g. a spin and coordinate, or a quasi-momentum and band index. The dummy index r is, to put it formally, the spectrum of $\hat{r}$, so it is a one-, two- or three-dimensional vector depending on the experimental configuration. The wave vector k is of course always three-dimensional. This has to be considered when discussing recoil effects. As already advertised, we perform a Rotating Wave Approximation in the Hamiltonian. In the Rotating Wave Approximation, the dipole term reduces to :

$$\hat{\mathcal{H}}_{sf} = -\int dr \int d\{i\} \langle 1|\hat{\sigma}^+(r)|2\rangle \hat{\Psi}^\dagger(1)\hat{\Psi}(2)\hat{\mathcal{E}}_\perp^+(r) + h.\,c.\,. \tag{12}$$

For later reference one should keep in mind that in writing the interaction in this way the polarization vectors in the operator for the electromagnetic field $\hat{\mathcal{E}}_\perp^\pm$ are always multiplied by *the* dipole moment $\mathbf{d}_{12}$ of the particles, and the operator $\hat{\sigma}^\pm(r)$ is given by $\hat{\sigma}^\pm \otimes \delta(r - \hat{r})$. Finally, we introduce in an obvious notation a spin operator $\hat{\mathcal{S}}(r)$.

3 The master equation

The coupling of the system to the electromagnetic field is eliminated in the Born-Markov approximation [7]. In the interaction representation with respect to $\hat{\mathcal{H}}_s + \hat{\mathcal{H}}_f$ this procedure yields the Liouville equation for the density operator of the system

$$\frac{d\hat{\varrho}_s(t)}{dt} = -\frac{1}{\hbar^2}\int_0^t d\tau Tr_f \left\{ [\hat{\mathcal{H}}_{sf}(t), [\hat{\mathcal{H}}_{sf}(\tau), \hat{\varrho}_{sf}(\tau)]] \right\}, \tag{13}$$

where the operators in $\hat{\mathcal{H}}_{sf}$ carry the time dependence of their free fields. The operator $\hat{\varrho}$ is the atomic density operator in Fock space. We assume as usual the factorization of $\hat{\varrho}(\tau)_{sf}$ into $\hat{\varrho}_f^{equi} \otimes \hat{\varrho}_s(\tau)$, thereby being left with free-field expectation values.

The integral over τ can be expanded to infinity because the decay of the correlations of the free field has a much shorter time scale than the dynamics of the system itself. We are then left with the half-sided Fourier transforms of the field correlation functions

$$\begin{aligned} \hbar^2(\Gamma(r_{12}) + i\Delta(r_{12})) &=: \int_0^\infty d\tau \langle \hat{\mathcal{E}}^-(r_1,t)\hat{\mathcal{E}}^+(r_2,t-\tau)\rangle e^{-i\omega_o\tau}, \\ \hbar^2(\gamma(r_{12}) + i\delta(r_{12})) &=: \int_0^\infty d\tau \langle \hat{\mathcal{E}}^+(r_2,t-\tau)\hat{\mathcal{E}}^-(r_1,t)\rangle e^{-i\omega_o\tau}, \end{aligned} \tag{14}$$

where $r_{12} = |\mathbf{r}_1 - \mathbf{r}_2|$. Note that the frequency ω_0 appears in these correlation functions because we left the interaction representation and must account for the fact that the operators $\hat{\mathcal{S}}(r_{1/2})$ are to be taken at different times t, τ. Here one basic assumption is made, namely that the "negative" frequency part of $\hat{\mathcal{S}}$, $\hat{\mathcal{S}}^+$ has only one narrow component, around ω_0, compared to the "flat" vacuum

The explicit form of the field correlation functions is

$$\Gamma(r_{12})/n(\omega_0) = \gamma(r_{12})/(n(\omega_0) + 1) =$$

$$\frac{\gamma_0}{2}\left\{j_0(k_0|r_{12}|) - \frac{1}{2}\left(1 - 3(e(r_{12})\cdot e(d_{12}))^2\right) j_2(k_0|r_{12}|)\right\}, \tag{15}$$

$$\left\{\begin{array}{c}\Delta(r_{12})\\ \delta(r_{12})\end{array}\right\} = \int_0^\infty d\omega\mathcal{P}\frac{1}{\omega-\omega_0}\left\{\begin{array}{c}n(\omega)\\ n(\omega)+1\end{array}\right\}\frac{\omega^3}{\omega_0^3}$$

$$\times\frac{\gamma_0}{2\pi}\left\{j_0(\omega|r_{12}|/c) - \frac{1}{2}\left(1 - 3(e(r_{12})\cdot e(d_{12}))^2\right) j_2(\omega|r_{12}|/c)\right\}. \tag{16}$$

Here j_i are the modified spherical Bessel functions of the first kind [8] and $e(\mathbf{x})$ is the unit vector in direction $\mathbf{x}$ and $\mathcal{P}(1/x)$ stands for the principal value.

We still need to carry out some reordering in order to identify the many-body contributions to the master equation as clearly as possible. This yields finally

$$\frac{d\hat{\varrho}}{dt} = -\frac{i}{\hbar}[\hat{\mathcal{H}},\hat{\varrho}]$$

$$-\frac{i}{\hbar}\int dr_1 dr_2 \frac{V_{dip}(r_{12})}{2}[\hat{\mathcal{S}}^+(r_1)\hat{\mathcal{S}}^-(r_2) + \hat{\mathcal{S}}^-(r_1)\hat{\mathcal{S}}^+(r_2),\hat{\varrho}]$$

$$+\frac{i}{\hbar}\int dr\frac{\hbar(\Delta(0)+\delta(0))}{2}[\hat{\Psi}^\dagger(r,e)\hat{\Psi}(r,e) - \hat{\Psi}^\dagger(r,g)\hat{\Psi}(r,g),\hat{\varrho}]$$

$$-\int dr_1 dr_2 \gamma(r_{12})$$

$$\times[\hat{\mathcal{S}}^+(r_1)\hat{\mathcal{S}}^-(r_2)\hat{\varrho} + \hat{\varrho}\hat{\mathcal{S}}^+(r_1)\hat{\mathcal{S}}^-(r_2) - 2\hat{\mathcal{S}}^-(r_2)\hat{\varrho}\hat{\mathcal{S}}^+(r_1)]$$

$$-\int dr_1 dr_2 \Gamma(r_{12})$$

$$\times[\hat{\mathcal{S}}^-(r_1)\hat{\mathcal{S}}^+(r_2)\hat{\varrho} + \hat{\varrho}\hat{\mathcal{S}}^-(r_1)\hat{\mathcal{S}}^+(r_2) - 2\hat{\mathcal{S}}^+(r_2)\hat{\varrho}\hat{\mathcal{S}}^-(r_1)], \tag{17}$$

where the term V_{dip} is the dipole force between excited atoms. Its explicit form is

$$\frac{V_{dip}(r_{12})}{2} := \left(\frac{\hbar\Delta(r_{12})}{2} - \frac{\hbar\delta(r_{12})}{2}\right)$$

$$= \frac{\hbar\gamma_0}{4}\left\{y_0(k_0|r_{12}|) - \frac{1}{2}\left(1 - 3(e(r_{12})\cdot e(d_{12}))^2\right) y_2(k_0|r_{12}|)\right\}. \tag{18}$$

Here y_i are the modified spherical Bessel functions of the second kind [8].

4 BBGKY hierarchy

Experiments usually measure expectation values of two-particle operators. The idea of a BBGKY-hierarchy [9] is to obtain equations of motion for reduced density matrices that contain only the expectation values of low-number particle operators. We proceed by defining a density operator in the N-particle subspace of the Fock space as

$$\hat{\varrho} = \int \frac{d^N\{i\}d^N\{i'\}}{N!}\rho(1,\ldots,N;1',\ldots,N') \prod_{i=N,\ldots,1} \hat{\Psi}^{\dagger}(i)|0\rangle\langle 0| \prod_{i=1',\ldots,N'} \hat{\Psi}(i'), \tag{19}$$

the norm being set by

$$Tr\hat{\varrho} = \int d^N\{i\}\rho(1,\ldots,N;1,\ldots,N) := 1. \tag{20}$$

With this definition, expectation values of operator products are in an obvious way related to reduced density matrices. We need only the two lowest ones,

$$\begin{aligned} \langle\hat{\Psi}^{\dagger}(b)\hat{\Psi}(a)\rangle &=: N\rho(a;b) \\ \langle\hat{\Psi}^{\dagger}(d)\hat{\Psi}^{\dagger}(c)\hat{\Psi}(a)\hat{\Psi}(b)\rangle &=: N(N-1)\rho(a,b;c,d). \end{aligned} \tag{21}$$

Our first goal is to derive an equation of motion for the single particle density matrix $\rho(a,b) =: \langle a|\hat{\rho}|b\rangle$. This equation clearly couples to the two-particle density matrix $\rho(a,b;c,d)$, and so on. In this way, we obtain a BBGKY-hierarchy, in complete analogy to classical physics. The new element here is the damping part in the basic N-particle equation that we start from to derive the hierarchy.

The general structure of the Liouville equation can be summarized in the form

$$\begin{aligned} \frac{d\hat{\varrho}}{dt} &= -\frac{i}{h}[\hat{\mathcal{H}}+\hat{\mathcal{V}},\hat{\varrho}] + \mathcal{L}\hat{\varrho}, \\ \hat{\mathcal{H}} &= \int d\{i\}\langle 1|\hat{H}|2\rangle\hat{\Psi}^{\dagger}(1)\hat{\Psi}(2), \\ \hat{\mathcal{V}} &= \int d\{i\}\frac{1}{2}\langle 1,2|\hat{V}|3,4\rangle\hat{\Psi}^{\dagger}(1)\hat{\Psi}^{\dagger}(2)\hat{\Psi}(4)\hat{\Psi}(3), \\ \mathcal{L}\hat{\varrho} &= -\oint \frac{d\Omega(k)}{4\pi}\frac{3\gamma_0}{4}n(\omega_0)[(\hat{\mathcal{S}}^+(k)\hat{\mathcal{S}}^-(k)+\hat{\mathcal{S}}^-(k)\hat{\mathcal{S}}^+(k))\hat{\varrho} \\ &+\hat{\varrho}(\hat{\mathcal{S}}^+(k)\hat{\mathcal{S}}^-(k)+\hat{\mathcal{S}}^-(k)\hat{\mathcal{S}}^+(k)) - 2(\hat{\mathcal{S}}^-(k)\hat{\varrho}\hat{\mathcal{S}}^+(k)+\hat{\mathcal{S}}^+(k)\hat{\varrho}\hat{\mathcal{S}}^-(k))] \\ &-\oint \frac{d\Omega(k)}{4\pi}\frac{3\gamma_0}{4}[\hat{\mathcal{S}}^+(k)\hat{\mathcal{S}}^-(k)\hat{\varrho}+\hat{\varrho}\hat{\mathcal{S}}^+(k)\hat{\mathcal{S}}^-(k) - 2\hat{\mathcal{S}}^-(k)\hat{\varrho}\hat{\mathcal{S}}^+(k)], \\ \hat{\mathcal{S}}(k) &= \int d\{i\}\langle 1|\hat{\sigma}^{\pm}(k)|2\rangle\hat{\Psi}^{\dagger}(1)\hat{\Psi}(2), \\ \hat{\sigma}^{\pm}(k) &= \sqrt{1-(e(k)\cdot e(d_{12}))^2}\int dr\hat{\sigma}^{\pm}(r)e^{\pm ik\cdot r}. \end{aligned} \tag{22}$$

For $\rho(a;b)$ this yields the equation of motion :

$$\frac{d\rho(a;b)}{N dt} = \frac{i}{\hbar}Tr\{[\hat{\mathcal{H}}, \hat{\Psi}^\dagger(b)\hat{\Psi}(a)]\hat{\varrho}\} + \frac{i}{\hbar}Tr\{[\hat{\mathcal{V}}, \hat{\Psi}^\dagger(b)\hat{\Psi}(a)]\hat{\varrho}\} + Tr\{\hat{\Psi}^\dagger(b)\hat{\Psi}(a)\mathcal{L}\hat{\varrho}\}. \tag{23}$$

We now successively discuss the three parts A, B and C on the right-hand side of this equation. The easiest part is of course the first contribution,

$$A = \frac{i}{\hbar}Tr\{[\hat{\mathcal{H}}, \hat{\Psi}^\dagger(b)\hat{\Psi}(a)]\hat{\varrho}] = -N\frac{i}{\hbar}\langle a|[\hat{H}, \hat{\rho}]|b\rangle, \tag{24}$$

where $\hat{\rho}$ denotes the single-particle density operator.

After some reordering of Eq. (17), we get for part B

$$B = \frac{i}{\hbar}Tr\{[\hat{\mathcal{V}}, \hat{\Psi}^\dagger(b)\hat{\Psi}(a)]\hat{\varrho}\} = -N(N-1)\frac{i}{\hbar}\langle a|Tr_2\{[\hat{V}, \hat{\rho}_2]\}|b\rangle, \tag{25}$$

where $\hat{V}$ stands for the dipole- and contact-interaction and $\hat{\rho}_2$ is the two-particle density matrix as defined in Eq. (21).

Finally, some reshuffling is also necessary for term C in order to separate one- and two-particle contributions. This gives

$$\begin{aligned} C = \; - \; & N\langle a| \{ \oint \frac{d\Omega(k)}{4\pi}\frac{3\gamma_0}{4} n(\omega_0) \{ \\ & (\hat{\sigma}^+(k)\hat{\sigma}^-(k) + \hat{\sigma}^+(k)\hat{\sigma}^-(k))\hat{\rho} + \hat{\rho}(\hat{\sigma}^+(k)\hat{\sigma}^-(k) + \hat{\sigma}^+(k)\hat{\sigma}^-(k)) \\ - \; & 2(\hat{\sigma}^-(k)\hat{\rho}\hat{\sigma}^+(k) + \hat{\sigma}^+(k)\hat{\rho}\sigma^-(k))\} \\ + \; & \oint \frac{d\Omega(k)}{4\pi}\frac{3\gamma_0}{4} \{\hat{\sigma}^+(k)\hat{\sigma}^-(k)\hat{\rho} + \hat{\rho}\hat{\sigma}^+(k)\hat{\sigma}^-(k) - 2\hat{\sigma}^-(k)\hat{\rho}\hat{\sigma}^+(k)\} \\ + \; & (N-1)\frac{i}{\hbar}Tr_2[\hat{V}_c, \hat{\rho}_2] \} |b\rangle. \end{aligned} \tag{26}$$

The imaginary potential $\hat{V}_c$ is

$$\begin{aligned} \hat{V}_c = \frac{i\hbar\gamma_0}{2} & \left\{ j_0(k_0|\hat{r}_{12}|) - \left(\frac{1}{2} - \frac{3(e(d_{12})\cdot e(r_{12}))^2}{2}\right) j_2(k_0|\hat{r}_{12}|)\right\} \\ & \times \{\hat{\sigma}^- \otimes \hat{\sigma}^+ - \hat{\sigma}^+ \otimes \hat{\sigma}^-\} \end{aligned} \tag{27}$$

and describes excitation and decay processes which are not locally confined. For instance, a particle that decays at x needs not to reappear at x in its lower electronic level. It should be noted that in Eq. (26) $\hat{V}_c$, is **not** to be taken as a symmetrized matrix element.

5 Hartree-Fock Approximation

The easiest way to close the BBGKY hierachy after the single particle density matrix self-consistently and to keep the influence of the particle statistics is the

Hartree-Fock approximation :

$$N(N-1)\rho(a,b;c,d) = N^2(\rho(a;c)\rho(b;d) - \eta\rho(a;d)\rho(b;c)), \tag{28}$$

where $\eta = \pm 1$ for fermions and bosons, respectively. We need to determine how the terms B and C of the preceding Section, which contain terms proportional to the two-particle density matrix, change under the Hartree-Fock approximation. First we look at the dipole interaction term. From Eq. (25) we see that the essential term is :

$$\begin{aligned} & \frac{1}{2}N(N-1)(\{a:1|\hat{V}|2:3\}\rho(2,3;b,1) - \rho(a,3;1,2)\{1:2|\hat{V}|b:3\}) \\ = \; & N^2\Big\{\langle a:1|\hat{V}|2:3\rangle(\rho(2;b)\rho(3;1) - \eta\rho(3;b)\rho(2;1)) \\ & -(\rho(a;1)\rho(3;2) - \eta\rho(a;2)\rho(3;1))\langle 1:2|\hat{V}|b:3\rangle\Big\}. \end{aligned} \tag{29}$$

Here, we have introduced the symmetrized matrix element $\{a:1|\hat{V}|2:3\} = \langle a:1|\hat{V}|2:3\rangle - \eta\langle a:1|\hat{V}|3:2\rangle$. The corresponding term in Eq. (26) may be expressed as

$$\begin{aligned} & \frac{1}{2}N(N-1)(\langle a:1|\hat{V}_c|2:3\rangle\rho(2,3;b,1) - \rho(a,3;1,2)\langle 1:2|\hat{V}_c|b:3\rangle) \\ = \; & \frac{1}{2}N^2\Big\{\langle a:1|\hat{V}_c|2:3\rangle(\rho(2;b)\rho(3;1) - \eta\rho(2;1)\rho(3;b)) \\ & -(\rho(a;1)\rho(3;2) - \eta\rho(a;2)\rho(3;1))\langle 1:2|\hat{V}_c|b:3\rangle\Big\}, \end{aligned} \tag{30}$$

which clearly shows that in the Hartree-Fock approximation, the interaction part of the Lindblad form gives a self-consistent **complex** potential.

We have now unveiled the full Liouvillian of our model and closed the chain of equations for the reduced density matrices by performing a Hartree-Fock approximation. In a previous publication [2], we presented and discussed some implications on the *coherent* part of the Hartree-Fock effective one-particle Liouvillian, neglecting all incoherent decay mechanisms. Internal consistency of the theory required that we also ignore the complex potential at this point in this work, since it has the same origin as the damping part in the effective single-particle Liouvillian. We are now in the process of numerically investigating the full Born-Markov-Hartree-Fock master equation of nonlinear atom optics. This work will be the object of a future publication.

This research is funded by the U.S. Office of Naval Research Contract No. N00014-91-J1205, by the National Sciences Foundation Grant PHY92-13762, and by the Joint Services Optics Program. Fruitful discussions with E. M. Wright and M. Wilkens are greatfully acknowledged.

References

[1] P. S. Julienne and F. H. Mies, *J. Opt. Soc. Am B* **6**, 2257 (1989).

[2] G. Lenz, P. Meystre and E. M. Wright, *Phys. Rev. Lett.* **71**, 3271 (1993).

[3] Weiping Zhang, D. F. Walls and B. C. Sanders, *Phys. Rev. Lett.* **72**, 60 (1994).

[4] C. Cohen-Tannoudji, J. Dupont-Roc and G. Grynberg, *Atom-Photon Interactions, Basic Processes and Applications* (Wiley-Interscience, New-York, 1992).

[5] C. Cohen-Tannoudji, J. Dupont-Roc and G. Grynberg, *Photons and Atoms, Introduction to Quantum Electrodynamics* (Wiley-Interscience, New-York, 1989).

[6] E. A. Power and T. Thirunamachandran *Proc. R. Soc. Lond. A* **372**, 265, (1980).

[7] R. Graham and F. Haake, *Springer Tracts in Modern Physics Vol. 66* (Springer Verlag, Berlin, 1973).

[8] M. Abramowitz and I. A. Stegun, *Handbook of Mathematical Functions* (Dover, New York, 1972).

[9] L. E. Reichl, *A Modern Course in Statistical Physiscs* (Uni. Texas Press, Austin, 1980).

Bragg Scattering of a Coherent Ultracold Atomic Beam in Nonlinear Atom Optics

W. Zhang[1] *and D.F. Walls*[2]

[1]School of Mathematics, Physics, Computing and Electronics, Macquarie University, New South Wales 2109, Australia
[2]Physics Department, University of Auckland, Auckland, New Zealand

Abstract. We present a vector quantum field theoretic description for nonlinear atom optics of a coherent ultracold atomic beam in a standing wave laser field. In the Bragg resonance regime, the standing wave laser acts as a nonlinear atomic beam splitter. The atomic nonlinearity is due to the dipole-dipole interaction between ultracold atoms which leads to self-phase and cross-phase modulation and two-wave mixing of the scattered atomic waves in the Bragg resonance regime.

1. Introduction

In past few years, rapid progress was made in laser cooling and trapping of neutral atoms[1-5]. By cooling techniques of neutral atoms, one can reduce thermal motion of an atom and achieve a long thermal de Broglie wavelength. Such cold atoms provide necessary tools to study optical effects of atoms in atom optics[6].

On the other hand, the combination of optical cooling and atom traps introduces the possibility of producing ultracold atomic sources where bosonic atoms are expected to lose their individual identities and degenerately condense into a macroscopic single quantum state by Bose-Einstein condensation (BEC). Such atomic sources provide possible path to generate coherent ultracold atomic beams which are expected to have properties similar to the role of lasers in conventional coherent optics.

Although there are many limitations to Bose-Einstein condensation in an optical trap[7-8], in principle it is believed that such a phase transition can be realized by combination of different cooling and trapping techniques, for example, evaporative cooling, combination of optical cooling and magnetic traps. In this paper, our purpose is to develop a theoretic description for a coherent ultracold atomic beam from a BEC atomic source and the interaction of such an atomic beam with a laser field. Due to condensation of many bosonic atoms in a single momentum eigenmode in the beam, the effect of quantum statistics would be of crucial importantance. In this case, the existing single-atom description is no longer valid for interaction of a coherent ultracold atomic beam with a light wave. In order to take quantum statistics into account, a vector quantum field theory is developed for a coherent ultracold atomic beam composed of identical bosonic atoms. A vector stochastic Schrodinger equation is derived to describe the interaction of the coherent atomic beam with a laser field. As an application of the vector quantum field theory, we study the Bragg scattering of an coherent ultracold atomic beam by a standing wave laser in atom optics. In the Bragg resonance regime[9], the standing wave laser acts as a nonlinear atomic beam splitter for the coherent atomic beam. The atomic nonlinearity originates from two-body "ultracold collisions" due to dipole-dipole interaction between atoms which

Springer Proceedings in Physics, Vol. 77 Quantum Optics VI
Editors: D.F. Walls · J.D. Harvey

leads to self-phase and cross-phase modulation and two-wave mixing of the scattered atomic waves. A numerical simulation shows that these nonlinear effects result in a fully different spatial scattering patterns of a coherent ultracold atomic beam from those of a single-atom beam.

2. Vector quantum field theory of a coherent ultracold atomic beam

In this section, we introduce a vector quantum field theoretic description for interaction of a coherent ultracold atomic beam with a laser beam. The coherent atomic beam is described as a vector quantum field with two components ψ_1 and ψ_2 corresponding to the internal ground state and excited state of the atoms, respectively. The two-component vector quantum field operators and their Hermitian conjugates satisfy the Bose commutation relations. The interaction of the coherent atomic beam with a laser field of complex amplitude $\vec{E}^{(\pm)}$ can be described by the nonlinear stochastic Schrödinger equations[10]

$$i\hbar\frac{\partial\psi_1}{\partial t}=-\frac{\hbar^2\nabla^2}{2m}\psi_1-\vec{\mu}\cdot\vec{E}^{(-)}\psi_2+$$

$$i\hbar\int d^3r'L(\vec{r}-\vec{r}')\psi_2^{+}(\vec{r}',t)\psi_1(\vec{r}',t)\psi_2(\vec{r},t)+G_1(\vec{r},t)\quad, \tag{1a}$$

$$i\hbar\frac{\partial\psi_2}{\partial t}=-\frac{\hbar^2\nabla^2}{2m}\psi_2-\hbar(\Delta+i\gamma/2)\psi_2-\vec{\mu}\cdot\vec{E}^{(+)}\psi_1$$

$$-i\hbar\int d^3r'L(\vec{r}-\vec{r}')^{*}\psi_1^{+}(\vec{r}',t)\psi_2(\vec{r}',t)\psi_1(\vec{r},t)+G_2(\vec{r},t)\quad, \tag{1b}$$

where $\gamma=4\,|\vec{\mu}\,|^2\,\omega_L^3/3\hbar c^3$ is the spontaneous emission rate of a single atom and $\Delta=\omega_L-\omega_a$ the detuning of the laser frequency from atomic transition frequency. The noise terms G_1 and G_2 describe the vacuum fluctuations and the nonlinear coefficient is[10]

$$L(\vec{r}-\vec{r}')=\gamma[K(\vec{r}-\vec{r}')/2-iW(\vec{r}-\vec{r}')]$$

$$=\frac{3\gamma}{4}[i\xi^2\sin^2\theta+(1-3\cos^2\theta)(\xi-i)]\frac{\exp(-i\xi)}{\xi^3}\quad, \tag{2}$$

where $\xi=|\vec{k}_L\cdot(\vec{r}-\vec{r}')\,|$ and θ is the angle between the dipole moment $\vec{\mu}$ and the relative coordinate $\vec{r}-\vec{r}'$. The real part $\gamma K(\vec{r}-\vec{r}')$ of the coefficient $L(\vec{r}-\vec{r}')$ accounts for the dissipation of atomic quantum field due to collisions induced by many-atom spontaneous emission. The imaginary part $-\gamma W(\vec{r}-\vec{r}')$ corresponds to the dipole-dipole interaction energy which is the result of vacuum photon exchanges between the ultracold atoms during their internal transitions[10]. In terms of equation (2), the nonlinear coefficient $L(\vec{r}-\vec{r}')$ has a sharp peak when the interatom distance is close to the atomic resonance wavelength. Hence it determines a long-range correlation of ultracold atoms in the range of atomic resonance wavelengths.

On the other hand, according to quantum electrodynamics, the complex amplitudes $\vec{E}^{(\pm)}$ of the quantized laser field in equations (1) satisfy the quantum propagation equations[11-12]

$$\nabla^2\vec{E}^{(\pm)} - \frac{1}{c^2}\frac{\partial^2\vec{E}^{(\pm)}}{\partial t^2} = \mu_0\frac{\partial^2\vec{P}^{(\pm)}}{\partial t^2} \quad , \tag{3}$$

where $\vec{P}^{(+)} = [\vec{P}^{(-)}]^+ = \vec{\mu}\psi_1^+(\vec{r},t)\psi_2(\vec{r},t)e^{-i\omega_L t}$ is the positive frequency part of the polarization operator of the coherent ultracold atomic beam in interaction picture.

In equations (1), there are several important terms which affect the propagation of the coherent atomic beam in a laser beam. There is the linear dissipation proportional to γ and the nonlinear dissipation proportional to $\gamma K(\vec{r}-\vec{r}')$ induced by spontaneous emission. The vacuum fluctuations described by the operators $G_j (j=1,2)$ result in the diffusion of the center of mass momentum of the atoms. The dissipation and diffusion cause a loss of atoms and destroy the coherence of the atomic beam. Similarly, in equation (3), the polarization of the atoms $\vec{P}^{(\pm)}$ results in absorption and dispersion of laser photons which are proportional to the density of the atomic beam. As a result, the absorption and dispersion of laser photons cause a spatial distortion of laser beam which depends on the spatial density distribution of the atomic beam[11]. To reduce these effects, a large laser detuning is required so that only few atoms are in the excited state. Then one can neglect the dissipation and fluctuations from spontaneous emission and the spatial distortion from absorption and dispersion.

3. Bragg scattering in nonlinear atomic optics

Equations (1) ,(2) and (3) give a general formalism for interaction of ultracold atoms with a laser field. In this section, we apply the general formalism to the diffraction of a coherent atomic beam by a standing wave laser. In terms of the arguments given in section 2, we choose a laser frequency which is detuned far from the atomic resonance. Hence, the dissipation and fluctuations due to spontaneous emission, and the absorption and dispersion of laser photons are ignored at this stage. On the other hand, in the off-resonance regime, the excited-state component of the atomic quantum field in equations (1) can be adiabatically eliminated. This finally leads to a nonlinear Schrödinger equation for propagation of the ground-state component of the coherent atomic beam[11]

$$i\hbar\frac{\partial\psi_1}{\partial t} = \left[-\frac{\hbar^2\nabla^2}{2m} + \frac{\hbar\Omega_0^2}{4\Delta}\exp\left(-\frac{x^2+z^2}{W_L^2}\right)\cos^2(k_L y)\right]\psi_1$$
$$-\hbar\chi\exp\left(-\frac{x^2+z^2}{W_L^2}\right)\cos^2(k_L y)\psi_1^+\psi_1\psi_1 \tag{4}$$

where $\chi = \frac{\Omega_0^2\gamma V_c}{2\Delta^2}$ defines the two-body collision rate of ultracold atoms in the adiabatic regime of atomic motion and $V_c = |\int W(\vec{r})d^3\vec{r}\,|$ gives the effective scattering volume of atoms due to vacuum photon exchange interaction. In the derivation of equation (4), we have used the sharpness of the long-range vacuum photon exchange potential $W(\vec{r}-\vec{r}')$ in the region of atomic resonance wavelength. The laser field is oriented along the y-axis and has a transverse Gaussian intensity profile in the x and z direction. The parameter $\Omega_0 = 2\vec{\mu}\cdot\vec{E}_0/\hbar$ defines the peak Rabi frequency.

The physics implicit in equation (4) is very clear. The standing wave laser induces a periodic atomic potential along the y-axis. The nonlinear term depends on the density of the atomic beam and the two-body collision rate χ which plays a similar role to that of a nonlinear optical susceptibility of a Kerr-type medium in conventional nonlinear optics. The density of the atomic beam takes the place of the intensity of light in conventional optics. By analogy with conventional nonlinear optics, equation (4) is similar to that of a light wave propagating in a nonlinear periodic dielectric medium. Therefore we see that atoms and photons exchange their role in conventional nonlinear optics. In other words, the light wave just acts as a nonlinear "medium" for a coherent ultracold atomic beam in nonlinear atom optics.

To simplify our discussions, we study a standard example in atom optics where the incident atomic beam propagates along the z-axis with momentum vector $(p_{x0}=0, p_{y0}=\hbar K_{0y}, p_{z0}=\hbar K_{0z})$ and kinetic energy E. In this example, the width of the incident atomic beam in the x direction is assumed to be wider than that in the y-direction and narrower than the width of the laser beam so that one can ignore the propagation of the atomic beam in the x-direction. The incident atomic beam is provided by an ultracold atomic source where there are a large number of bosonic atoms condensed in a macroscopic quantum state. In this case, the mean field approximation is applicable[12] and the atomic quantum field operator for the ultracold atomic beam in equation (4) can be replaced by a c-number macroscopic wave function $\Phi(\vec{r})$.

When the atomic beam passes through the standing wave laser beam, the periodic structure results in diffraction of the incident atomic beam into many components in the y direction. However, if the momentum component of the incident atomic beam in the y-direction is arranged to match one single-photon recoil momentum, one can have the Bragg resonance. In this case, the standing wave laser just acts as an atomic beam splitter. Fig.1 is the schematic diagram for the diffraction of an atomic beam by a standing wave laser in the Bragg resonance condition.

In this paper, we assume the Bragg resonance condition satisfied. Then the atomic macroscopic wave function can be expanded in terms of the Bragg scattering waves as follows[11]

$$\Phi(\vec{r}) = \left(\Phi_0(y,z)e^{-ik_L y} + \Phi_1(y,z)e^{ik_L y}\right)e^{iK_{0z}z - iEt/\hbar} \quad . \tag{5}$$

Substituting the above equation into (4), we have the coupled wave equation for the Bragg scattering waves $\Phi_0(y,z)$ and $\Phi_1(y,z)$

$$i\left(\frac{\partial}{\partial\tau} - v_R\frac{\partial}{\partial y}\right)\Phi_0 = v(\tau)\Phi_0 + g(\tau)\Phi_1 + 2c(\tau)(\Phi_0^+\Phi_0 + 2\Phi_1^+\Phi_1)\Phi_0$$

$$+c(\tau)(2\Phi_0^+\Phi_0 + \Phi_1^+\Phi_1)\Phi_1 + c(\tau)\Phi_1^+\Phi_0^2 \quad , \tag{6a}$$

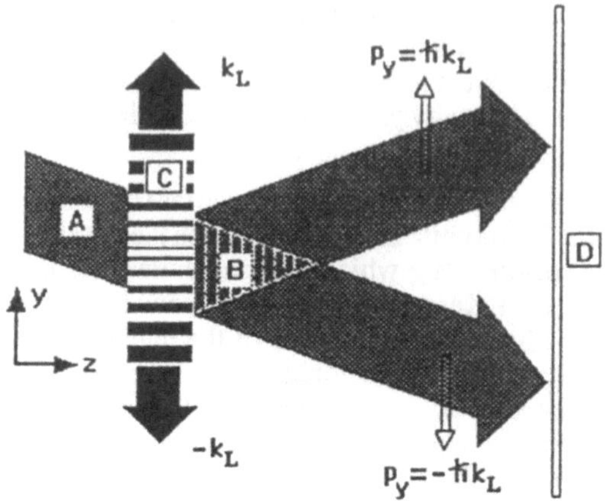

Fig. 1. The schematic diagram for Bragg scattering of an atomic beam by a standing wave laser beam: A denotes the incident atomic beam, B the interference region, C the standing wave laser and D the detection screen.

$$i\left(\frac{\partial}{\partial\tau}+v_R\frac{\partial}{\partial y}\right)\Phi_1 = v(\tau)\Phi_1 + g(\tau)\Phi_0 + 2c(\tau)(\Phi_1^+\Phi_1 + 2\Phi_0^+\Phi_0)\Phi_1$$

$$+c(\tau)(2\Phi_1^+\Phi_1 + \Phi_0^+\Phi_0)\Phi_0 + c(\tau)\Phi_0^+\Phi_1^2 \quad , \tag{6b}$$

where $\tau = v_g^{-1}z$ defines the effective time variable, $v_g = \frac{\hbar K_{0z}}{m}$ is the group velocity of the atomic beam in the z direction, $v_R = \frac{\hbar k_L}{m}$ the single-photon recoil velocity, $v(\tau) = 2g(\tau) = |\Omega_0|^2 \exp(-\tau^2/\tau_0^2)/8\Delta$ the frequency shift for the two Bragg scattering waves. The nonlinear coupling coefficient is denoted as $c(\tau) = -\frac{\chi}{4}\exp\left(-\frac{\tau^2}{\tau_0^2}\right) = -\beta g(\tau)$ with $\beta = \frac{2\chi V_c}{\Delta}$. The flight time of atoms through the laser beam with width W_L is $\tau_0 = W_L/v_g$.

Mathematically, equations (6) describe two-wave mixing in a nonlinear coupler. The well-known examples for a nonlinear coupler are the nonlinear beam splitter in conventional nonlinear optics and the optical coupler in nonlinear guidedwave optics. By analogy with nonlinear optics, the nonlinearity in equations (6) has a similar form to a Kerr-type optical nonlinearity and induces self-phase modulation (SPM) and cross-phase modulation (XPM) of atomic waves[13]. Optical self-phase modulation and cross-phase modulation are well-known nonlinear phenomena in nonlinear optics. To study the atomic version of these nonlinear phenomena in atom optics, we numerically simulate the propagation of an incident atomic beam in the Bragg resonance regime in the next section.

4. Numerical simulation for nonlinear Bragg scattering

In this section, we give the numerical analysis for the diffraction of an incident coherent ultracold atomic beam in the Bragg resonance regime. The incident atomic beam is assumed to have a Gaussian density profile with wave vector matching the single photon recoil momentum

$$\Phi_{in}(y,-\infty)=\sqrt{\rho_0}\exp\left(-\frac{y^2}{2w_y^2}-ik_L y\right), \tag{7}$$

where w_y is the beam width in the y direction and $\rho_0=J_0/v_g$ is the density of the atomic beam which is provided by an ultracold atomic source localized at $z=-\infty$ (see Fig.1). The atomic source continuously releases the ultracold atomic flow with the rate J_0 which has the equivalent meaning to the light intensity in conventional optics. In terms of equations (5) and (7), we have the initial conditions for equations (6)

$$\Phi_0(y,-\infty)=\sqrt{\rho_0}\exp\left(-\frac{y^2}{2w_y^2}\right), \tag{8a}$$

$$\Phi_1(y,-\infty)=0\ . \tag{8b}$$

Under conditions (8), we numerically simulate the spatial propagation of the incident atomic beam in terms of equations (5) and (6). The spatial density distributions are shown in Fig.2 and Fig.3. For comparisons, Fig.2 is the result for a low-density atomic beam with a negligible nonlinearity. The laser parameters in Fig.2 is chosen to satisfy the condition $g_0\equiv\int_{-\infty}^{\infty}g(\tau)d\tau=\pi/4$ so that the incident atomic beam splits into two components with identical density profiles. In terms of the results in Fig.2, when the incident atomic beam passes through the laser region, the interaction of atoms with the laser beam results in the splitting of the initial atomic beam into two coherent beams, respectively, with group velocities $\pm v_R$ in the y direction. The splitting of the

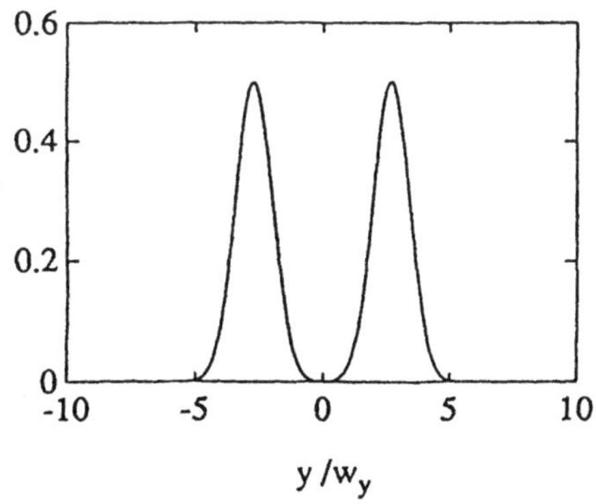

Fig. 2. The density distributions of the scattered atomic beams in Bragg resonance regime with a weak atomic nonlinearity $\beta\rho_0=0.001$. The coupling strength is $g_0=\pi/4$.

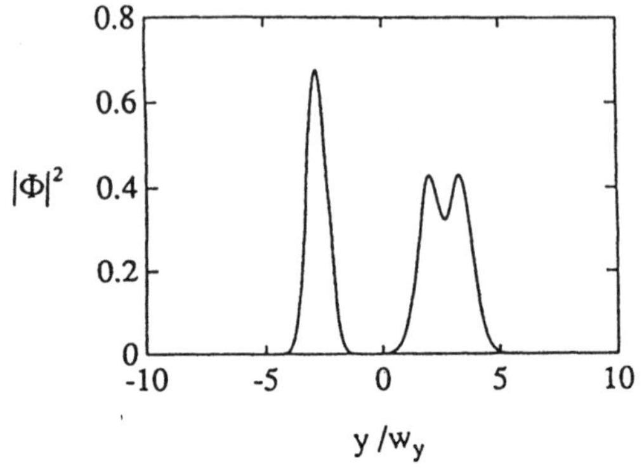

Fig. 3. The density distributions of the scattered atomic beams in Bragg resonance regime with a strong atomic nonlinearity $\beta\rho_0 = 0.3$. The coupling strength is $g_0 = \pi$.

atomic beam is due to the exchange of atoms between two diffraction waves. Such an exchange means that some atoms in the incident beam change their momentum in the y-direction. The change of atomic momentum is caused by the mechanical effect induced by the laser beam. The number of exchanged atoms between two diffraction waves depends on the linear coupling strength g_0.

However, for a coherent ultracold atomic beam with high density, the nonlinearity is important and the diffraction of atoms will be affected by it. Fig.3 shows the density distributions of the diffracted atomic waves for an incident coherent ultracold atomic beam. We find that including the atomic nonlinearity, the incident atomic beam coherently splits into two components and one of the split components exhibits two-peak structure. To understand the phenomena, we seek for the approximately analytical solutions of equations (6) by neglecting the spatial propagation of the beam in y direction. After a simple derivation, we have the approximate solutions of equations (6)

$$|\Phi_0(y,z)|^2 = |\Phi_{in}(y,-\infty)|^2 \cos^2(\theta(\tau) + \Delta\theta(y,\tau)) \tag{9a}$$

$$|\Phi_1(y,z)|^2 = |\Phi_{in}(y,-\infty)|^2 \sin^2(\theta(\tau) + \Delta\theta(y,\tau)) \tag{9b}$$

where $\theta(\tau) = \int_{-\infty}^{\tau} g(\tau)d\tau = \frac{g_0}{2}\left(1 + erf\left(\frac{\tau}{\tau_0}\right)\right)$ is the phase shift due to the linear coupling. The nonlinearity induces an additional phase shift $\Delta\theta(y,\tau) = -\beta\theta(\tau)\{|\Phi_{in}(y,-\infty)|^2 + \frac{1}{2}[\Phi_0^*(y,z)\Phi_1(y,z) + \Phi_1^*(y,z)\Phi_0(y,z)]\}$. In the far-field region $z \to \infty$, the phase shift due to linear coupling is a constant $\theta(\infty) = g_0$ and the nonlinear phase shift $\Delta\theta(y,\infty)$ is still a function of the spatial coordinate y and depends on the transverse distributions of the atomic beams. Hence the nonlinear phase shift is nonuniform in the y direction. Such a nonuniform phase shift modulates the spatial distributions of the split atomic beams and causes the multiple-peak structures in the density profiles. As an example, we consider the case shown in Fig.3

where $\theta(\infty)=g_0=\pi$, the far-field atomic densities of the diffracted beams have the form

$$|\Phi_0(y,z)|^2=|\Phi_{in}|^2\cos^2(\Delta\theta(y,\infty)), \tag{10a}$$

$$|\Phi_1(y,z)|^2=|\Phi_{in}|^2\sin^2(\Delta\theta(y,\infty)). \tag{10b}$$

For negligible nonlinearity, we have $|\Phi_0(y,\infty)|^2=|\Phi_{in}(y,-\infty)|^2$ and $|\Phi_1(y,\infty)|^2=0$. Hence the incident beam doesn't split in this case. However, it is evident that a large nonlinearity causes splitting of the incident beam and also causes a nonuniform spatial modulation of the split beams in terms of equations (10). Hence the nonuniform phase modulation due to atomic nonlinearity plays an important role in the propagation of a coherent ultracold atomic beams.

5. Conclusions

In this paper, we develop a systematic theory for the interaction of a coherent ultracold atomic beam with a laser field in the framework of the vector quantum field theory. A vector stochastic nonlinear Schrödinger equation for ultracold atoms is derived. These equations form the basis to study many-body quantum statistics and atomic nonlinearity in atom optics. As a straightforward application of the theory, we study the diffraction of an coherent ultracold atomic beam by a standing wave laser. We find that the vacuum photon exchanges between ultracold atoms in the beam induces an effective atomic nonlinearity for the atomic waves which is analogous to an optical Kerr-type nonlinearity for coherent light waves. Such an atomic nonlinearity can result in self-phase modulation and cross-phase modulation and two-wave mixing of atomic waves. In the Bragg resonance regime, we simulate the propagation of a coherent ultracold atomic beam. The atomic nonlinearity results in a spatial modulation of atomic density profiles for the Bragg scattering waves.

In addition, we wish to emphasize that the observation of the nonlinear diffraction of atomic beams requires that the incident atomic beam should be prepared in a coherent ultracold beam. In the languages of quantum statistical theory, the condition for a coherent ultracold atomic beam is that the single-mode degeneracy of bosonic atoms in the beam or the "brightness" of the atomic beam, $N_v=\rho_0\lambda_{dB}^3$, must be larger than one. Here $\rho_0=J_0/v_g$ is the peak density of the atomic beam and $\lambda_{dB}=\sqrt{2\pi\hbar^2/mk_BT}$ is its thermal de Broglie wavelength which determines the spatial coherence of the atomic beam. Under this condition, the thermal de Broglie wavelength λ_{dB} must be comparable with the average interatom distance $\rho_0^{-1/3}$. For vacuum photon exchange interaction which has a characteristic interatom distance close to the atomic resonance wavelength, the thermal de Broglie wavelength must achieve the order of the atomic resonance wavelength, ie., $\lambda_{dB}\sim\lambda_L$. In this case, the density of the atomic beam must meet the criterion $\rho_0\lambda_L^3>1$.

Finally we point out that although we only discuss the nonlinear diffraction in this paper, the quantum field theory developed here provides a general method which could be applied for wide problems in atom optics, quantum optics and interaction of light wave with condensed matter. For example, the atomic nonlinearity could result in generation of atomic soliton[12] and self-focussing of atomic beams[14] in laser beams. It is not an exaggeration that almost all nonlinear optical phenomena could find their versions in nonlinear atom optics in terms of the general formalism constructed in this paper. Particularly, the absorption and dispersion of laser photons lead to a complicated form of atomic nonlinearity which would induce new nonlinear phenomena in atom optics when absorption and dispersion are not negligible. In addition, the quantum field theory and the stochastic nonlinear Schrodinger equations present a simple way to study the ultracold collisions of atoms in a laser field, quantum statistics in ultracold regime and the critical condition for Bose-Einstein condensation in laser-cooled neutral atomic gases. These open a new window onto future research in laser cooling and atom optics.

Acknowledgements

We thank Dr B. C. Sanders for his helpful comments on the manuscript. WZ thanks the support of the Australian Research Council for Australian Research Fellow. DFW thanks the support of New Zealand Foundation of Research Science and Technology, the New Zealand Lotteries Grants Board, and the University of Auckland Research Committee.

References:

1. S. Chu, J. E. Bjorkholm, A. Ashkin and A. Cable, Phys. Rev. Lett. 57, 314 (1986)
2. M. Prentiss, A. Cable, J. E. Bjorkholm, S. Chu, E. Raab and D. E. Pritchard, Opt. Lett. 13, 452 (1988)
3. D. Sesko, T. Walker, C. Monroe, A. Gallagher and C. Wieman, Phys. Rev. Lett. 63, 961 (1989)
4. P. Lett, P. Jessen, W. Phillips, S. Rolston, C. Westbrook and P. Gould, Phys. Rev. Lett. 67, 2139 (1991)
5. C. Monroe, W. Swann, H. Robinson and C. Wieman, Phys. Rev. Lett. 65 1571 (1990)
6. See special issue on atom interferometry, Appl. Phys B 54 (1992)
7. T. Walker, D. Sesko, and C. Wieman, Phys. Rev. Lett. 64, 408 (1990)
8. D. Sesko, T. Walker, and C. Wieman, J. Opt. Soc. Am. B 8, 946 (1991)
9. P. Martin, B. Oldaker, A. Miklich and D. Pritchard Phys. Rev. Lett. 60, 515 (1988)
10. Weiping Zhang Phys. Lett. A 176, 53 (1993)
11. Weiping Zhang and D. F. Walls, Phys. Rev. A. (in press, 1994)
12. Weiping Zhang, D. F. Walls and B. C. Sanders, Phys. Rev. Lett. 72, 60 (1993)
13. Weiping Zhang, Proceedings of The Eighth Conference of The Australian Optical Society (Sydeny, 3-5 February) pT38 (1993)
14. Weiping Zhang, D. F. Walls and B. C. Sanders, (to be published).

Atom Optics in Quantized Light Fields

D.S. Krähmer[1], *A.M. Herkommer*[1], *E. Mayr*[1], *V.M. Akulin*[1,*], *I.Sh. Averbukh*[2], *T. van Leeuwen*[3], *V.P. Yakovlev*[1,**], *and W. Schleich*[1,***]

[1]Abteilung für Quantenphysik, Universität Ulm, D-89069 Ulm, Germany
[2]Department of Chemical Physics, Weizmann Institute of Science, Rehovot 76100, Israel
[3]Eindhoven University of Technology, Department of Physics, 5600 MB Eindhoven, The Netherlands

We consider the deflection and focusing of atoms in a quantized light field. We study the influence of spontaneous emission on the deflection pattern and propose a method to create narrow atomic wave packets. A possible experiment is suggested.

1 Introduction

In a majority of matter-light interactions it suffices to describe the electromagnetic field classically, even when the light is contained within an optical cavity. This is a consequence of the high average number of photons in the resonator and the fact that the light tends to be in a coherent state — the state closest to a classical state. The recent progress in creating experimentally non-classical states of the radiation field [1] allows to approach new questions: Direct observation of the quantization of the electromagnetic *energy* in a cavity would constitute a direct demonstration of the concept of the photon [2]. Determining the number of photons in a cavity also directly touches on the subject of quantum measurement: the cavity, usually assumed to be in a coherent state before the measurement, is forced in a "photon number state" by the measurement. The new subject of atom optics [3] combined with the physics of quantized light fields is a unique instrument to study in detail these fundamental questions. The dipole force acting on the atoms when passing through the cavity depends on the intensity of the light and hence on the number of photons in the cavity. Thus, the deflection pattern contains information on the field state in the cavity [4] — this is one important feature of this problem. A joint measurement [5, 6] of atomic momentum and field state after the scattering of the atom providing information about the initial field state is another important aspect. To resolve the discreteness of the light field via a quantum lens [7, 8]

*Also at: Moscow Institute of Physics and Technology, Dolgoprudny, Moscovskaja Obl., Russian Federation
**Also at: Moscow Engineering Physics Institute, Moscow 115409 and International Institute of Physics, Moscow, Russia
***Also at: Max-Planck-Institut für Quantenoptik, D-85748 Garching bei München, Germany

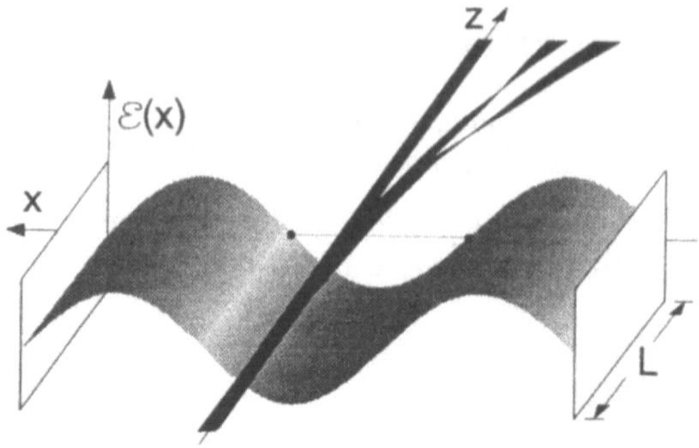

Figure 1: Schematic illustration of the atomic deflection and focusing. Different Fock states deflect the atomic beam at different angles and focus the beam at different points.

constitutes another interesting possibility offered by this marriage of atom optics and cavity quantum electrodynamics [9]. In the present paper we briefly review and highlight our work on this topic.

Figure 1 shows a schematic illustration of our model. An atomic beam with an intensity profile $f^2(x)$ propagating in the z direction enters a cavity with a single mode of a quantized electromagnetic field. The field is inhomogeneous in the x direction and therefore scatters the atoms. For the sake of simplicity the system is assumed to be uniform in y direction. At the entrance of the cavity, at $z = -L$, the state vector $|\Psi(z = -L)\rangle$ of the combined system is the direct product of the atomic state $|g\rangle$, the field state $\sum_n w_n|n\rangle$, and the state $\int dx\, f(x)|x\rangle$ of the atomic transverse motion, and $w_n \equiv \langle n|\psi\rangle$ denotes the probability amplitude of the photon number state $|n\rangle$ in the state $|\psi\rangle$ of the field. Hence the state vector $|\Psi\rangle$ reads

$$|\Psi\rangle = \sum_n w_n|n\rangle \otimes \int_{-\infty}^{\infty} dx\, f(x)\,|x\rangle \otimes |g\rangle. \tag{1}$$

We take the longitudinal kinetic energy $Mv_z^2/2$ to be large compared to the atom-light interaction energy and, hence, the interaction with the light field does not change considerably the velocity v_z of the atom. Therefore, the z coordinate plays the role of time $t = (z + L)/v_z$ in the evolution of the state vector corresponding to the transverse motion and the field. Consequently the dynamics of the atom-field system follows from the Hamiltonian

$$\hat{H} = \frac{\hat{p}_x^2}{2M} + \hat{H}_{\text{int}},$$

where $\hat{p}_x^2/(2M)$ denotes the kinetic energy of the transverse motion of the atom and $\hat{H}_{\text{int}}$ accounts for the interaction between the atom and the field.

We use this model with the resonant Jaynes-Cummings interaction Hamiltonian to consider in Sec. 2 the deflection of atoms by a quantized light field.

Here we also take into account the influence of spontaneous emission. In Sec. 3 we discuss the focusing of an atomic wave by a standing light field. In Sec. 4 we suggest a new method for creating a set of narrow wave packets for atoms. We conclude by presenting in Sec. 5 a proposal for a possible experiment probing the discreteness of the electromagnetic field.

2 Photon Statistics from Atomic Deflection

2.1 Atomic Deflection by a Resonant Quantum Field

In this section we investigate the deflection [10] of an atomic wave by a quantized electromagnetic light field [11, 4]. We show that a joint measurement of the atomic momentum and an appropriate field variable allows us to reconstruct the original photon statistics. We furthermore demonstrate that the momentum distribution of atoms scattered at the node of the standing wave also follows the photon statistics of the field.

Consider the deflection of a two-level atom by a standing, resonant quantized light field. We describe the interaction of the field and the atom by the resonant Jaynes-Cummings Hamiltonian

$$\hat{H}_{\text{int}} = g(x)\left(\hat{\sigma}_+\hat{a} + \hat{\sigma}_-\hat{a}^\dagger\right), \tag{2}$$

where the coupling $g(x) = \mu\mathcal{E}_0 \sin(kx)$ is determined by the dipole moment μ, the electric field per photon $\mathcal{E}_0$, and the spatial mode function $\sin(kx)$ of the standing light field of wavenumber $k = 2\pi/\lambda$. The annihilation and creation operators of the field we denote by $\hat{a}$ and $\hat{a}^\dagger$, and $\hat{\sigma}_+$ and $\hat{\sigma}_-$ are the Pauli matrices. In the Raman-Nath-regime [12] we can ignore the transverse kinetic energy $\hat{p}_x^2/(2M)$, assuming that the transverse displacement during the interaction time τ is small compared to the wavelength λ.

The quantum state after the interaction contains all information about the complete system of atom and field

$$|\Psi(\tau)\rangle \equiv e^{-i\hat{H}\tau/\hbar}\,|\Psi(t=0)\rangle \tag{3}$$

$$= \int_{-\infty}^{\infty} dx\, f(x) \sum_{n=0}^{\infty} w_n \{\cos\left[\kappa\sqrt{n}\sin(kx)\right]|n\rangle|g\rangle \tag{4}$$

$$-i\sin\left[\kappa\sqrt{n}\sin(kx)\right]|n-1\rangle|e\rangle\}|x\rangle,$$

where we have defined the resonant interaction parameter $\kappa = \mu\mathcal{E}_0\tau/\hbar$. Note that the interaction of the atom with the field has created a highly entangled state. As a consequence, a measurement of the atomic variables provides some information about the field state.

In order to gain insight into this interplay between atomic and field dynamics we first consider a rectangular transverse distribution $f(x)$ of the atom which is constant over many wavelengths of the standing light field. We can

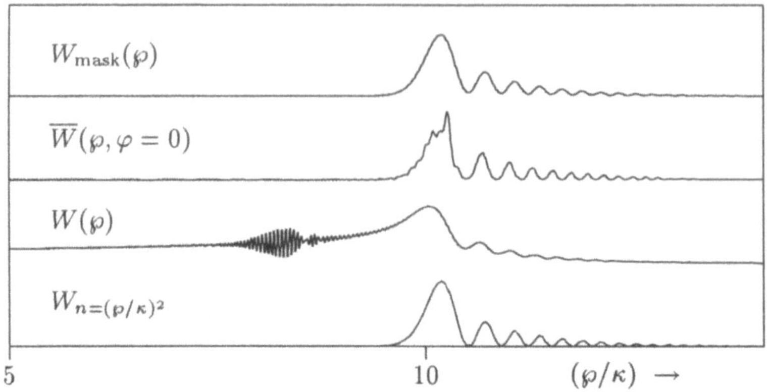

Figure 2: The photon statistics of a squeezed displaced state of squeezing parameter $s = 50$ and displacement parameter $\alpha = 10$ (lower curve) and its readout via the momentum distribution of deflected atoms with interaction parameter $\kappa = 110$. The curve $\overline{W}(\wp, \varphi = 0)$ corresponds to a joint measurement of the momentum of the atom and the field phase, Eq. (6), whereas the distribution $W(\wp)$, given by Eq. (5), ignores the field phase. The top curve, $W_{\rm mask}(\wp)$, Eq. (9), gives the momentum distribution of atoms filtered by a slit of width $d = \lambda/10$ placed at a node of the standing wave. The joint measurement strategy gives an adequate readout, while ignoring the field phase results in an less effective readout as well as in additional rapid oscillations. The momentum distribution obtained by the mask shows an exact image of the photon statistics.

then perform the Fourier transform of Eq. (3) and calculate the momentum distribution $W(\wp)$ of the atom

$$W(\wp) = \sum_{\ell=-\infty}^{\infty} \delta(\wp - \ell) \sum_{n=0}^{\infty} |w_n|^2 J_\wp^2(\kappa\sqrt{n}) \; . \tag{5}$$

Here we have scaled the momentum $\wp = p/(\hbar k)$ in units of photon momenta. Note that the spatial periodicity of the standing wave has led to quantization of the atomic momentum in units of $\hbar k$. Due to sharp maxima of the Bessel functions at $\wp = \kappa\sqrt{n}$ there is a mapping of the photon statistics onto the momentum distribution of the atoms as shown in Fig. 2 and discussed in Ref. [13].

So far we have ignored all the information stored in the field by tracing over the field as well as the internal atomic variables. We can considerably improve the readout of the photon statistics by making use of the correlation between atomic and field variables. In other words we can replace the simple measurement of the atomic momentum by a joint measurement strategy [5] on atom and field. To get insight into this we calculate the conditional probability amplitude to find the atom with momentum $\wp$ provided the field is in a given reference state. In Ref. [6] we show that for real coefficients w_n it is the London phase state corresponding to zero phase, $|\varphi = 0\rangle = (2\pi)^{-1/2} \sum_{n=0}^{\infty} |n\rangle$, which allows the best read out of the photon statistics. After projecting the reference state $|\varphi = 0\rangle$ onto the state vector $|\Psi(\tau)\rangle$, Eq. (3), we obtain for the conditional probability to find the momentum $\wp$ given the phase of the field is $\varphi = 0$

$$\overline{W}(\wp, \varphi = 0) = N \sum_{\ell=-\infty}^{\infty} \delta(\wp - \ell) \left| \sum_{n=0}^{\infty} w_n J_{\wp}(\kappa\sqrt{n}) \right|^2 , \tag{6}$$

where N accounts for the normalization. Note that in contrast to Eq. (5) the probability $\overline{W}(\wp, \varphi = 0)$, Eq. (6), does contain quantum mechanical interference terms $w_n w_m^*$.

Figure 2 shows the momentum distributions $\overline{W}(\wp, \varphi = 0)$, Eq. (6), and $W(\wp)$, Eq. (5), for the case of a highly squeezed and displaced state [14] and demonstrates the importance of the interference terms: We note that $\overline{W}$ follows precisely the photon statistics $|w_n|^2$, shown in the lower curve of Fig. 2, while only a few features of $|w_n|^2$ survive in W. This is a consequence of the mathematical fact, that the kernel of the sum Eq. (6), $J_{\wp}(\kappa\sqrt{n})$, acts more like a delta function than the kernel $J_{\wp}^2(\kappa\sqrt{n})$ in Eq. (5), which extends over a broader region. The physical meaning of the conditional probability $\overline{W}(\wp, \varphi = 0)$ is, that from all scattering events we select only those atoms which have not altered the phase of the initial field state and created a minimum of disturbance of the field [15]. These are apparently the atoms which have passed through the nodes of the field. We illustrate this by calculating the conditional position distribution $\overline{W}(x, \varphi = 0)$ to find the atom at position x, given the phase of the field is $\varphi = 0$. From Eq. (3) we obtain

$$\overline{W}(x, \varphi = 0) = N|f(x)|^2 \left| \sum_{n=0}^{\infty} w_n e^{i\kappa\sqrt{n}\sin(kx)} \right|^2 . \tag{7}$$

Figure 3 shows this probability for the same parameters as given in Fig. 2 and proves that the conditional position distribution $\overline{W}(x)$ is indeed sharply peaked at the nodes of the light field. To gain even more insight we can approximately perform the summation in Eq. (7) for the case of a coherent state and find

$$\overline{W}(x, \varphi = 0) \cong N|f(x)|^2 \, e^{-\kappa^2 \sin^2(kx)}, \tag{8}$$

which is obviously a function sharply peaked at the nodes. In this sense the joint measurement procedure plays the role of a spatial filter. A mechanical slit or mask placed in front of a node of the electromagnetic field works in a similar way and therefore gives an adequate readout of the photon statistics. Indeed, when we take the width of the atomic wave packet to be much smaller than the wavelength λ we can replace the sinusoidal potential in $g(x)$, Eqs. (2) and (3), by its linear expansion $\sin(kx) \cong kx$. In this case we find for the momentum distribution of the scattered atom

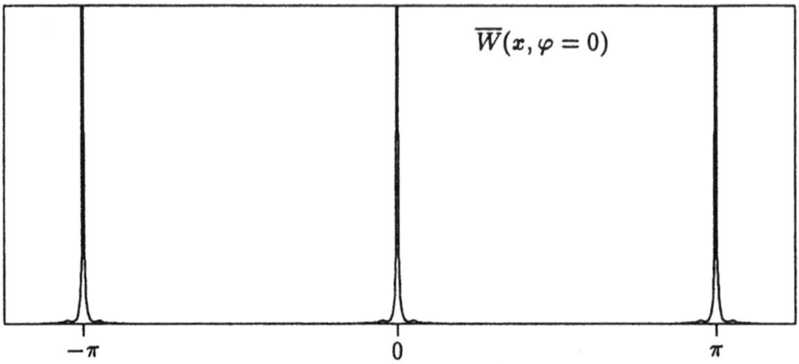

Figure 3: The conditional position distribution $\overline{W}(x, \varphi = 0)$, Eq. (7) for the same parameters as given in Fig. 2. The position distribution is sharply peaked at the nodes of the light field.

$$W_{\text{mask}}(\wp) = \sum_{n=0}^{\infty} |w_n|^2 \frac{1}{2} \left\{ W^0(\wp - \kappa\sqrt{n}) + W^0(\wp + \kappa\sqrt{n}) \right\} , \tag{9}$$

where $W^0(\wp)$ is the momentum distribution of the atom as it enters the light field. The top curve of Fig. 2 shows this momentum distribution for atoms sent through a slit of width $d \ll \lambda$. It reproduces precisely the photon statistics. In order to achieve an adequate readout of the photon statistics the acquired momentum $\wp = \kappa\sqrt{n}$ has to be larger than the original momentum spread of $W^0(\wp)$, determined by the diffraction at the slit. This implies that the width d of the slit cannot be too small as to avoid too large diffraction at the slit. We conclude by noting that this photon-distribution-from-node-scattering method is not restricted to the on-resonance Hamiltonian but of course could also be applied to the off-resonance case, see section 3.

2.2 Influence of Spontaneous Emission

So far we have only considered the coherent evolution of the combined quantum system and have neglected spontaneous emission. In this section we take spontaneous emission [16] into account and present an analytical expression for the momentum distribution [17] of an atom scattered at a node. This allows to monitor the dynamics of the relaxating quantum system "atom plus field" on a macroscopic scale.

The equation of motion for the density matrix of the combined system of atom and resonator field mode is given by

$$\frac{\partial \hat{\rho}}{\partial t} = \frac{1}{i\hbar} \left[\hat{H}, \hat{\rho} \right] + \hat{L}_{\text{irr}} \hat{\rho} , \tag{10}$$

where the Liouville operator $\hat{L}_{\text{irr}}$ describes the irreversible spontaneous relaxation of the atom due to the interaction with the photon bath, and the coherent part is governed by the Hamiltonian Eq. (2). The relaxating part $\hat{L}_{\text{irr}}\hat{\rho}$ we describe [11] by

$$\hat{L}_{\text{irr}}\hat{\rho} = \gamma\left(2\hat{\sigma}_-\hat{\rho}\hat{\sigma}_+ - \hat{\sigma}_+\hat{\sigma}_-\hat{\rho} - \hat{\rho}\hat{\sigma}_+\hat{\sigma}_-\right), \tag{11}$$

where γ is the spontaneous emission rate of the atom into modes other than the cavity mode.

Let us take the case of an atomic wave packet of width much smaller than the wavelength λ, that is $kx \ll 1$, passing through the node of the light field. This assumption allows us to replace the sinusoidal potential in $g(x)$, Eqs. (2) and (3), by its linear expansion $\sin(kx) \cong kx$. In Ref. [18] we find a regime where we can monitor the coherent and relaxating dynamics in the momentum distribution of the scattered atom: We take the momentum acquired by the atom during the coherence time to be larger than the spread of the initial wave packet. Moreover we consider interaction times τ such that the typical number of spontaneously emitted photons exceeds unity but still remains smaller than the average number of photons in the field. In this case we find for the momentum distribution $W(\wp)$

$$W(\wp) = \sum_{n=0}^{\infty} |w_n|^2 \left[\mathcal{R}_n^{(\mathrm{sc})}(\wp) + \mathcal{R}_n^{(\mathrm{sp})}(\wp)\right]. \tag{12}$$

Here the scattering part of the kernel reads

$$\mathcal{R}_n^{(\mathrm{sc})}(\wp) = \frac{1}{2}e^{-\gamma\tau/2}\left[W^0(\wp-\kappa\sqrt{n})+W^0(\wp+\kappa\sqrt{n})\right], \tag{13}$$

and the spreading part is

$$\mathcal{R}_n^{(\mathrm{sp})}(\wp) = \int_{-1}^{1} dz\, W^0(\wp-\kappa\sqrt{n}z)\mathcal{D}(z;\gamma\tau), \tag{14}$$

with

$$\mathcal{D}(z;\gamma\tau) = \frac{\gamma\tau}{4}e^{-\gamma\tau/2}\left[\mathrm{I}_0\left(\tfrac{\gamma\tau}{2}\sqrt{1-z^2}\right) + \frac{\mathrm{I}_1\left(\frac{\gamma\tau}{2}\sqrt{1-z^2}\right)}{\sqrt{1-z^2}}\right]. \tag{15}$$

Here I_0 and I_1 denote the modified Bessel functions.

The term $\mathcal{R}_n^{(\mathrm{sc})}$ corresponds to atoms which have not lost their coherence with the field and therefore are deflected according to the gradient of the optical potential inside the resonator. This gradient leads to displaced images of the initial momentum distribution centered at $|\wp| = \kappa\sqrt{n}$ and weighted with the photon statistics $|w_n|^2$. It allows us to read off the photon statistics from the momentum distribution of the atom and thereby to monitor the coherent dynamics. For small interaction times $\gamma\tau \cong 1$ this diffractive part is dominating and in the limit $\gamma \to 0$ we obtain the result Eq. (9) of the previous section.

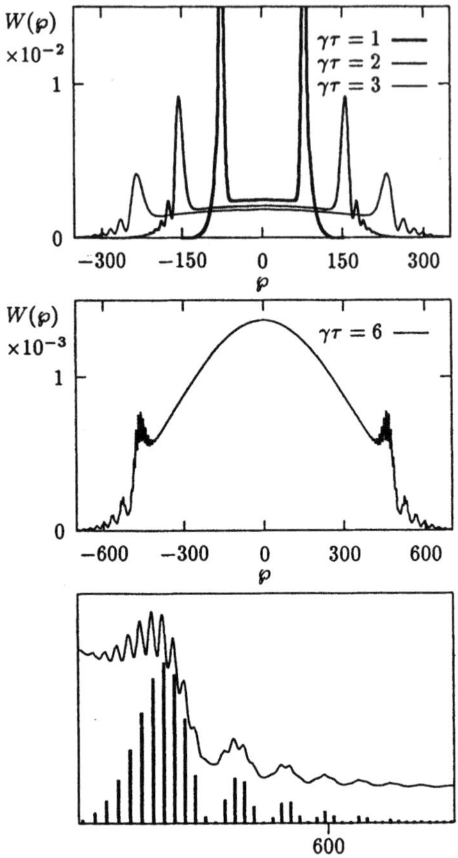

Figure 4: Momentum distributions of an atom passing through a node of a squeezed displaced light field for different interaction times. The distribution consists of a diffusive part corresponding to relaxating dynamics and a diffractive part, which reflects the coherent dynamics and images the field photon statistics. On the bottom we magnify the edge of the momentum distribution and compare it to the photon statistics of the squeezed light field. Note that even the discrete structure of the photon statistics can be resolved. Here we have chosen the interaction parameter $\kappa/(\gamma\tau) = 15$, the squeezing parameter $s = 20$ and the displacement parameter $\alpha = 5$. The initial wave packet has a width $\Delta\wp = 4$.

The term $\mathcal{R}_n^{(\mathrm{sp})}$ represents atoms which have lost their coherence due to spontaneous emission. As a result the force changes randomly sign and the atoms perform a random walk in momentum space. These atoms are therefore deflected by an amount $\delta\wp = \kappa\sqrt{n}z$, where z is the difference of the relative times spent in the symmetric and antisymmetric dressed state and hence $|z| < 1$. Note that in the limit $\gamma\tau \gg 1$ the distribution $\mathcal{D}(z)$ given by Eq. (15) reduces to a Gaussian $\mathcal{D}(z;\gamma\tau) \cong \sqrt{\gamma\tau/(4\pi)}\exp\{-\gamma\tau z^2/4\}$. Therefore the probability to find the change of the momentum $\delta\wp$ is given by the Gaussian distribution $\exp\{-\delta\wp^2\gamma\tau/(4n\kappa^2)\}$ of Brownian motion.

Figure 4 shows the evolution of the momentum distribution for an initially Gaussian wave packet centered at a node of the light field. The quantum field is in an amplitude squeezed state. As the interaction time increases we note a transition from the diffractive to the diffusive regime [19]. The figure also shows that the photon statistics is reproduced at the edge of the diffusive background for sufficiently large interaction times. Note, that this image of the photon statistics is obtained for times much longer than the spontaneous decay time. This is due to the fact that despite the area under the diffractive part is decaying exponentially, one can still trace it at the edge of the smooth diffusive background, because the diffusive part is by itself exponentially small there.

We conclude by noting that the analytical expression for the momentum distribution of the deflected atom allows us to monitor the transition of the dynamics of the system from the coherent evolution to the relaxating regime.

3 Focusing of Atomic Waves

In the preceding section we have shown that the individual Fock states building up the field state deflect the atoms into different directions. In the present section we show that they also provide some focusing which allows us to resolve the individual photon states.

We take the frequency of the field to be strongly detuned from the atomic transitions. In this case the atom remains in the initial internal state, and we can describe the interaction by the effective Hamiltonian [20]

$$\hat{H}_{\text{eff}} = g(x)\,\hat{a}^\dagger\hat{a}, \tag{16}$$

where the coupling $g(x) = \alpha\,\mathcal{E}_0^2(x)$ is the atomic linear susceptibility α multiplied by the "square of the electric field per photon."

In the cavity we neglect the kinetic energy $\hat{H}_{\text{free}} = \hat{p}_x^2/(2M)$ of the transverse motion of the atom compared to $\hat{H}_{\text{eff}}$. Therefore the state vector of the system at the exit of the cavity at $z = 0$ reads

$$|\Psi(z=0)\rangle = \exp\left[-\frac{i}{\hbar}\hat{H}_{\text{eff}}\frac{L}{v_z}\right]|\Psi(z=-L)\rangle.$$

Outside of the cavity the dynamics of the atom in x direction is governed by $\hat{H}_{\text{free}}$ and, hence, the state vector at a position z is given by

$$|\Psi(z)\rangle = \exp\left[-\frac{i}{\hbar}\hat{H}_{\text{free}}\frac{z}{v_z}\right]|\Psi(z=0)\rangle.$$

From this expression we obtain the probability

$$W(x,z)=\sum_{n=0}^{\infty}\left|\langle x,n|\Psi(z)\rangle\right|^2=\sum_{n=0}^{\infty}|w_n|^2\,R_n(x,z)\,, \tag{17}$$

to find the particle at the point with the coordinates x and z. We evaluate the intensities R_n of the atomic partial wave corresponding to the nth Fock state for a Gaussian profile of the atomic beam of width $d \ll \lambda$ centered at $x = 0$ and find [7]

$$R_n(x,z)=\frac{1}{\sqrt{\pi}\,D_n(z)}\exp\left\{-\left[\frac{nz+Nx}{ND_n(z)}\right]^2\right\}.$$

Here

$$D_n(z)\equiv\left[\left(\frac{\hbar z}{dMv_z}\right)^2+d^2\left(1-\frac{n}{N}\frac{g_2}{g_1}z\right)^2\right]^{1/2}$$

and

$$N\equiv\frac{Mv_z^2}{g_1L},$$

where $g_j = \left.\frac{\partial^j g}{\partial x^j}\right|_{x=0}$ for $j = 0,\ 1,\ 2$.

Hence the nth partial wave with intensity $R_n(x,z)$ is a Gaussian beam propagating in the x-z plane under an angle

$$\theta_n = -\arctan(n/N)$$

with respect to the z-axis. Positive g_1 result in a positive value for N and thus in negative deflection angles θ_n. Negative g_1 result in positive deflection angles. Note, that in the framework of our approximation $H_{\text{free}} \ll H_{\text{eff}} \ll Mv_z^2/2$ we always have $N \gg n$ and hence $|\theta_n| \ll 1$.

For positive g_2 the width D_n of the partial beam reaches its minimum for positive z and thus the nth partial wave focuses at the point

$$\begin{aligned} z_n &= \mathcal{F}_n \equiv \frac{g_1}{g_2}\frac{N}{n}=\frac{1}{n}\frac{Mv_z^2}{g_2L}, \\ x_n &= \mathcal{F}_n\tan\theta_n = -g_1/g_2. \end{aligned}$$

Hence all foci lie along a straight line parallel to the direction of incident atoms. The net probability $W(x,z)$, Eq. (17), for an arbitrary field state is a superposition of partial Gaussian beams weighted with the photon statistics $|w_n|^2$. This results in a multi-focal structure for $W(x,z)$ as shown in Fig. 5.

In order to resolve the contributions from different Fock states the widths of the foci have to be narrower than their spacings. A velocity spread of the atoms translates itself into an additional spread of the foci. Therefore we need an atomic beam of good monochromaticity. Furthermore, in order to guarantee the applicability of our approach based on the nondemolition Hamiltonian, Eq. (16), we have to fulfill further conditions concerning transitions to the

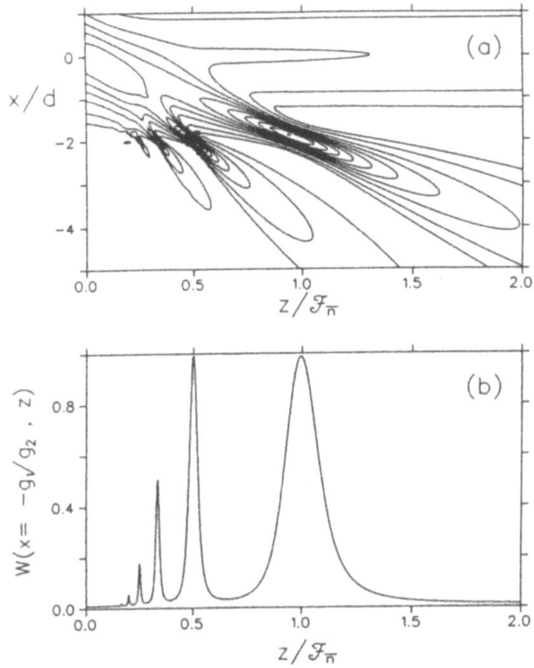

Figure 5: Contour plot (a) for probability $W(x, z)$ of finding the atom at the point with coordinates x and z. The Gaussian atomic beam centered at $x = 0$ leaves the cavity at $z = 0$. The cavity field is in a coherent state of average number of photons $\bar{n} = 1$. The undeflected and unfocused partial wave R_0 associated with the cavity vacuum state represents the profile of the incident beam. The deflected partial waves R_1, R_2, R_3 and R_4 associated with the photon states $n = 1, 2, 3$ and 4 in the coherent state of the field focus along the line $x = -g_1/g_2$. The intensity of atoms along this line is shown in (b). Parameters used are: $g_1/(g_2 d) = 2$; $(g_2 d^2/\hbar)(L/v_z) = 3\pi/2$.

upper level and the influence of spontaneous emission. For a detailed discussion of all these points we refer to Ref. [7] and note that the experimental facilities necessary for this approach are presently available.

We conclude this section by noting that a discussion of this quantum lens in terms of the Wigner function can be found in Ref. [8].

4 Narrow Wave Packets from Spontaneous Emission

Spontaneous emission is detrimental to the coherence of the atomic wave. Consequently one tries to reduce spontaneous emission by detuning the light field from the atomic transition frequency. In this section we make use of spon-

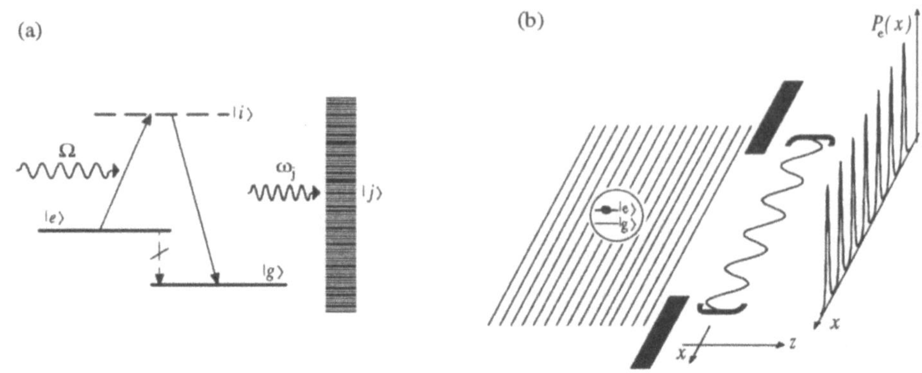

Figure 6: (a): Illustration of the Anti-Stokes Raman Process. Transitions from the excited state $|e\rangle$ to the ground state $|g\rangle$ are induced by the standing light field. The intermediate state $|i\rangle$ spontaneously decays due to the coupling of the atom-field system to a photon bath. (b): Schematic illustration of the light-induced grating for atomic waves. A plane wave of atoms in the excited state $|e\rangle$ interacts with a standing light field. After the interaction the position distribution $P_e(x)$ is periodically modulated due to the mode function $\sin x$.

taneous processes to transfer a plane atomic wave into a set of narrow wave packets located at the nodes of a standing light field. For this purpose we consider atomic de Broglie waves incident perpendicular to the light field. Before the interaction the atoms are in the excited state $|e\rangle$. We take this atomic state to be metastable, that is the dipole transition to the ground state $|g\rangle$ is forbidden. The only possibility to make a transition to the ground state is via Raman scattering during the interaction with the standing light field (see Fig. 6a). The spatial modulation of the light field, described by the mode function $s(x)$, creates a modulation of the atomic population. At the nodes of the standing light field no Anti-Stokes scattering processes are possible because the electric field and hence the atom-field coupling vanishes. In contrast, the probability of a transition to the ground state while flying through the anti-nodes is very high. Hence we expect a spatial hole burning in the population of excited atoms as shown in Fig. 6b.

The Liouville equation describing the translational motion as well as the internal dynamics of the atom reads

$$\dot{\hat{\rho}} = \frac{1}{i\hbar}\left[\hat{p}_x^2/(2M)\,,\,\hat{\rho}\right] - \gamma\left(\hat{s}^2\hat{\sigma}_+\hat{\sigma}_-\hat{\rho} + \hat{\rho}\hat{\sigma}_+\hat{\sigma}_-\hat{s}^2 - 2\hat{s}\hat{\sigma}_-\hat{\rho}\hat{\sigma}_+\hat{s}\right). \tag{18}$$

The mode function $\hat{s} = s(\hat{x})$ contains the position operator and is therefore treated quantum mechanically. In order to calculate the position and momentum distribution of atoms in the excited state we perform the Raman-Nath approximation, that is we neglect the kinetic energy term in Eq. (18) during the atom-field interaction.

We consider the atom before the interaction with the light field as a plane wave with the transverse width L. As mode function we take for simplicity $s(x) = \sin(kx)$. For the position distribution of an atom in the excited state after the interaction time t we find

$$P_e(x,t) = \frac{1}{L} e^{-2\gamma t \sin^2(kx)}, \tag{19}$$

that is we actually have a spatial modulation of the atomic population. For large interaction parameter γt we obtain sharp peaks for $P_e(x,t)$ around the nodes of the standing light field.

5 Proposal for an Experiment

In this section we propose an experiment dealing with the deflection of atoms in a quantized light field. In Sec. 2.1 we have used a mechanical mask to send the atom only through the nodes of the standing light field. Since it is difficult to produce and handle such a mechanical mask suitable for the optical regime we suggest to replace it by an effective *self-induced mask* based on the selective detection of atoms which have not spontaneously emitted photons. This modification also allows us to obtain *nondemolition* measurements of the photon number. An analysis of experimental requirements shows that the experiment is feasible. An overview of the proposed experimental set-up is shown in Fig. 7.

A collimated beam of metastable atoms passes through a resonant optical cavity. No mechanical mask is used. The atomic and experimental parameters are chosen such, that spontaneous emissions will occur in the cavity. The diffraction of the atoms leaving the cavity is determined using a position-sensitive metastable atom detector.

The search for a suitable atomic transition has resulted in the choice of metastable helium, using the $2s^3S_1 \rightarrow 2p^3P_2$ two-level transition at a wavelength $\lambda = 1083$ nm. The optimal axial beam velocity for the experiment is around 300 m/s. Hence laser-slowing of the beam is necessary. Diode-laser pumped LNA lasers will be constructed and used both for the slowing/collimation and for the excitation of the cavity. The choice of a slow helium beam is partly motivated by the large single-photon recoil velocity. This makes the required angular resolution, corresponding to a fraction of the recoil velocity, relatively easy to obtain. This resolution is needed to select those atoms, which have not undergone spontaneous emission.

The detector selects only those atoms, which have not spontaneously emitted any photon. We obtain this selectivity in the following way: The momentum change due to the coherent part of the interaction with the light (absorption and stimulated emission processes only), is always a multiple of the photon momentum $\pm\hbar\vec{k}$, where $\vec{k}$ denotes the wave vector of the cavity mode at the waist. In contrast, a spontaneous emission leads to a random momen-

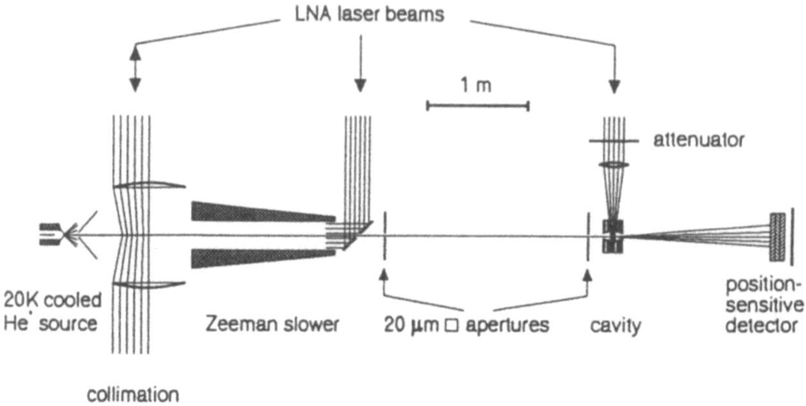

Figure 7: Overview of the proposed set-up. The metastable helium beam is produced by a supersonic discharge source, cooled with 20 K helium gas from a cryocompressor to obtain a low initial axial velocity. Transverse cooling in two dimensions (to increase the centerline intensity) and slowing/cooling to an axial velocity of 300 m/s using the Zeeman slowing scheme is provided for by 1.08 μm LNA laser light. After collimation, the atoms pass through the cavity and are diffracted by the light field. The position-sensitive detector consists of a triple multichannel plate electron multiplier with a resistive anode.

tum change, causing the atom to deviate from the discrete one-dimensional grid with $\hbar k$ periodicity in momentum space. When the initial collimation of the atomic beam and the resolution of the detector are good enough to allow a resolution in atomic momentum change $\Delta p \ll \hbar k$, efficient discrimination against atoms which have spontaneously emitted one or more photons can be obtained. For light atoms as helium or neon, momentum resolution far below the single-photon recoil can be easily obtained when using slow atomic beams as produced by laser cooling.

Atoms passing through the cavity at an antinode of the standing wave, experience the maximum rate of spontaneous emission and hence have the smallest probability of being detected. For atoms passing at the nodes, the rate is zero, and the probability of detection is unity (assuming 100% detection efficiency). Hence the *detected* atoms act, as if they have passed a mask transmitting the atoms near the nodes of the standing wave.

We conclude by noting that in contrast to the mechanical mask, this *self-induced mask* is automatically aligned to the nodes. As it coincides with the cavity light field, diffraction in propagation from the mask to the interaction region does not play a role. Moreover, the detected atoms have not changed the number of photons in the cavity. Hence, although the complete interaction Hamiltonian does not represent a quantum nondemolition measurement scheme, we have selected nondemolition measurements by selective detection of atoms.

Acknowledgments.

One of us (W. S.) would like to express his appreciation to J. Harvey and D. F. Walls for the invitation to participate at a most splendid conference. I. Sh. A. thanks the University of Ulm for its kind hospitality, A. M. H. acknowledges the support of the Deutsche Forschungsgemeinschaft and V. P. Y. is grateful to the Heraeus Foundation for a stipend.

References

[1] G. Rempe, R. J. Thompson, R. J. Brecha, W. D. Lee, and H. J. Kimble, Phys. Rev. Lett. **67**, 1727 (1991).

[2] M. J. Holland, D. F. Walls, and P. Zoller, Phys. Rev. Lett. **67**, 1716 (1991).

[3] For a review of the field of atom optics, see the special issue Appl. Phys. B **54** (1992), and S. Stenholm, in Lectures at the CXVIII International School of Physics *Laser manipulation of atoms and ions*, E. Arimondo, W. D. Phillips, F. Strumia (eds.), Proceedings of the International School of Physics "Enrico Fermi", Course CXVIII, Varenna, 1991.

[4] P. Meystre, E. Schumacher, and S. Stenholm, Opt. Commun. **73**, 443 (1989).

[5] For a joint measurement strategy aimed to localize an atomic beam in a QND situation see P. Storey, M. Collett, and D. Walls, Phys. Rev. Lett. **68**, 472 (1992).

[6] A. M. Herkommer, V. M. Akulin, and W. P. Schleich, Phys. Rev. Lett. **69**, 3298 (1992).

[7] I. Sh. Averbukh, V. M. Akulin, and W. P. Schleich, Phys. Rev. Lett. **72**, 437 (1994)

[8] E. Mayr, D. Krähmer, A. M. Herkommer, V. M. Akulin, W. P. Schleich, and I. Sh. Averbukh, Acta Phys. Pol. A **85**, 483 (1994).

[9] S. Haroche and D. Kleppner, Physics Today, January 1989, pp. 24–30 and references therein.

[10] A. Aspect, E. Arimondo, R. Kaiser, N. Vansteenkiste, and C. Cohen-Tannoudji, Phys. Rev. Lett. **61**, 862 (1988).

[11] A. P. Kasantsev, G. I. Surdutovich, V. P. Yakovlev, *Mechanical Action of Light on Atoms*, (World Scientific, Singapore, 1990).

[12] C. W. Raman and N. S. Nath, Proc. Ind. Acad. Sci. **2**, 406 (1933).

[13] V. M. Akulin, Fam Le Kien, and W. P. Schleich, Phys. Rev. A **44**, R1462 (1991).

[14] For the photon statistics of a squeezed state, see for example W. P. Schleich, D. F. Walls, and J. A. Wheeler, Phys. Rev. A **38**, 1177 (1989).

[15] This is closely related to the concept of a least disturbing measurement, in which an atomic beam probes the nodes of a ring resonator; see for example M. O. Scully, W. P. Schleich, and H. Walther, Phys. Rev. A, to be published.

[16] For the experiment see P. L. Gould, P. J. Martin, G. A. Ruff, R. E. Stoner, J. L. Picque, D. E. Pritchard, Phys. Rev. A **43**, 585 (1991).

[17] For the discussion of the motion of a spontaneously emitting atom in a *classical* field, see C. Tanguy, S. Reynaud and C. Cohen-Tannoudji, J. Phys. B **17**, 4623 (1984); see also the review by C. Cohen-Tannoudji in *Fundamental systems in quantum optics*, edited by J. Dalibard, J. M. Raimond and J. Zinn-Justin (North-Holland, Amsterdam 1992).

[18] A. M. Herkommer, V. M. Akulin, and W. P. Schleich, to be published in Phys. Rev. A **49** in April 1994.

[19] S. M. Tan and D. F. Walls, Phys. Rev. A **44**, R2779 (1991).

[20] See, for example, M. Brune, S. Haroche, V. Lefevre, J. M. Raimond and N. Zagury, Phys. Rev. Lett. **65**, 976 (1990).

Quantum Nonlinear Dynamics in Atom Optics

S. Dyrting, W.Y. Chen, and G.J. Milburn

Department of Physics, University of Queensland, St. Lucia, Australia 4072

Abstract. We consider examples of quantum motion in integrable and non-integrable systems in which the center-of-mass motion of an atom interacts with an optical field mediated by the coupling to internal electronic states. In particular we consider the formation of superposition states and fractional revivals of center-of-mass motion for an atom moving in a standing wave and an atom bouncing on an evanescent wave. As examples of non-integrable motion we consider the same systems but with periodically modulated optical potentials.

Recent advances in laser cooling and trapping enable us to probe the quantum nonlinear motion of atoms in optical potentials. Indeed the isolation from fluctuations is so good that the dynamics is largely Hamiltonian and it is increasingly likely that quantum coherence effects will become experimentally accessible. In this lecture I will consider the different quantum and classical descriptions of nonlinear dynamics for both integrable and non-integrable systems. I will only consider single atom dynamics with bounded motion. Due to the long interaction times considered here the kinetic energy term, omitted in the Raman-Nath approximation, must be included. The examples of integrable systems are; an atom moving in a standing wave in the high detuning limit, and an atom bouncing vertically from an evanescent wave at a flat dielectric surface. As examples of non-integrable systems we take the same models but include a modulation of the light intensity. This corresponds to a periodically modulated potential for the center-of-mass motion of the atoms.

A nonlinear oscillatory system is characterised by the fact that the period of motion is a function of the initial energy. The nonlinear frequency is defined by,

$$\omega(H) = \left(\frac{\partial I(H)}{\partial H}\right)^{-1}, \tag{1}$$

where $I(H)$ is the action on a curve of constant energy H. Points on different energy surfaces in phase space rotate at different rates. This leads to a rotational shearing of a localised phase-space density, which eventually becomes uniformly spread around the elliptic fixed point defining the periodic motion. The result for the mean position and momentum is a decay to zero on a time scale of the nonlinear period, while the variance in position and momentum saturate at a constant value determined by the radius of the annular phase-space density resulting from the shearing motion.

The quantum description however follows the classical only for short times. At some time the quantum mean values return arbitrarily closely to their initial values, as the initial state partially recurs. An even more surprising quantum effect is the appearance of coherent superpositions of copies of the initial states at intermediate times.

A two-level atom interacting with a standing wave, but detuned by a large amount Δ from the cavity resonance frequency, will experience a spatially varying force, while remaining in its initial state. The center of mass motion is then determined by the effective Hamiltonian

$$H = \frac{p^2}{2} - F\cos q \quad , \tag{2}$$

where we are using scaled variables defined by $\tau = \omega t$, $p = \frac{2kp_x}{m\omega}$, $q = 2kx$, $F = \frac{\hbar\Omega^2 k^2}{2m\Delta\omega^2}$ where ω is a convenient frequency scale, Ω is the Rabi frequency, m is the atomic mass and k is the wave number of the standing wave. In these variables the commutation relations for position and momentum are $[q,p] = iK$, where $K = \frac{4\hbar k^2}{m\omega}$ is a parameter which determines how 'quantum' the system is (i.e. very small K correspnds to the semiclassical regime). This is the Hamiltonian for a nonlinear pendulum, although the position variable is not an angle variable in this case. The nonlinear frequency as a function of the energy H starts at the linear oscillator value near $H = -F$, varies slowly as H is increased until the energy approaches the top of the cosine potential where it drops rapidly to zero, corresponding to infinite period on the separatrix. In figure 1 we plot the mean and variance in momentum, versus the number of classical nonlinear periods, for an initial gaussian centred on $\langle p\rangle = 0$; $\langle q\rangle = -1.5$; with variances $\sigma_q = 0.18$; $\sigma_p = 0.33$ and the parameters $F = 1.2$; $K = 0.24$. The features discussed above are clearly evident. In contrast, the quantum dynamics plotted in Figure 2 , show near revivals of initial values at $n = 26$ and $n = 52$. In addition the variance shows a rapid variation around $n = 13$. This feature is characterisitic of the formation of a superposition state, and is more evident if the contours of the Q-function are plotted[1].

Averbukh and Perelman[2] have discussed the formation of such superposition states in nonlinear oscillators and we use their results here. The time scale for revivals is determined by

$$T_{REV} = \frac{4\pi}{\omega(H)}\left(K\left|\frac{\partial\omega(H)}{\partial H}\right|\right)^{-1}. \tag{3}$$

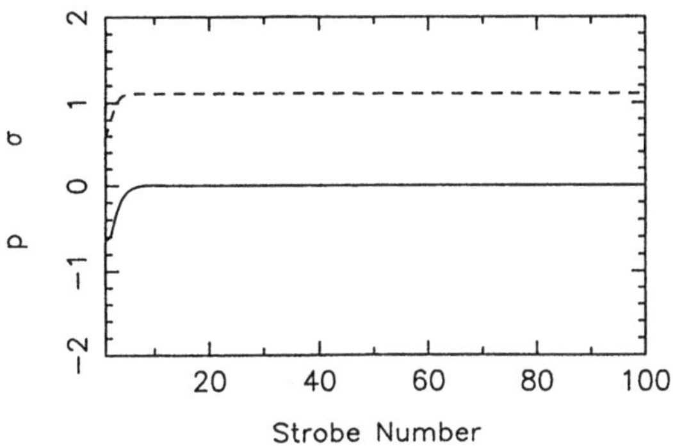

Figure 1. Classical mean and variance of momentum, for an initial Gaussian; $\langle p \rangle = 0$; $\langle q \rangle = -1.5$; $\sigma_q = 0.18$; $\sigma_p = 0.33$.

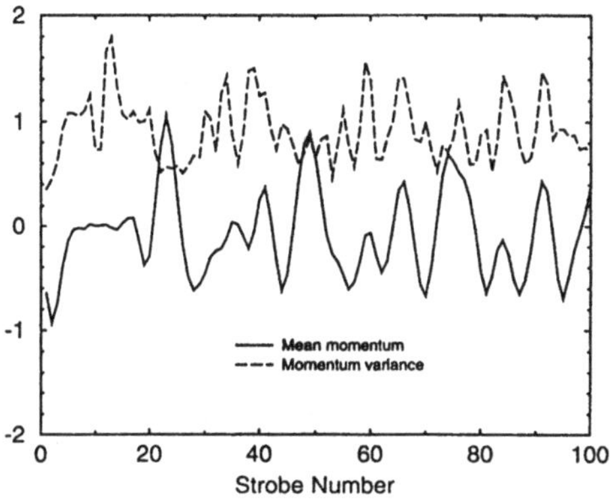

Figure 2. Quantum mean and variance of momentum, for the same values as Figure 1.

For strongly nonlinear systems the variation of $\omega(H)$ with frequency is large and the time scale for revivals is short, likewise if the system is strongly classical (large K).

In this example we find $T_{REV} = 52 \times$ *classical period* which agrees well with the numerical results. What does a fractional revival mean in practice? Once the atom leaves the standing wave, the momentum distribution at the end of the interaction determines the position distribution far from the interaction region. If the classical result were correct, we would find a broad spread of atoms transverse to the cavity. The quantum momentum distribution, at $n = 13$, is shown in Figure 3. Two distinct peaks are evident, as expected for this superposition state. This would lead to two distinct beams of atoms leaving the cavity, in contrast to a broad, classical distribution.

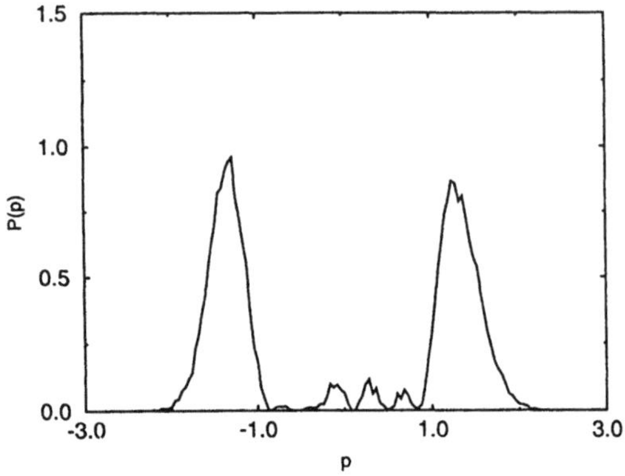

Figure 3. Momentum distribution at a fractional revival.

For an experiment with ytterbium to see a fractional revival would require an interaction length of the order of $10mm$ with an atomic velocity of $92m/s$ and a collimation of the order of $0.01\mu m$[1]. Current progress in atomic cooling, and beam shaping are likely to reach these, admittedly difficult, parameters in the near future. However spontaneous emission is an important consideration.

It is well known that quantum coherence effects such as fractional revivals are extrordinarly sensitive to damping or simply noise. To address this issue we use the approximate master equation derived by Dyrting[3]. This is an expansion in the small parameters, Ω/Δ and γ/Δ. Taking matrix elements of the total density operator in the electronic state basis, $(|a\rangle\ ,|b\rangle)$ where a lables the ground state and b labels the excited state, we have

$$\frac{d\hat{\rho}_a}{dt} = \frac{-i}{K}\left[\frac{\hat{p}^2}{2}+\frac{2F}{|\nu|^2}\sin^2\frac{\hat{q}}{2},\hat{\rho}_a\right]+\Gamma\mathcal{N}\hat{\rho}_b \tag{4}$$

$$-\frac{\alpha\Gamma}{2}\left[\{\sin^2\frac{\hat{q}}{2},\hat{\rho}_a\}-2\sin\frac{\hat{q}}{2}\ \hat{\rho}_b\ \sin\frac{\hat{q}}{2}\right], \tag{5}$$

$$\frac{d\hat{\rho}_b}{dt} = \frac{-i}{K}\left[\frac{\hat{p}^2}{2}-\frac{2F}{|\nu|^2}\sin^2\frac{\hat{q}}{2},\hat{\rho}_b\right]-\Gamma\hat{\rho}_b \tag{6}$$

$$-\frac{\alpha\Gamma}{2}\left[\{\sin^2\frac{\hat{q}}{2},\hat{\rho}_b\}-2\sin\frac{\hat{q}}{2}\ \hat{\rho}_a\ \sin\frac{\hat{q}}{2}\right], \tag{7}$$

where $\Gamma = \gamma/\omega$; $\alpha = \frac{\Omega^2}{4\Delta^2|\nu|^2}$; $\nu = 1 - i\frac{\gamma}{2\Delta}$. We choose the values $\Gamma = 1.525$; $\alpha\Gamma = 0.0003$; $\frac{\gamma}{2\Delta} = 0.00003$.

Using the theory of quantum trajectories[4], these equations may be interpreted as a series of quantum jumps of the centre-of-mass wavefunction as the atom makes stimulated and spontaneous transitions between the internal states. Between jumps the centre-of-mass wavefunction satisfies a

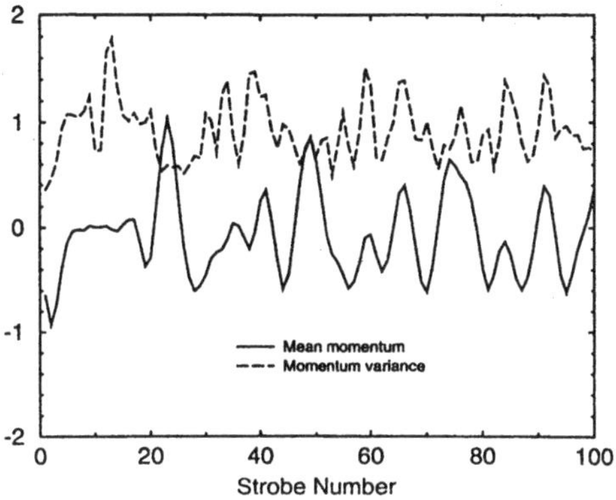

Figure 4. The mean and variance of momentum for an atom

non-hermitian 'Hamiltonian' evolution, which depends on the internal state;

$$iK\frac{d|\psi_{a,b}\rangle}{dt} = H_{+,-}|\psi_{a,b}\rangle \,, \tag{8}$$

where

$$H_{+,-} = \frac{p^2}{2} \pm \frac{2F}{1 \pm i\frac{\gamma}{2\Delta}} \sin^2\frac{q}{2} \,. \tag{9}$$

When the internal statemakes a stimulated jump from gound (excited) to excited (ground), the wave function just after the jump is given in terms of the state just before the jump,

$$|\psi^{(+)}\rangle = \frac{\sin\frac{q}{2}|\psi^{(-)}\rangle}{\sqrt{\langle\psi^{(-)}|\sin\frac{q}{2}|\psi^{(-)}\rangle}} \,. \tag{10}$$

Spontaneous emission is described by the super-operator $\mathcal{N}$, where

$$\mathcal{N}\rho = \int d\mathbf{n}\phi(\mathbf{n})\exp(i\mathbf{n}\frac{\hat{q}}{2})\ \rho\ \exp(-i\mathbf{n}\frac{\hat{q}}{2}). \tag{11}$$

The internal state jumps to the ground state at a constant rate Γ and after the jump the wavefunction is $|\psi^{(+)}\rangle = \exp(-i\bar{\mathbf{p}}.\mathbf{q}/K)|\psi^{(-)}\rangle$. The momentum $\bar{p} = K\mathbf{n}$ acquired by the atom is due to recoil and is itself a random variable given by the probablity distribution ϕ. The result for the moments of the momentum, computed by averaging 1000 trajectories,with an average number of emission events up to $1/4T_{REV}$ equal to 0.03, are shown in figure 4 . Previous results suggest that for ytterbium one can expect on average, about 0.02 spontaneous emission events during the atomic transit time. It is clear from figure 4 , that at this level fractional revivals should be clearly visible.

Another atom optical system which can exhibit fractional revivals was proposed by Wallis et al.[5], in which a falling atom is reflected from a surface

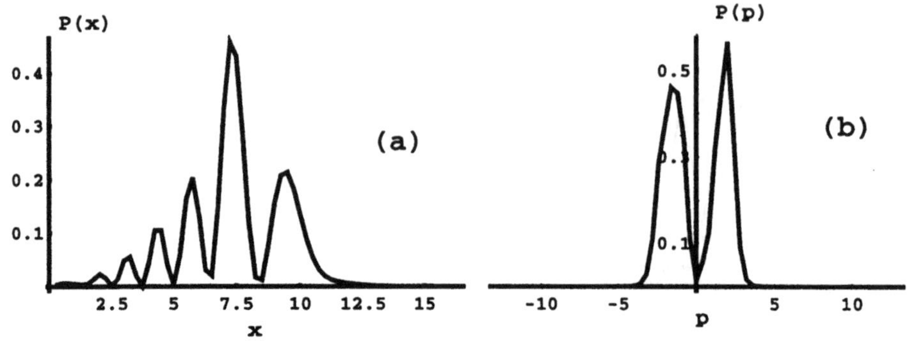

Figure 5. Position (a) and momentum (b) probability distribution for the atomic bouncer at one quarter of a revival time.

by an optical evanescent wave. Consider the simpler case of an atom bouncing elastically from a flat plate. In this case the nonlinear period is just the time taken for the atom to return to its initial position after a bounce.

The Hamiltonian between bounces is just $H = \frac{p^2}{2} + \mu q$ where we are using the scaled variables $q = z/\beta$; $p = \frac{p_z}{m\beta\omega}$ where the position scale is $\beta = \left(\frac{\hbar^2}{2m^2 g}\right)^{1/3}$ the frequency scale is $\omega = 2\pi mg\beta/\hbar$ and the scaled acceleration is $\mu = \frac{g}{\beta\omega^2}$. The scale for the canonical commutation relations, K, is given by $K = \frac{\hbar}{m\beta^2\omega}$. The nonlinear frequency is then given by $\omega(H) = \frac{\pi\mu}{\sqrt{2H}}$, which has a much slower variation with H than the case for the standing wave considered above.

In MKS units the fractional revival time is given by $\frac{16H^2}{\pi m g^2 \hbar}$ To illustrate the formation of a superposition state at a quarter of the revival time we consider an initial Gaussian with a variance of β^2, which is approximately equal to the spread in position for the Airy function ground state. The Gaussian is initially located at $q = 10$ in which case the classical period in scaled time is about 6.3, while the revival time is about 41 times this classical period. In figure 5 we plot the position and momentum distribution at one quarter of the revival time. The position distribution shows inteference fringes while, as expected, the momentum distribution shows a superposition of two states of oppositely directed momentum. Such a state could possibly be detected by looking for Doppler shifted resonances with a weak probe, or perhaps the position density fringes could be probed directly.

The two systems considered above are nonlinear but integrable. The easiest way to produce atom optical systems which are non-integrable is to add periodic time dependent modulations to the optical potentials. Consider first the case of a modulated standing wave

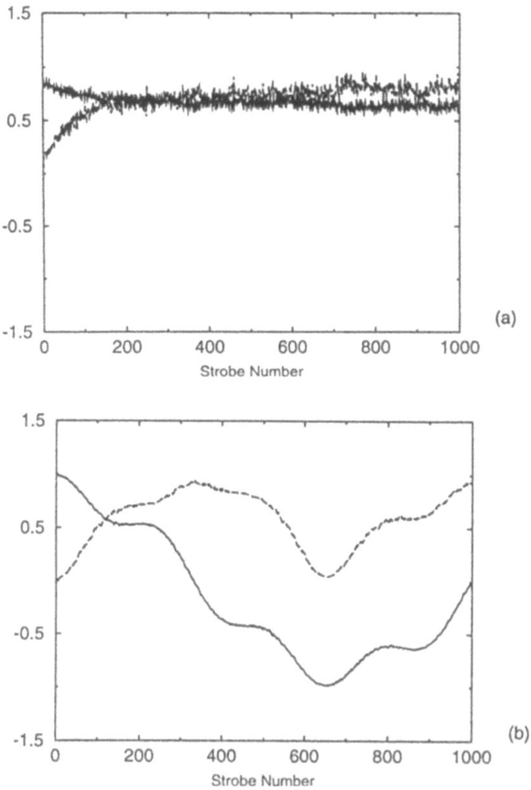

Figure 6. Classical(a) and quantum (b) mean and variance of momentum for an atom in a modulated standing wave. Solid $\langle p \rangle$ dashed $Var(p)$.

$$H(t) = \frac{p^2}{2} - F(1 - 2\epsilon \cos \Omega t) \cos q \tag{12}$$

We have previously given a classsical analysis of this system[6]. The motion inside the separatrix is characterised by chaos and first and second order resonances. A classical distribution located well within a second order resonance zone cannot penetrate the surrounding KAM tori. Thus the mean values simply oscillate around values corresponding to the co-ordinates of the resonance, see figure 6a . The quantum motion however is quite different and can exhibit tunneling between the classical second order resonances, see figure 6b .

In a similar way we can consider modulations of the intensity of the evanescent wave in the atomic bouncer. This has the effect of moving the surface from which the atom is bouncing. A particle bouncing on a moving plate has long been known as an example of chaotic motion, but now there is the possibility of investigating the motion in the mesocopic quantum domain.

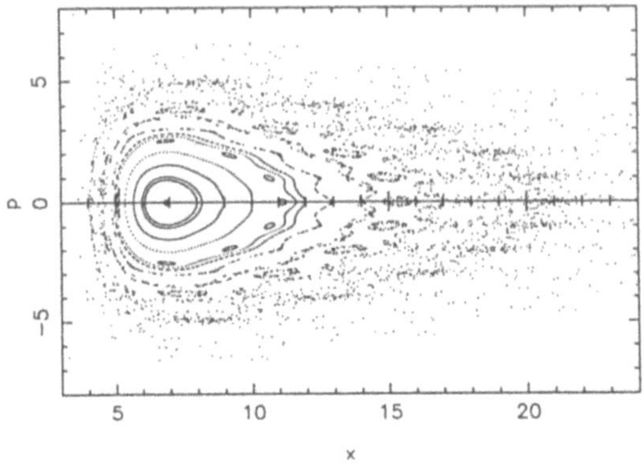

Figure 7. Classical Poincare section for an atom bouncing on a modulated evanescent field.

In this case it is easiest to take the full exponential potential and thus we consider the Hamiltonian

$$H(t) = \frac{p^2}{2} + \lambda q + (\kappa + \epsilon \cos \Omega t)e^{-q}. \tag{13}$$

Typicallly, for realistic atomic systems we expect $0.01 < \lambda < 1$; $100 < \kappa < 1000$. A typical classical stroboscopic phase-portrait is shown in figure 7 for $\lambda = 1.0; \kappa = 1000; \epsilon = 100$. Damping of course would turn multiple period fixed points into attractors and thus permit mode locking of the atomic oscillations. We have not yet completed the quantum analysis of this system, however a related system has been studied by Dembinski et al.[7]. They show that in the chaotic domain the bouncing atom absorbs energy diffusely while the quantum motion prevents this diffusive growth in the energy. In the atom optical case such an energy diffusion would appear as a heating effect and thus a quantum coherent suppression of chaotic diffusion would appear as a suppression of this classical heating. We are currently investigating how realistic such a prospect is for the current state of experiments in this area. A number of other interesting possibilites come to mind in time-periodic potentials. In such problems the stationary states, which characterise the quantum dynamics of time independent oscillators, are replaced by Floquet states, eigenstates of the time evolution over one period. Floquet states are often more localised in phase-space than stationary states and may thus be a lot easier to load from a cooled ball of atoms. It is also of interest to ask if multi-atom cooperative effects can be seen when a large number of atoms are placed in a single Floquet state.

References

[1] S.Dyrting and G.J.Milburn, Phys. Rev. **47**,R2484, (1993).

[2] I.Sh.Averbukh and N.F.Perelman, Phys. Lett. **139**, 449 (1989).

[3] S. Dyrting and G.J.Milburn, to appear Phys. Rev A. (1994).

[4] H.J.Carmichael, *An Open Systems Approach to Quantum Optics*, lecture Notes in Physics, 18, (Springer, berlin 1993);J.Dalibard, Y.Castin, and K.Molmer, Phys. Rev. Lett. **65**, 580, (1992); H.M.Wiseman and G.J.Milburn, Phys. Rev. A**47**,1652 (1993).

[5] H.Wallis, J.Dalibard and C. Cohen-Tannoudji, Z. Phys. B**54**,407 (1992).

[6] S.Dyrting, G.J.Milburn and C.A.Holmes, Phys. Rev E **48**, 969, (1993).

[7] S.T.Dembinski, A.J.Makowski and P.Peplowski, Phys. Rev. letts. **70**, 1093 (1993).

Non-Classical States of Motion and Quantum Collapse and Revival in an Ion Trap

J.I. Cirac[1], *R. Blatt*[2], *A.S. Parkins*[3], *and P. Zoller*[3]

[1]Departamento de Fisica Aplicada, Universidad de Castilla-La Mancha, Campus Unversitario s/n, E-13071 Ciudad Real, Spain

[2]Institut für Laser-Physik, Universität Hamburg, Jungiusstrasse 9, D-20355 Hamburg, Germany

[3]Joint Institute for Laboratory Astrophysics, University of Colorado and National Institute of Standards and Technology, Boulder, CO 80309-0440, USA

A novel scheme for the preparation of non–classical states of motion in a three–dimensional harmonic ion trap is proposed. The technique is based on adiabatic passage along dressed energy levels of the strongly coupled ion + trap system by varying the laser frequency. For an observation of these non–classical states we propose to measure quantum collapse and revival in the population inversion of a single trapped ion.

1. Introduction

Recent experiments on laser cooling and trapping [1] have reported the observation of quantum effects related to the center–of–mass motion of ions and atoms. In ion traps, sideband cooling to the vibrational ground state of the trapping potential has been demonstrated [2], and for neutral atoms quantized motion in optical potentials has been observed in spectroscopy of optical molasses [3]. An intriguing new perspective is the possibility to prepare *nonclassical states of motion* of cold atoms and ions [4, 5, 6]. This is interesting from the point of view of high resolution spectroscopy with cold atoms and ions. In addition this opens a new route for experiments to test fundamental questions in quantum mechanics, such as the formation and decay of macroscopic superposition states. We have proposed [5] the generation of Fock states of motion of trapped ions (energy eigenstates of the trap) by observation of quantum jumps in three–level systems placed at the node of a standing laser wave. In a similar configuration, squeezed states of motion can be prepared as a "dark state" of the atom in a multichromatic laser wave [6]. For an ion located at the node of a standing light wave there exists an interesting analogy between the ion in the trap and the Jaynes–Cummings model (JCM) of cavity quantum electrodynamics (CQED) [5, 6].

Springer Proceedings in Physics, Vol. 77 Quantum Optics VI
Editors: D.F. Walls · J.D. Harvey

2. 3–D Fock states of the ion motion

The Hamiltonian describing the motion of a trapped ion is given by [7]

$$H_{tp} = \sum_{i=x,y,z} \left[\frac{P_i^2}{2m} + \frac{1}{2} m \nu_i^2 R_i^2\right] = \sum_{i=x,y,z} \hbar\nu_i \left[a_i^\dagger a_i + \frac{1}{2}\right], \tag{1}$$

where $a_i = [m\nu_i/(2\hbar)]^{(1/2)}[R_i + iP_i/(m\nu_i)]$ is the usual annihilation operator along the direction i ($\vec{R}$ and $\vec{P}$ are position and momentum operators, and m is the ion mass), and ν_i is the frequency of the ion motion in the i direction. Denoting $|n_i\rangle_i$ ($i = x, y, z$) the Fock state (eigenstate) of the harmonic oscillator with energy $\hbar\nu_i(n_i+1/2)$, we seek to prepare (i) Fock states in three dimensions, i.e.

$$|\Psi\rangle = |n_x, n_y, n_z\rangle \equiv |n_x\rangle_x |n_y\rangle_y |n_z\rangle_z, \tag{2}$$

and (ii) linear superpositions of these states, such as

$$|\Psi\rangle = \alpha|n_x, n_y, n_z\rangle + \beta|n_x, n_y + 1, n_z\rangle \ . \tag{3}$$

To prepare these states, the ion is assumed to interact with three standing waves of frequencies ω_x, ω_y, and ω_z along the directions x, y, and z, respectively. These laser beams excite an internal transition $|g\rangle \rightarrow |e\rangle$ of the ion modeled here by a two–level system with transition frequency ω_0. For the preparation of the nonclassical states (2) and (3) we propose to use a slow variation of the laser frequency in such a way that the motion of the ion is left in one of the states (2) or (3), whereas the internal atomic state is the ground state $|g\rangle$. This adiabatic passage is coherent, and we will assume that there is no spontaneous emission during the preparation. This is the case, for example, when the lasers excite an electric–dipole forbidden transition [2]. After the frequency chirp cycle the ion will be left in its internal ground state, therefore the prepared motional state persists due to the absence of spontaneous emission.

The Hamiltonian describing the motion of the ion in the presence of the laser beams is

$$H = H_{tp} + H_I + H_{int}, \tag{4}$$

where H_{tp} is given in (1), $H_I = \omega_0\sigma_z/2$ ($\hbar = 1$) is the free Hamiltonian describing the two–level system, and

$$H_{int} = \sum_{i=x,y,z} \frac{\Omega_i}{2} \sin(k_i R_i + \phi_i)(\sigma_+ e^{-i\omega_i t} + h.c.)$$

gives the interaction with the laser. Here, $\sigma_\pm$, σ_z are the usual spin–1/2 operators describing the two–level system, Ω_i is the peak Rabi frequency of the laser standing wave in the direction $i = x, y, z$; $k_i = \omega_i/c$, and ϕ_i depends on

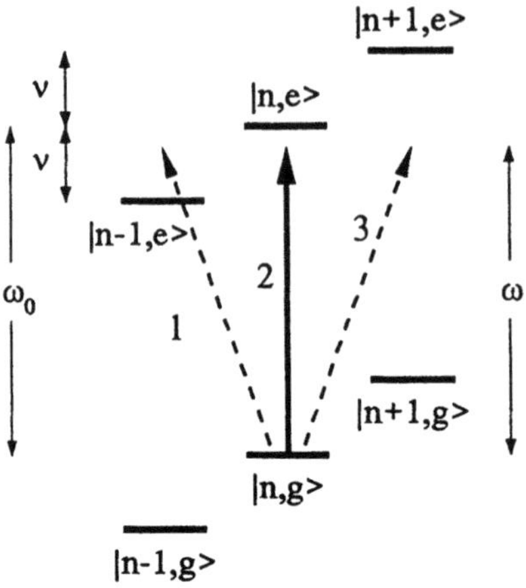

Figure 1: Energy levels for a two–level ion moving in a harmonic potential. For $\omega \simeq \omega_0$, transitions from $|n,g\rangle \rightarrow |n,e\rangle$ are close to resonance (2), whereas transitions with $|n,g\rangle \rightarrow |n \pm 1, e\rangle$ are very far off resonance (1) and (3). For $\omega \simeq \omega_0 - \nu$ transitions from $|n,g\rangle \rightarrow |n-1,e\rangle$ are close to resonance (1) whereas all other transitions are very far off resonance. For $\omega \simeq \omega_0 + \nu$ only transitions from $|n,g\rangle \rightarrow |n+1,e\rangle$ are close to resonance (3).

the relative phase between each of the lasers forming the standing wave in this direction.

In the following we will consider that at any given time only one standing wave (denoted by i) interacts with the ion. In this case, in a frame rotating with the laser frequency ω_i, the Hamiltonian is

$$H = H_{tp} - \frac{(\omega_i - \omega_0)}{2}\sigma_z + \frac{\Omega_i}{2}\sin(k_i R_i + \phi_i)(\sigma_+ + h.c.)\,. \tag{5}$$

In the absence of spontaneous emission and considering interaction with a single laser only, the motion in the three spatial dimensions decouples, which allows us to study the Hamiltonian H in one dimension. For simplicity we will suppress the subscript i in the following. In contrast to the standard theory of ion cooling [8, 9], we will not assume in the following that we are in the LDL, i.e. we do not assume that the ion is localized in a region small compared to the optical wavelength [8]. This allows us to consider the preparation of motional states larger than the wavelength of the light.

The bare Hamiltonian ($\Omega = 0$) shows degeneracies whenever the laser detuning is a multiple of the trap frequency, $\omega - \omega_0 = k\nu$ ($k = 0, \pm 1, \ldots$), i.e. the laser is tuned to one of the "motional sidebands" corresponding to a degeneracy between $|n,g\rangle$ and $|n+k,e\rangle$. In the presence of the laser these degeneracies become avoided crossings, and for sufficiently weak laser excitation these avoided crossings will be isolated (nonoverlapping). For example, when the laser frequency is close to the two–level transition resonance ($|\omega - \omega_0| \ll \nu$) transitions changing the harmonic oscillator quantum number n are off–resonance ($k = 0$) and can be neglected (arrows 1 and 3 in Fig. 1).

Figure 2 is a numerical example of a dressed energy spectrum. In this figure we have plotted the exact eigenvalues of H (calculated numerically) as a function of the laser frequency ω, for $\eta = 0.5$, $\Omega = 0.3\nu$, and $\phi = \pi/4$. The dotted line represents the energies of the "bare" states. For the laser frequencies far below the two–level resonance (point "C" in Fig. 2) no transitions between states $|g\rangle$ and $|e\rangle$ take place, since they are very far off resonance. Hence, the eigenstates of H are those of the free part $H_{tp} + H_I$, i. e. $|n, g\rangle$ and $|n, e\rangle$. For increasing values of the laser frequency, states $|n, g\rangle$ couple to states $|n-1, e\rangle$ via Hamiltonian H^{JC}, and therefore the eigenstates (or dressed states) are linear combinations of these two states.

The ability to change adiabatically the laser frequency offers the possibility to generate Fock states of the atomic motion by adiabatically following one of the dressed states. To illustrate this, let us assume that the initial state of the ion is in $|0, g\rangle$. This is the vibrational ground state in which the ion is left after sideband cooling. The laser frequency is initially set to a certain value between ω_0 and $\omega_0 + \nu$ (denoted by "A" in Fig. 2), so that all transitions are off resonance. The corresponding energy level is denoted "1" in the energy diagram of Fig. 2. When the laser frequency is decreased adiabatically, the state of the ion changes according to the dressed energy diagram of Fig. 2. Once the frequency ω reaches "C", the state of the ion has evolved to $|1, g\rangle$ (point "2" in the figure). Now the laser is switched off and on again, but with the original frequency "A". Thus, the state of the ion corresponds to the point

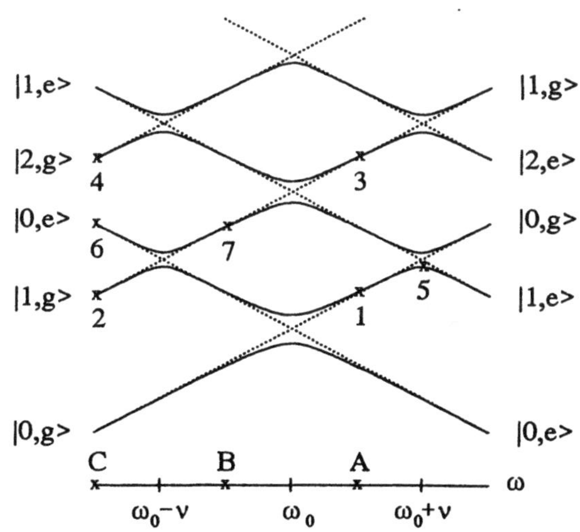

Figure 2: Energy level diagram for a two–level ion interacting with laser light, as a function of the laser frequency ω. Dotted lines correspond to the bare states.

"3" in the figure. By decreasing adiabatically the laser frequency again the ion ends up in the state $|2,g\rangle$ (point "4" in the figure). By repeating this cycle n times, the state of the ion becomes $|n,g\rangle$, i.e. a Fock state with n quanta is prepared. Note that after any cycle, the internal structure of the ion is left in its ground state $|g\rangle$, and therefore the Fock state is not modified due to spontaneous emission from the electric dipole forbidden transition.

The conditions for the adiabatic following can be easily estimated for a laser frequency varying linearly with time. In this case one finds $\bar{\Omega}T \gg \nu/\bar{\Omega}$, where T is the time duration of each cycle, and $\bar{\Omega}$ is the effective Rabi frequency for transitions $|n,g\rangle \to |n-1,e\rangle, |n,e\rangle$ which for the sake of simplicity, we assume to be similar. For example, for a trap frequency of $(2\pi)1$ MHz, and an effective Rabi frequency of $(2\pi)50$ kHz, adiabaticity condition implies interaction times $T \gg 60$ μs. These numbers are within reach of current experiments with trapped ions. It is worth emphasizing the fact that under low excitation conditions [10] and when the adiabaticity condition is satisfied, this procedure for preparing Fock states is insensitive to the specific values of the parameters involved in the process. Furthermore, it can be easily generalized to other laser configurations, such as traveling wave excitations.

General linear combinations of neighboring Fock states can be prepared following the above procedure, with the initial state of the ion in the superposition state $(\alpha|g\rangle + \beta|e\rangle)|0\rangle$. As can be deduced following the level diagram of Fig. 2,

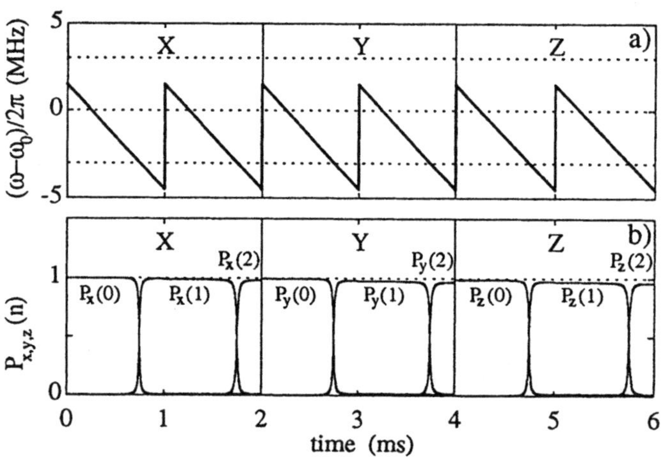

Figure 3: Preparation of the 3D Fock state $|2\rangle_x|2\rangle_y|2\rangle_z$ by adiabatic variation of the laser detuning. (a) Variation of the laser frequency as a function of time (X, Y, and Z indicate the direction of propagation of the laser that is on during each time interval; these time intervals are separated by vertical lines in the figure. (b) Populations of the states $|n\rangle_x$, $|n\rangle_y$, and $|n\rangle_z$ ($n = 0,1,2$) vs time. Note that we have plotted the populations only during the time intervals in which they change due to the presence of the laser.

after one cycle the state of the ion becomes $(\alpha|1\rangle + \beta|0\rangle)|g\rangle$. After n cycles, the state of the ion will be $(\alpha|n\rangle + \beta|n-1\rangle)|g\rangle$.

Fock states in two and three dimensions may be prepared as follows. We assume that the ion has been cooled via sideband cooling to its lowest state along the three directions, i.e. the initial state of the ion is $|0,0,0,g\rangle$. Applying lasers along one direction, for example x, and with the procedure described above, the ion ends up in the state $|n_x,0,0,g\rangle$. Next, the lasers along x are switched off, and those along y switched on. By adiabatic change of the laser frequency the state $|n_x,n_y,0,g\rangle$ is prepared. Finally, proceeding in the same manner with the z direction, one can prepare state (2). Obviously, by combining this procedure with that described in the previous paragraph one is able to generate state (3).

As a numerical example, we have plotted in Fig. 3 the time evolution for a pulse sequence that leads to the preparation $|0,0,0\rangle \rightarrow |2,2,2\rangle$ by an adiabatic variation of the laser detuning. These curves were calculated by solving the full time–dependent Schrödinger equation using a truncated basis of oscillator eigenstates. The frequencies of the lasers along the three directions as a function of time are plotted in the upper part of the Fig. 3(a). Figure 3(b) shows the populations $|n\rangle_x$, $|n\rangle_y$, and $|n\rangle_z$ $(n = 0,1,2)$ of the trap states.

3. Quantum collapse and revival

Collapses and revivals are features of the quantized motion of the ion which are directly related to the discreteness of the vibrational states in the trap [11]. Experimental observation of these collapses and revivals provides a novel means of measuring the statistics of the quantum motion of the ion [12], and thus of detecting non–classical states of motion (such as Fock states and squeezed states of motion [5, 6]).

For an observation of the collapse and revival phenomenon we propose to employ a single ion trapped in a harmonic trap (e.g., a Paul trap) and an internal level scheme as indicated in Fig.4 (a). The two–level transition $|g\rangle \leftrightarrow |r\rangle$ used to observe the collapse and revival is a forbidden transition (e.g., an electric quadrupole transition), for which the spontaneous decay time (Γ_r^{-1}) is long compared to the observation time τ and the trap period ν^{-1}. It is assumed that this transition is driven by a very strong standing wave field in such a way that the effective coupling constant leads to several Rabi oscillations during the interaction time. The third level $|e\rangle$ is used for optical cooling of the ion (initial preparation), and for a measurement of the population inversion (on the $|g\rangle$ - $|r\rangle$ transition) as explained below.

In a first step, one prepares the ion in a given state of motion, which serves as the initial harmonic oscillator state for interaction with the two–level system $(|g\rangle \leftrightarrow |r\rangle)$. For example, after initial cooling on the transition $|g\rangle \leftrightarrow |e\rangle$ a Fock state of motion could be prepared via adiabatic passage on the transition $|g\rangle \leftrightarrow |r\rangle$ as outlined in the previous section.

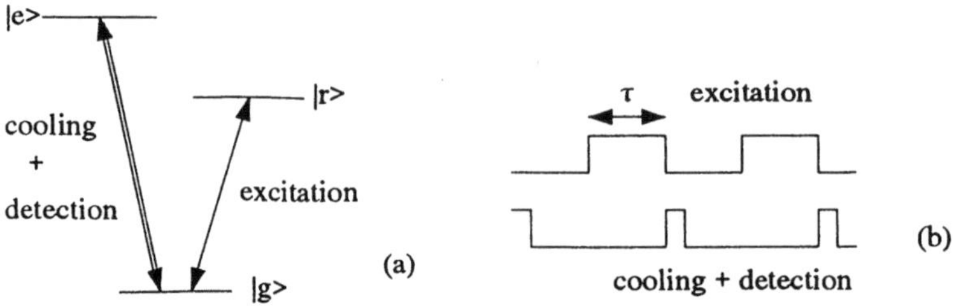

Figure 4: (a) Three–level ion with a strong transition $|g\rangle \leftrightarrow |e\rangle$ for laser cooling and weak transition $|g\rangle \leftrightarrow |r\rangle$ which serves as the effective two–level system for the observation of collapse and revival phenomena. (b) The lasers on the strong and weak transition are alternately switched on and off. After a certain interaction time τ on the weak transition, interaction on the strong transition serves to measure the ground state population.

Collapse and revival in the three–level atom shown in Fig.4 (a) can then be observed by exciting the narrow transition $|g\rangle \leftrightarrow |r\rangle$ in the following way (cf. Fig.4 (b)): (i) after the ion is laser cooled and an initial state distribution $\rho(t_i)$ is prepared, the cooling radiation is turned off, (ii) then laser radiation on the narrow (two–level) transition is applied for a certain time τ with the ion placed at the node of the standing wave laser field, and (iii) at the end of this time interval τ the laser exciting the strong transition is turned on again; the absence or presence of fluorescence from the strong transition will indicate whether or not the ion has made a transition (quantum jump) to the excited metastable level $|r\rangle$ [13, 14, 15]. We repeat the same sequence in order to determine the probability $P_r(\tau)$ that after the interaction time τ the ion is in the state $|r\rangle$ (i.e., to determine the population inversion $\langle\sigma_z(\tau)\rangle = 1 - 2P_r(\tau)$). Repeating this measurement for different interaction times τ, we build up the population inversion as a function of time τ. Depending on the initial state distribution, this population inversion exhibits different quantum collapses and revivals which provide a signature of the statistical distribution of the initially prepared state $\rho(t_i)$.

Here we present a discussion and numerical results for collapse and revivals of the population inversion of a trapped ion given the harmonic oscillator to be in a Fock state ($|n_0\rangle$), i.e. we start with the initial population distribution

$$P_n^{Fock} = \delta_{n,n_0}. \tag{6}$$

In Fig. 5 we plot our results for a Fock state $|n_0 = 4\rangle$. The different curves indicate the behavior for various laser bandwidths. The solid line shows the sinusoidal behavior as is expected for a Fock state, and a measurement of its period allows the determination of the quantum number n_0. As expected, for larger laser bandwidth, the inversion damps out as a function of time.

The appearance of collapse and revival depends sensitively on the initial state distribution and demonstrates that the proposed measurement of collapse

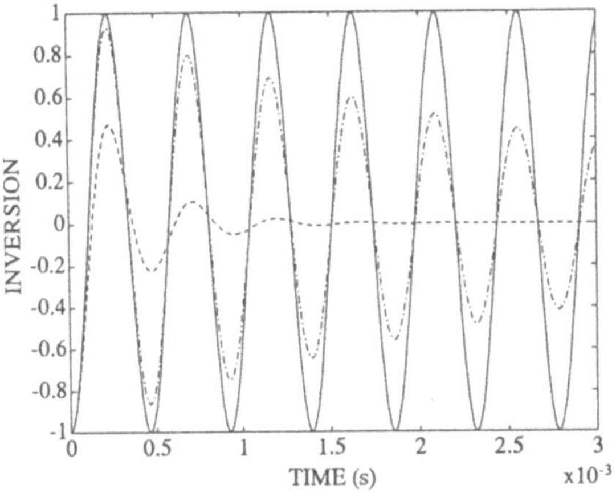

Figure 5: Population inversion as a function of time for an initial Fock state $n = 4$ and laser bandwidth $\Gamma' = 2\pi \times 0, 100, 1000\ \mathrm{s}^{-1}$ (solid, dashed–dotted, dashed line, respectively). The Rabi frequency is $\Omega = 2\pi \times 10^5$.

and revival phenomena is a new tool to analyze the state distribution in an ion trap.

Finally, it should be noted that for an experiment with a single trapped ion the localization at the node of the standing wave is only required to suppress unwanted transitions $|g,n\rangle \to |e,n\rangle$ which may appear at the exact resonance frequency. We remember that placing an ion at the node of the standing wave field allows interaction only with the motional sidebands corresponding to transitions $|g,n\rangle \to |e,n\pm 1\rangle$. For a highly forbidden transition, such as, e.g. that used in the quantum jump experiments [15], a traveling wave detuned to the lower sideband (i.e. $\omega_L = \omega_0 - \nu$) would be sufficient for the observation of these phenomena. Thus, an experimental realization of the proposed technique seems to be readily feasible with single trapped Ba^+, Ca^+, Sr^+, Hg^+, In^+ and Yb^+ ions.

4. Conclusion

In summary, we have proposed a novel scheme for the preparation of a variety of three–dimensional nonclassical states of motion in a trap. This technique is based on the excitation of an electric dipole forbidden transition by different lasers in an optical molasses configuration, in such a way that during the preparation no spontaneous emission takes place. Nonclassical states can be produced by adiabatically varying the frequency of the lasers. The scheme presented here is valid beyond the Lamb–Dicke limit, and is not sensitively dependent on the specific values of the laser parameters, trap frequencies, etc.

Furthermore, after the state of motion is prepared, the internal structure of the ion is left in its ground state, and therefore the nonclassical state persists. It is expected that the observation of the quantum collapse and revival phenomenon in appropriate three–level ions will allow the observation of such non–classical states.

References

[1] S. Chu and C. Wieman (Eds.), special issue of JOSA B **6** (11) (1989); S. L. Gilbert and C. E. Wieman, Opt. Photonics News **4**, 8 (1993).

[2] F. Diedrich, J. C. Bergquist, W. M. Itano, and D. J. Wineland, Phys. Rev. Lett. **62**, 483 (1989).

[3] P. Verkerk *et al.*, Phys. Rev. Lett. **68**, 3861 (1992); P. S. Jessen *et al.*, Phys. Rev. Lett. **69**, 49 (1992); A. Hemmerich and T. W. Hansch, Phys. Rev. Lett. **70**, 410 (1993); G. Grynberg *et al.*, Phys. Rev. Lett. **70**, 2249 (1993).

[4] D.J. Heinzen and D.J. Wineland, Phys. Rev. A **42**, 2977 (1990).

[5] J. I. Cirac, R. Blatt, A. S. Parkins, and P. Zoller, Phys. Rev. Lett. **70**, 762 (1993).

[6] J. I. Cirac, A. S. Parkins, R. Blatt, and P. Zoller, Phys. Rev. Lett. **70**, 556 (1993).

[7] J. I. Cirac, R. Blatt, and P. Zoller, Phys. Rev. A 49, in press

[8] See, for example, S. Stenholm, Rev. Mod. Phys. **58**, 699 (1986).

[9] J. I. Cirac, R. Blatt, P. Zoller, and W. D. Phillips, Phys. Rev. A **46**, 2668 (1992).

[10] By low excitation we mean that transitions only take place under close–to–resonance excitation. For transitions $|n\rangle \rightarrow |n\rangle, |n \pm 1\rangle$ this is the case for $\Omega \sin(\phi), \Omega\eta\cos(\phi) \ll \nu$.

[11] C. A. Blockley, D. F. Walls, and H. Risken, Europhys. Lett. **17**, 509 (1992).

[12] J. I. Cirac, R. Blatt, A. S. Parkins, and P. Zoller, Phys. Rev. A 49 (2), (1994)

[13] W. Nagourney, J. Sandberg, and H. Dehmelt, Phys. Rev. Lett. **56**, 2797 (1986).

[14] Th. Sauter, W. Neuhauser, R. Blatt, and P. E. Toschek, Phys. Rev. Lett. **57**, 1696 (1986).

[15] J. C. Bergquist et al., Phys. Rev. Lett. **57**, 1699 (1986).

Part III

Quantum Measurement and Quantum Noise

Quantum Non-Demolition Measurements and Back-Action-Induced Squeezing in Atomic-Three-Level Systems

P. Grangier[1], *J.-Ph. Poizat*[1], *P. Grelu*[1], *F. Castelli*[2], *L.A. Lugiato*[2], *and A. Sinatra*[2]

[1]Institut d'Optique, B.P. 147, F-91403 Orsay Cedex, France
[2]Dipartimento di Fisica dell'Università, via Celoria 16, I-20133 Milano, Italy

This paper deals with two topics :

(a) the implementation of a quantum non-demolition (QND) measurement in the optical domain, using the two-photon non-linearity of three-level atoms placed inside a doubly-resonant optical cavity.

(b) the experimental observation and theoretical interpretation of a squeezing effect which occurs in the same set-up, when the optical cavity is detuned from the exciting laser beams. The origin of this effect can be attributed to the transfer of the intensity fluctuations of one beam to the phase fluctuations of the other one, (similar to the information transfer which occurs in the QND measurement), followed by an optical feedback onto the intensity of the initial beam. This feedback effect is obtained through mixing of the phase and intensity quadratures, due to the detuning of the optical cavity which contains the nonlinear medium.

Both experiments show that the information obtained by the QND measurement can be used to correct the intensity fluctuations of the signal beam by a build-in mechanism, which does not require any use of external electronics.

1 Introduction

A quantum measurement usually perturbs the quantity which is measured, adding back-action noise to the system under study. The purpose of Quantum Non-Demolition (QND) measurements is to leave the observed quantity unperturbed, while adding the back-action noise into another complementary observable [1]. A travelling wave optical QND scheme usually involves the coupling of two field modes: a *signal* mode, which is to be measured, and a *meter* mode, which is directly detected and yields some informations about the signal mode. Throughout this paper, the measured quantity will be the intensity, measured via optical crossed Kerr effect, and the back-action noise will therefore appear on the phase [2]. A perfect Kerr effect QND measurement device should leave the intensity quadrature component of the signal mode unchanged, and turn the phase quadrature component of the meter mode into some kind of macroscopic copy of the signal [3]. If a good QND measurement is achieved, the conditional variance of the output signal beam, given the measurement read out, is smaller than the shot-noise limit [4]. Therefore, the information from the meter beam can be used to realize a quantum noise reduction feedback or feedforward scheme.

We want here to investigate the situation where an input coherent beam is turned into an intensity squeezed beam, via a measurement of its quantum intensity noise followed by an optical feedback. During recent years, various methods have been implemented in order to produce squeezed light [5]. Several proposals have also been made in order to squeeze the quantum intensity fluctuations of an intense coherent field, achieving thus a so-called quantum noise eater (see for example [6][7][8]). A possible way, which has been theoretically proposed but not yet implemented, is to achieve a quantum non-demolition (QND) measurement [1] of the intensity fluctuations of a laser beam, and then to use the result of this measurement to correct the initial beam [9]. A quantum feedforward scheme has already been realized using twin beams generated by an optical parametric oscillator [10]. In that case, there is no input beam to be corrected, but the sub-shot-noise correlations between the two output beams allow one to reduce the intensity fluctuation of one of them, using the information given by the direct detection of the other one [11].

In this paper, we want to demonstrate an implementation of intensity noise reduction of a laser beam, using a scheme related to a QND measurement. The QND effect uses two-photon crossed Kerr effect, obtained in three-level atoms in a cascade configuration contained in an optical cavity [12][13][14]. In our experiment, the feedback is performed optically by detuning the optical cavity from resonance. In this situation, the input to output transformation of the signal beam is no more a QND one, since the coherent input signal is actually squeezed at the output. We will show that this squeezing effect results from an all-optical "measurement and feedback" scheme, which is basically due to the Kerr coupling between the two beams. This effect is therefore different from the "bright squeezing" scheme, based on electromagnetically induced transparency, which was considered in ref. [8] using an Λ configuration for three-level atoms.

2 Description of the experiment

The set-up used for this experiment (fig. 1) has been described in ref. [14] for realizing QND measurements. The non-linear coupling of the signal and meter fields is obtained in a beam of three-level atoms in a ladder configuration ($3s_{1/2} - 3p_{3/2} - 3d_{5/2}$ levels of sodium

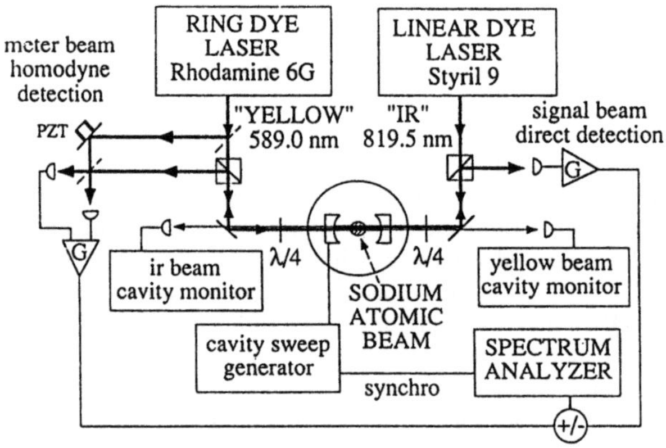

Figure 1. Experimental set-up. Two lasers at 589.0 nm (*meter* beam, labelled 1) and 819.5 nm (*signal* beam, labelled 2) are tuned close to the resonances of the $3s_{1/2} - 3p_{3/2} - 3d_{5/2}$ cascade in a sodium atomic beam. The quarter-wave plates are used to polarize circularly the two beams in the interaction region, and also to separate the input and output beams. The intracavity intensities are monitored through the leaks of the high reflector mirror at each wavelength.

atoms). The atomic beam provides a density of 5.10^{11}atoms/cm^3, over an interaction length of 1 cm, with a Doppler width of about 200 MHz (FWHM). Two *cw* electronically stabilized dye lasers are tuned around 589 nm (meter beam) for the lower transition and 819.5 nm (signal beam) for the upper one. The non-linearity is enhanced by placing the non-linear medium in a doubly resonant optical cavity (5.2 cm long, 5 cm mirror radius) with a good finesse (100 for each wavelength) and low losses. One of the mirrors is a high reflector at 589 nm ($t^2 = 7.10^{-4}$) and is the input-output mirror for the 819.5 nm beam ($t^2 = 7.10^{-2}$) while the other one is high reflector at 819.5 nm ($t^2 = 5.10^{-4}$) and is the input-output mirror for the 589 nm beam ($t^2 = 7.10^{-2}$), so that the cavity is single ended at each wavelength. The total losses are characterized by the on-resonance cavity reflection coefficient which is 0.95 for the 819.5 nm beam and 0.90 for the 589 nm laser beam. One of the mirror is piezoelectrically mounted, allowing the length of the cavity to be adjusted. In order to have a phase-sensitive low-intensity detection, a homodyne detection with local oscillator power of $800\mu W$ and fringe visibility of 0.92 is used on the meter beam, while the signal output is directly detected. The detectors are $p-i-n$ silicon photodiodes with quantum efficiency of 0.78 at 589 nm and 0.93 at 819.5 nm followed by low-noise preamplifiers.

We use the *ghost transition* configuration of ref [15] (see also ref [16]). The meter beam is tuned around the lower atomic transition and has a very small power ($5\mu W$) compared to the signal beam (about $300\mu W$) which is tuned around the upper transition. In this configuration, most of the atomic population remains in the ground state, and the absorption of the signal beam is very small. This guarantees a nearly perfect non-demolition interaction while the coupling between the amplitude of the signal and the phase of the meter beam can be quite large [15].

The experimental procedure for operating in the QND regime is the following. The cavity length being scanned, the detuning and eventually the power of the lasers are

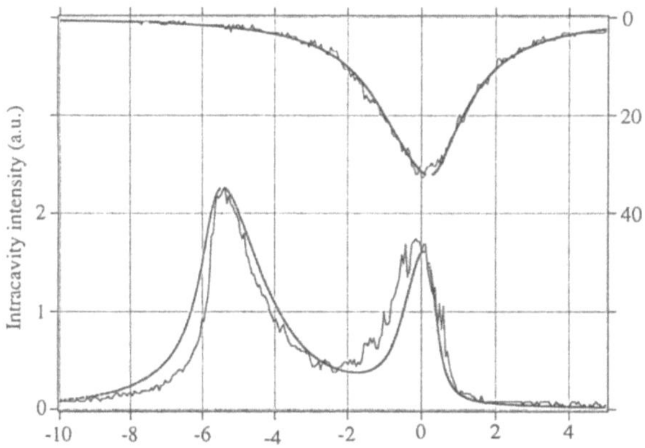

Figure 2. Behaviour of the intracavity mean intensities during a cavity scan (experimental result and theoretical fit). The lower trace is the *meter* intensity, and the upper trace is the *signal* intensity, inverted and rescaled for clarity. The left peak in the lower trace corresponds to the resonance of the meter field in absence of the signal field. In this plot, and in all the relevant following plots, the horizontal axis show the cavity detuning of the signal field, normalized to its cavity damping rate, and the origin on the cavity scan axis corresponds to the resonance of the IR signal beam alone.

adjusted in order to find a good working point, which is obtained when the cavity is resonant at both wavelengths with a reasonably strong non-linear coupling between the two fields. A typical situation is shown in fig. 2: the meter beam is "attracted" to resonance by the strong signal beam due to the non-linear phase shift induced by the signal field. The latter is itself not perturbed by the meter beam (of much weaker intensity) therefore avoiding two-photon bistability [17]. A typical detuning of the meter beam from the lower transition is -1.5 GHz , while the signal beam is tuned at about +0.5 GHz of the upper atomic transition. All noise measurements presented below are done at 10 MHz analysis frequency. Similar results could be obtained between 5 and 15 MHz, these limits being respectively due to excess noise and cavity roll-off.

When the system is operating in the QND regime, the transfer of the signal-to-noise ratios (SNR) from SNR_s^{in} to SNR_m^{out} is measured to be $T_m = 0.4$, and the transfer to

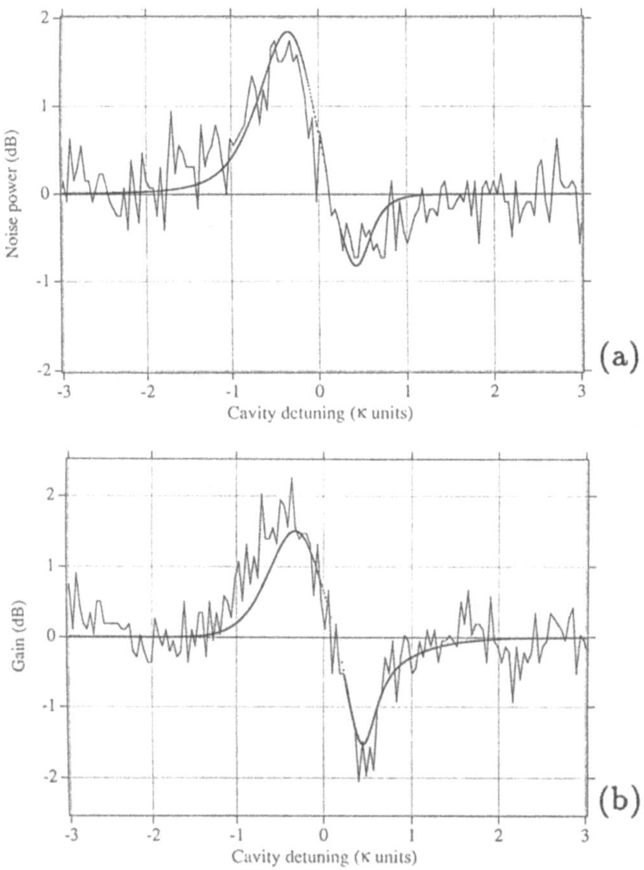

Figure 3. (a) Experimental result and theoretical fit for the noise of the signal beam in dB (noise analysis frequency 10 MHz). The shotnoise level is at 0 dB, and it is obtained when the meter beam is turned off (we checked that the input signal beam is indeed shot-noise limited). (b) Same curves, showing now a classical modulation at 10 MHz. Optical losses including detector quantum efficiency are 0.84 on the IR beam. In these two graphs, both experimental conditions and fit parameters are the same as in Fig.2.

SNR_s^{out} is $T_s = 0.9$. As a consequence, the sum of the signal and measurement SNR transfers is $T_s + T_m = 1.3$. A value of $T_s + T_m$ larger than one is the indication of operation in the quantum regime, and a perfect QND device would have $T_s = T_m = 1$ [13][14][18]. Moreover, the measured conditional variance [4] of the output signal beam, knowing the result of the meter read out, was found equal to $V_{s|m} = 0.85$, which is clearly below the classical limit of one.

Let us now look at the quantum intensity noise of the signal beam, when the cavity is scanned in the vicinity of the double resonance. A typical result is shown on fig. 3(a), which displays the signal beam intensity noise at 10 MHz : it is clear that the noise is below the shot-noise level (SNL) on one side of the peak, equal to the SNL at resonance (QND effect, see Section IV), and above the SNL on the other side. A similar effect can be observed on a classical amplitude modulation at 10 MHz, which is applied by an electro-optical modulator on the signal beam (fig. 3(b)). Therefore, both the signal quantum noise and classical modulation can be deamplified by the device detuned from the exact QND condition. Since the signal beam is exciting the upper transition of the three-level atoms, obviously no effect can occur without the "meter" beam exciting the lower transition. Therefore, the interpretation of this effect must take into account the coupling between the two beams in the non-linear system.

The fits, which are in good agreement with the experimental results, are performed by a computer code which calculates the mean values of the intra-cavity intensities, and the quantum noise properties of the system (see Section 5). The atomic medium is described by a cascade three level configuration taking into account spontaneous emission for arbitrary parameters [15][19][20].

3 The crossed Kerr effect model

In order to get a physical insight into this effect, we will first consider a simple non-linear coupling between the two beams through a phenomenological crossed Kerr effect [3], that ignores the atomic nature of the non-linear medium. We have shown in previous papers [3][15][17] that the crossed Kerr model is indeed a good first order approximation for two-photon dispersion in atomic three-level systems. This model will be used to explain the basic idea of the "measurement and feedback" effect, and to give a first approach to the optimization of the parameters. This shall give an intuitive idea of the phenomenon, and allows one to interpret simply the effect as a result of the mixing of the quadrature components due to the cavity detuning [21].

We consider a crossed-Kerr medium located in a good finesse doubly resonant optical cavity. As derived in ref.[22], it can be described using a set of four Quantum Langevin equations, for the annihilation and creation operators a_j, $a_j^\dagger$ ($j = 1, 2$) of the two relevant cavity modes, taking into account the cavity decay rates κ_j of the two fields, the empty cavity detunings ϕ_j, and the crossed Kerr coupling coefficient f .

Assuming that the quantum fluctuations are much smaller than the stationary mean values of the fields, the latter are obtained by setting to zero the time derivatives, and by neglecting the quantum fluctuations and correlations (semi-classical approximation). This leads to a set of 4 coupled non-linear equations, for $j = 1, 2$ and conjugates, which read :

$$\langle a_j \rangle = \frac{-\sqrt{2\kappa_j}}{\kappa_j(1 + i\phi_j) - if\,|\langle a_{3-j}\rangle|^2}\langle a_j^{in}\rangle \,. \tag{1}$$

where the superscripts *in* are for the input modes to the cavity. Except for the height of the peaks, which will depend on atomic absorption, this model accounts for the main features of the behaviour of the mean fields [17].

In order to deal with the quantum fluctuations of the fields, we use a linearized analysis of this system, which leads to a set of equations in Fourier space. An important parameter in these equations is the non-linear cavity detuning $\alpha_j = \phi_j - \mu_j$, which takes into account the non-linear phase-shifts $\mu_j = f \mid \langle a_{3-j} \rangle \mid^2 / \kappa_j$. The non-linear phase-shifts determine also the coupling between the two field fluctuations, which depends on the coupling coefficient $\mu = \sqrt{\mu_1 \mu_2}$.

As it is often the case with Kerr-type non-linear configuration, this system is bistable or unstable for some range of parameters [17][23]. The stability condition is that the real part of the eigenvalues of the time evolution matrix be strictly negative. A simple form of this condition can be obtained using the Routh-Hurwitz criterion, which yields the double inequality :

$$4\mu^2\alpha_1\alpha_2 < \left(1+\alpha_1^2\right)\left(1+\alpha_2^2\right) \tag{2}$$

$$\text{and} \quad -4\mu^2\alpha_1\alpha_2 < \frac{1}{\kappa_1\kappa_2}\left((\kappa_1+\kappa_2)^2 + \frac{\left(\kappa_1^2\alpha_1^2-\kappa_2^2\alpha_2^2\right)^2}{(\kappa_1+\kappa_2)^2} + 2\left(\kappa_1^2\alpha_1^2+\kappa_2^2\alpha_2^2\right)\right). \tag{3}$$

At the boundary of the region defined by eq.(2), one meets a steady-state instability, related to the turning point of the hysteresis cycle of the intracavity field as a function of any control parameter, e.g. the cavity length. The boundary of the region defined by eq.(3) corresponds to an Hopf instability, which leads to the onset of spontaneous oscillations of the intracavity intensities. In fact the squeezing of the signal beam occurs when $\alpha_1\alpha_2 > 0$, so that only the first inequality will be relevant in the following of this paper, and for simplicity we will limit ourselves to the situation where both α_1 and α_2 are positive. For simplicity also, we will assume that $\kappa_1 = \kappa_2 = \kappa$, which is close to the experimental situation. It can be predicted by solving eq.(2) for α_1 or α_2, the system is always stable when $\mu < 1$, and the unstable area becomes larger as the non-linear coupling increases.

4 Theoretical interpretation

Using the previous model, one can derive input-output relations for the quantum fluctuations of the two fields. As usual, it is convenient to use the quadrature components

$$\begin{aligned} p_j(\omega) &= \frac{\langle a_j \rangle^*}{|\langle a_j \rangle|}\delta a_j(\omega) + \frac{\langle a_j \rangle}{|\langle a_j \rangle|}\delta a_j^{\dagger}(\omega) \\ q_j(\omega) &= -i\left(\frac{\langle a_j \rangle^*}{|\langle a_j \rangle|}\delta a_j(\omega) - \frac{\langle a_j \rangle}{|\langle a_j \rangle|}\delta a_j^{\dagger}(\omega)\right) \end{aligned} \tag{4}$$

which correspond respectively to intensity and phase fluctuations. The general input-output expressions are rather tedious, so we will first look at some physically interesting special cases. For $\alpha_1 = \alpha_2 = \omega = 0$, one obtains the usual QND coupling at zero frequency ($\omega = 0$ will be omitted in the equations) [3]

$$\begin{aligned} p_j^{out} &= p_j^{in} \\ q_j^{out} &= q_j^{in} + 4\mu p_{3-j}^{in} \end{aligned} \tag{5}$$

Therefore, the intensity fluctuations of each field are not changed, but they can be read out on the phase fluctuations of the other field. The excess phase noise on the signal beam is entirely due to the meter beam, and it appears as the "back-action" of the measurement.

Let us now detune the meter field (field 1), keeping $\alpha_2 = \omega = 0$. As the QND case presented above, this configuration is always stable (see eq.2 and 3). The input-output relation for the signal field (field 2) becomes :

$$\begin{aligned} p_2^{out} &= p_2^{in} \\ q_2^{out} &= q_2^{in} + \frac{4\mu}{\sqrt{1+\alpha_1^2}} r_1^{in} + \frac{8\mu^2\alpha_1}{1+\alpha_1^2} p_2^{in} \end{aligned} \tag{6}$$

where r_1^{in} is a quadrature component of the input meter field, which depends on α_1. This transformation is still QND for the signal intensity, but the back-action noise now includes a term which depends on p_2^{in}, and where the coupling coefficient μ appears squared, evidencing a "back-coupling" effect [21]. This term is due to the mixing of the intensity and phase of the meter, which transfers the information about p_2 from the meter phase to the meter intensity, and then to the signal phase via the crossed Kerr effect. A consequence of this effect is that a mixed quadrature component of the signal (combination of p_2^{out} and q_2^{out}) will be squeezed, as it can be easily checked from the above expressions.

Finally, if one wants the signal intensity itself to be squeezed, one has to detune also the signal beam, destroying therefore the QND condition, and turning it into a "back-action-induced squeezing" condition. If the back-coupling from p_2^{in} to q_2^{out} is large, only a small detuning will be enough to bring back and substract the information from the intensity quadrature. However, two things should be noted at that point. First, the intensity fluctuations of the signal at zero frequency will always remain at the shot-noise level for the purely dispersive medium considered; a related effect for usual (one-beam) Kerr effect was described in Ref.[7]. Therefore, it will be necessary to consider non-zero frequency noise when looking for intensity squeezing. Second, the system can become unstable if both α_1 and α_2 are different from zero (see eq.2), and these instabilities will play a role in the optimization procedure. As a matter of fact, some combination of the four quadrature components has to diverge when approaching an instability threshold, and since the two-beam system remains in a minimum uncertainty state, another orthogonal combination has to be squeezed in the same limit [24]. For our purpose, this squeezed combination should include mostly the signal intensity.

Although there is no intensity squeezing at zero frequency, it turns out that the intensity squeezing and best squeezing are very close at non-zero frequencies, as it is illustrated below in fig. 4. It is therefore convenient to use a simple optimization procedure on the best squeezing at zero frequency in order to find a good value of the intensity squeezing at non-zero frequency [7]. The best squeezing for the signal field is obtained when the detuning α_1 of the meter beam is much larger than the detuning α_2 of the signal beam, in order to have an efficient mixing of the meter quadratures. The intensity and best squeezing spectra, for $\mu = 2$ and $\alpha_{1,2}$ obtained using this optimization procedure are shown in fig. 4. There is in principle no lower limit for the degree of squeezing with the Kerr model, provided the coupling coefficient μ is large enough.

As a conclusion of this part, the effect described here can be observed typically when the two fields intensities and detunings are such that the state of the system comes close to an instability turning point, with very different values (but same signs) of the detunings. The *meter* field must have a large detuning, insuring large mixing of the intensity and phase quadratures and good back-coupling to the signal phase, while the *signal* field must have a small detuning, insuring an optimum feedback onto the output

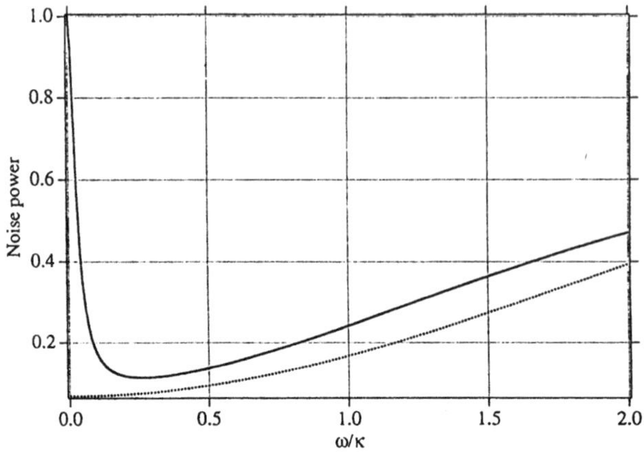

Figure 4. Noise spectra for the best squeezing (dotted line) and for the intensity squeezing (solid line) for $\mu = 2$ with an optimized choice of the detunings ($\alpha_1 = 3.75$ and $\alpha_2 = 0.25$). For this figure, as well as for other noise spectras, we have taken $\kappa_1 = \kappa_2$.

intensity quadrature. Quite large effects are obtained for relatively small values of the non-linearity parameter (typically 10 dB noise reduction for $\mu = 2$), and the squeezing will increase arbitrarily with μ. We will now discuss the corresponding results using a full three-level model for the crossed Kerr non-linearity.

5 Full three level model

As it was said above, our experiment can be modeled by atoms with three energy levels in a ladder configuration, coupled to two laser fields injected in a ring cavity. The behaviour of the mean values of the normalized atomic and intracavity fields variables is described by twelve Bloch-Langevin equations[17]. The stationary solutions are found by taking the temporal derivatives of the variables equal to zero, and solving the non linear system thus obtained. In order to check the stability of each stationary solution, we have also studied the full set of time- dependent differential equations with a Runge-Kutta method for numerical integration, obtaining qualitative and quantitative agreement with experimental observations. In this paper, we consider only the case of stable stationary solutions, disregarding parameter values wich lead to spontaneous oscillations (oscillations or chaos were studied in [17]). The quantum noise is calculated for each cavity length, by analyzing the set of dynamical equations linearized around the stationary solution. A second computer code calculates the relevant noise spectra for arbitrary values of the parameters [15][19][20].

These results are in good agreement with the experimental data (Figs. 2 and 3), provided that one takes into account the extra losses in the optics and the detection quantum efficiency. An optimized result for the squeezing of the intensity is shown on fig. 5 with parameters inspired by the Kerr analysis presented above. We have also displayed on this figure the theoretical curve obtained using the Kerr model for corresponding parameters. The main difference between these two results is the large excess noise given by the full three level atomic model on the antisqueezed side of the resonance, that is

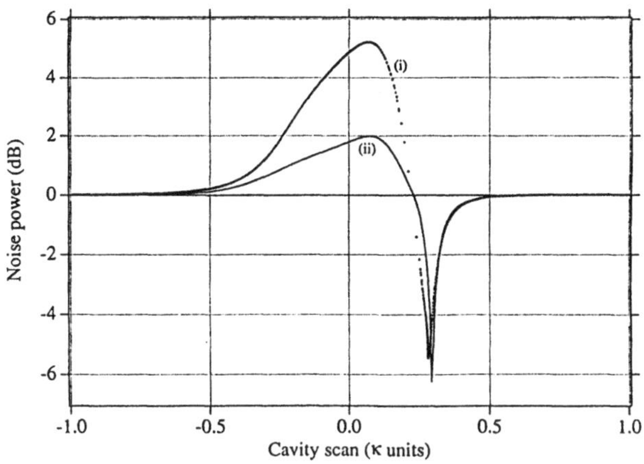

Figure 5. Intensity squeezing during a cavity scan with optimized parameters for the full three-level model (trace (i)). Trace (ii) is the result given by the Kerr model, for corresponding values of the parameters.

not predicted by the parametric Kerr model. This fact is not surprising since it is a general feature of spontaneous emission, which is not included in the Kerr model, to give rise to excess noise. However a reasonably good agreement is obtained on the degree of squeezing and the location of the peak. The agreement is also quite satisfactory for the general shape of the intracavity fields with respect to the cavity detuning (not represented here). These theoretical predictions show clearly that large effects should be observable for real atoms in an optimized experiment.

6 Conclusion

As a conclusion, we have described a situation where squeezing appears as a by-product of a QND measurement, through direct optical feedback from the meter beam to the signal beam. The effect has been clearly observed experimentally, and it could in principle be quite large in an optimized experimental configuration. Beyond its interest in relating the field of QND measurements and the field of squeeezing, this method could be useful if the two-photon non-linear effects, which are used here, turned out to be less sensitive to defects of the non-linear medium (e.g., collisions...) than their one-photon counterparts.

Acknowledgements

This work was completed in the framework of the EC Contract ESPRIT III BRA 6934. Support from the CNR contract number 92.01369.02 is also acknowledged. Institut d'Optique Théorique et Appliquée is URA 14 of the Centre National de la Recherche Scientifique.

References

[1] C. M. Caves, K. S. Thorne, R. W. P. Drever, V. D. Sandberg, and M. Zimmermann, Rev. Mod. Phys. **52**, 341 (1980).

[2] G. J. Milburn and D. F. Walls, Phys. Rev. A **28**, 2055 (1983); N. Imoto, H. A. Haus, and Y. Yamamoto, Phys. Rev. A **32**, 2287 (1985).

[3] P. Grangier, J. F. Roch and S. Reynaud, Opt. Commun. **72**, 387 (1989).

[4] M. J. Holland, M. J. Collett and D. F. Walls, M. D. Levenson, Phys. Rev. A **42**, 2995 (1990).

[5] Special issue on squeezed light, J. Opt. Soc. Am. B, **4** 1450 (1987), H.J. Kimble and D.F. Walls eds.; Special issue on squeezed light, J. Mod. Optics, **34** 709 (1987), P.L. Knight and R. Loudon eds.; Special issue: *Quantum noise reduction in optical system - Experiments*, Appl. Phys. B **55** 189-303 (1992), E. Giacobino and C. Fabre eds.

[6] M. D. Reid and D. F. Walls, Phys. Rev. A **32**, 396 (1985); M. J. Collett and D. F. Walls, *Quantum Optics V*, Springer Proceedings in Physics, Vol 41 (Springer Verlag, Heidelberg 1989).

[7] S. Reynaud, C. Fabre, E. Giacobino and A. Heidmann, Phys. Rev. A **40** 1140 (1989).

[8] K.M. Gheri and D. F. Walls, Phys. Rev. A **46** 6793 (1992).

[9] Y. Yamamoto, N. Imoto, and S. Machida, Phys. Rev. A **33**, 3243 (1986).

[10] A. Heidmann, R. J. Horowicz, S. Reynaud, E. Giacobino, C. Fabre, and G. Camy, Phys. Rev. Lett. **59** 2555 (1987); T. Debuisschert, S. Reynaud, A. Heidmann, E. Giacobino, and C. Fabre, Quantum Opt. **1**, 3 (1989); J. Mertz, T. Debuisschert, A. Heidmann, C. Fabre, and E. Giacobino, Opt. Lett. **16** 1234 (1991).

[11] J. Mertz, A. Heidmann, C. Fabre, E. Giacobino, and S. Reynaud, Phys. Rev. Lett. **64**, 2897 (1990).

[12] P. Grangier, J. F. Roch, and G. Roger, Phys. Rev. Lett. **66**, 1418 (1991).

[13] J. F. Roch, G. Roger, P. Grangier, J. M. Courty, and S. Reynaud, Appl. Phys. B **55**, 291 (1992).

[14] J. Ph. Poizat and P. Grangier, Phys. Rev. Lett. **70**, 271 (1993).

[15] K.M. Gheri, P. Grangier, J. Ph. Poizat, and D. F. Walls, Phys. Rev. A. **46**, 4276 (1992).

[16] H. A. Bachor and P. T. H. Fisk, Applied Phys. B **49**, 291 (1989).

[17] P. Grangier, J. F. Roch, G. Roger, L. A. Lugiato, E. M. Pessina, G. Scandroglio, and P. Galatola, Phys. Rev. A. **46**, 2735 (1992)

[18] P. Grangier, J. M. Courty, and S. Reynaud, Opt. Commun. **89**, 99 (1992).

[19] J. M. Courty, P. Grangier, L. Hilico, et S. Reynaud, Opt. Commun. **83**, 251 (1991).

[20] J. Ph. Poizat, M. J. Collett, and D. F. Walls, Phys. Rev. A **45**, 5171 (1992).

[21] P. Grangier, and J. F. Roch, Opt. Commun. **83**, 269 (1991).

[22] C. A. Blockley and D. F. Walls, Optics Comm. **79**, 241 (1990); J. P. Poizat, M. J. Collett, and D. F. Walls Opt. Commun. **84**, 409 (1991).

[23] P.Grangier and J.F. Roch, Quantum Optics **1**, 17 (1989).

[24] M. J. Collett and D. F. Walls, Phys. Rev. A **32**, 2887 (1985).

Competing $\chi^{(2)}$ Nonlinearities in a Cavity

M.A.M. Marte

Institut für Theoretische Physik, Universität Innsbruck,
A-6020 Innsbruck, Austria

The interaction between two cavity–resonant nonlinear processes in an intracavity $\chi^{(2)}$ medium is studied, namely doubly resonant second harmonic generation and resonant nondegenerate down conversion. A closed set of semiclassical equations of motion for appropriately chosen quantities of the full system is derived and an explicit stationary solution is given. The most striking feature of this solution is a stable plateau in the second harmonic, i.e. the field of frequency $\omega_2 = 2\omega_1$ becomes independent of the pump field $\mathcal{E}_1$, provided $\mathcal{E}_1$ is larger than a certain threshold value. It is demonstrated that undamped selfpulsing can only prevail, if the ratio of the rates of second harmonic generation over parametric down conversion exceeds a value which depends solely on the relative damping rates of second harmonic and fundamental.

I. THE MODEL

Second harmonic generation (SHG) and nondegenerate down conversion in an optical parametric oscillator (NDOPO) have been among the most widely used systems for generating squeezed states of light. Consequently their behaviour has been studied very thoroughly during the past decade (cf. [1] for an overview or e.g. [2] - [6] for SHG and e.g. [7] - [8] for the NDOPO).

Here I will present a semiclassical theory for a situation, where both $\chi^{(2)}$ processes are resonant inside the cavity; the cavity resonance condition and the phase matching condition in the crystal choose appropriate sideband modes with frequency $\omega_\pm = \omega_1 \pm \Delta$ (in the special case $\Delta = 0$ the two modes differ in polarization).

As we will see, the interaction between the two nonlinear processes leads to a new – and in fact rather simple – solution of combined oscillations of all modes. Indeed, such combined oscillations have already been observed in a monolithic total internal reflection ring resonator [9] (MOTIRR) at the University of Konstanz [10], [11].

The situation graphically depicted in Fig.1, can be described by the model Hamlitonian

Springer Proceedings in Physics, Vol. 77 Quantum Optics VI
Editors: D.F. Walls · J.D. Harvey

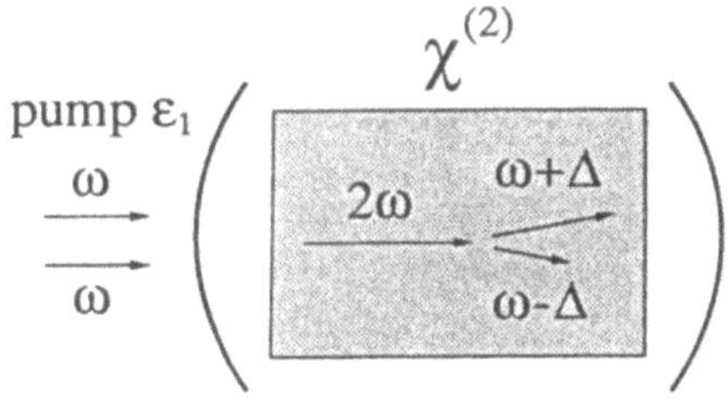

FIG. 1. Schematic diagram of the nonlinear processes in the intracavity $\chi^{(2)}$ medium: the pump field $\mathcal{E}_1$ at frequency ω is frequency doubled and the second harmonic is subsequently down converted to a pair of photons at frequencies $\omega_\pm = \omega \pm \Delta$ (and vice versa).

$$\begin{aligned} H = & \hbar\omega_1 a_1^\dagger a_1 + \hbar\omega_2 a_2^\dagger a_2 + \hbar\omega_+ a_+^\dagger a_+ + \hbar\omega_- a_-^\dagger a_- \\ & + i\hbar\frac{\kappa}{2}\left(a_1^{\dagger 2} a_2 - a_1^2 a_2^\dagger\right) + i\hbar\chi\left(a_2 a_+^\dagger a_-^\dagger - a_2^\dagger a_+ a_-\right) \\ & + i\hbar\left(\mathcal{E}_1(t)\, a_1^\dagger - \mathcal{E}_1^*(t)\, a_1\right) , \end{aligned} \tag{1}$$

which has the following interpretation: the crystal converts two 'photons' of a strong pump field of frequency ω_1 (called fundamental) into one 'photon' of frequency $\omega_2 = 2\omega_1$ and vice versa (doubly resonant SHG). The coupling constant for this process will be denoted by $\kappa \equiv \chi^{(2)}(\omega_1 = \omega_2 - \omega_1)$. At the same time nondegenerate down conversion processes are possible, which change one 'photon' of frequency $\omega_2 = 2\omega_1$ into a pair of two 'photons' of frequency $\omega_+ = \omega_1 + \Delta$ and $\omega_- = \omega_1 - \Delta$, respectively, and vice versa. The coupling constant for the NDOPO will be denoted by $\chi \equiv \chi^{(2)}(\omega_+ = \omega_2 - \omega_-) = \chi^{(2)}(\omega_- = \omega_2 - \omega_+)$. For simplicity the coupling constants have been chosen real.

A semiclassical theory can be derived be calculating the Heisenberg equations of motion for the operators $a_i \in \{a_1, a_1^\dagger, a_2, a_2^\dagger, a_+, a_+^\dagger, a_-, a_-^\dagger\}$ according to $i\hbar\, \dot{a}_i = [a_i, H]$ and replacing the operators by c-number amplitudes α_i. Assuming that the pump field $\mathcal{E}_1$ is in resonance with mode a_1, i.e. oscillates with frequency ω_1, $\mathcal{E}_1(t) = \mathcal{E}_1 e^{-i\omega_1 t}$, it is straightforward to derive the following set of coupled differential equations

$$\begin{aligned} \dot{\alpha}_1 &= -\gamma_1\alpha_1 + \kappa\,\alpha_1^*\,\alpha_2 + \mathcal{E}_1 \quad , \\ \dot{\alpha}_2 &= -\gamma_2\alpha_2 - \frac{\kappa}{2}\alpha_1^2 - \chi\alpha_+\alpha_- \quad , \\ \dot{\alpha}_+ &= -\gamma_+\alpha_+ - i\Delta\alpha_+ + \chi\alpha_2\alpha_-^* \ , \\ \dot{\alpha}_- &= -\gamma_-\alpha_- + i\Delta\alpha_- + \chi\alpha_2\alpha_+^* \end{aligned} \tag{2}$$

(and the four conjugate complex equations), where effective damping rates γ_i for the various modes have been included.

Due to the phase diffusion in the NDOPO [1], [7], the OPO amplitudes $\alpha_\pm$ have only the trivial stationary solution $\alpha_\pm^0 = 0$. Thus, in order the study the influence of the NDOPO semiclassically, one has to go to higher order moments such as the intensities $I_\pm = \alpha_\pm^* \alpha_\pm$ and the sideband coherence $U = \alpha_+ \alpha_-$, which can have a nonzero stationary value (this corresponds to intensity and phase variables [12]).

The *closed* set of equations for the amplitudes of the fundamental α_1, the second harmonic α_2, the intensities in the two sidebands of the NDOPO $I_\pm$, and the sideband coherence U reads

$$
\begin{aligned}
\dot{\alpha}_1 &= -\gamma_1\, \alpha_1 \;+\; \kappa\, \alpha_1^*\, \alpha_2 \;+\; \mathcal{E}_1 \;, \\
\dot{\alpha}_2 &= -\gamma_2\, \alpha_2 \;-\; \frac{\kappa}{2}\, \alpha_1^2 - \chi\, U \;, \\
\dot{U} &= -(\gamma_+ + \gamma_-)\, U \;+\; \chi\, \alpha_2\, (I_+ \;+\; I_-) \;, \\
\dot{I}_\pm &= -2\gamma_\pm\, I_\pm \;+\; \chi\, (\alpha_2 U^* \;+\; \alpha_2^* U) \;,
\end{aligned}
\tag{3}
$$

(and three corresponding c.c. equations).

II. THE SEMICLASSICAL SOLUTION

It can easily be verified that the expressions

$$
\begin{aligned}
\alpha_1^0 &= \frac{\mathcal{E}_1}{\gamma_1 \left(1 \;+\; r\, \frac{\sqrt{\gamma_+\, \gamma_-}}{\gamma_1}\right)} \;, \\
\alpha_2^0 &= -\; \frac{\sqrt{\gamma_+\, \gamma_-}}{\chi} \;, \\
U^0 &= -\; \frac{1}{2}\, r\, (\alpha_1^0)^2 - \frac{\gamma_2}{\chi}\, \alpha_2^0 \;, \\
I_+^0 &= \sqrt{\frac{\gamma_-}{\gamma_+}}\, (-\, U^0) \;, \\
I_-^0 &= \sqrt{\frac{\gamma_+}{\gamma_-}}\, (-\, U^0) \;,
\end{aligned}
\tag{4}
$$

satisfy Eq.(3) as a stationary solution. Here $r = \kappa/\chi$ has been introduced to denote the relative coupling strength of the two nonlinear processes.

Above, the pump field has been chosen real and positive ($\mathcal{E}_1 > 0$); consequently the stationary amplitudes are also real: the fundamental is in phase with the pump field, i.e. positive, whereas the second harmonic is π out of phase, i.e. negative. Since the two intensities I_+^0 and I_-^0 are proportional to $-U^0$, this solution is only physically reasonable, if $U^0 < 0$, that is if the pump field $\mathcal{E}_1$ exceeds the threshold

$$
\mathcal{E}_{thresh} = \frac{\gamma_1}{\chi}\, (1 + r\, \frac{\bar{\gamma}_1}{\gamma_1})\, \sqrt{\frac{2\, \gamma_2\, \bar{\gamma}_1}{r}} \;,
\tag{5}
$$

(with $\bar{\gamma}_1 = \sqrt{\gamma_+ \gamma_-}$).

The most striking feature of the above solution is the 'plateau' in the amplitude of the second harmonic: for $\mathcal{E}_1 > \mathcal{E}_{thresh}$, α_2 is *independent* of the pump field (except for the condition $\mathcal{E}_1 > \mathcal{E}_{thresh}$), in contrast to the fundamental mode which grows linearly with the pump field $\mathcal{E}_1$. The reason for this is that for $\chi \neq 0$, light of frequency $\omega_2 = 2\omega_1$ which is produced by SHG is subsequently 'drained' again by the parametric down conversion process; thus it serves only as an intermediate step in the balanced equilibrium of SHG and NDOPO.

In the following, for the sake of simplicity, I will assume $\gamma_+ \approx \gamma_- \approx \gamma_1$, which is a good approximation, if the sidebands of the NDOPO are not too far away from the frequency ω_1 of the fundamental, thus experiencing more or less the same losses. In this case $I_+ = I_-$, and the number of equations to form a closed system can be reduced by one. In terms of real variables, defined by $\alpha_j = a_j + ib_j$, $U = u + iv$, $I = I_+ + I_-$, the solution reads

$$\begin{aligned}
a_1^0 &= \frac{\mathcal{E}_1}{\gamma_1 (1 + r)} \quad \text{and} \quad b_1^0 = 0 \ , \\
a_2^0 &= -\frac{\gamma_1}{\chi} \quad \text{and} \quad b_2^0 = 0 \ , \\
u^0 &= -\frac{1}{2}\, r\, (a_1^0)^2 - \frac{\gamma_2}{\chi}\, a_2^0 \quad \text{and} \quad v^0 = 0 \ , \\
I^0 &= -2\, u^0 \ .
\end{aligned} \tag{6}$$

III. STABILITY REGIMES AND PARAMETERS OF THE PROBLEM

In order to investigate the stability of this solution, one can calculate the eigenvalues of the matrix A^0 of the linearized equations of motion for $\delta a_i = a_i - a_i^0, \delta b_i = b_i - b_i^0, \delta u = u - u^0, \delta v = v - v^0, \delta I = I - I^0$

$$\left(\delta\dot{a}_1, \delta\dot{b}_1, \delta\dot{a}_2, \delta\dot{b}_2, \delta\dot{u}, \delta\dot{v}, \delta\dot{I}\right)^T = A^0\, (\delta a_1, \delta b_1, \delta a_2, \delta b_2, \delta u, \delta v, \delta I)^T \ . \tag{7}$$

As a main result one finds that two qualitatively different regimes exist and that they can be classified by the relative size of two intuitive parameters

$$\begin{aligned}
r &\equiv \frac{\kappa}{\chi} \ , \\
G &\equiv \frac{(\gamma_1 + \gamma_2)}{\gamma_1} \ ,
\end{aligned} \tag{8}$$

i.e. the ratio of the strengths of the nonlinear couplings on the one hand, and a dimensionless parameter depending solely on the damping rates on the other hand.

The relative size of r and G is of importance, since it determines whether the threshold Eq.(5), which with $\gamma_+ \approx \gamma_- \approx \gamma_1$ simplifies to $\mathcal{E}_{thresh} = \frac{\gamma_1}{\chi}(1 + r)\sqrt{\frac{2\gamma_1\gamma_2}{r}}$ lies below or above the selfpulsing threshold of SHG

$$\mathcal{E}_{SP}^{\chi=0} = \frac{2\gamma_1 + \gamma_2}{\kappa} \sqrt{2\gamma_2 (\gamma_1 + \gamma_2)} \tag{9}$$

derived by Drummond, McNeil and Walls [2]. (Here I have introduced the superscript $\chi = 0$ as a reminder that this threshold was derived for SHG alone.) It is easily shown that

$$\mathcal{E}_{SP}^{\chi=0} = \mathcal{E}_{thresh} \qquad \Longleftrightarrow \qquad \frac{\kappa}{\chi} = \frac{\gamma_1 + \gamma_2}{\gamma_1} . \tag{10}$$

(The fact that Eq.(10) contains only *ratios* of coupling strengths and decay rates is due the invariance of the equations of motion under a scaling

$$(\gamma_i, \chi, \kappa, \mathcal{E}_1, \alpha_i, U, I_\pm, t) \mapsto (s\,\gamma_i, \chi, \kappa, s^2\mathcal{E}_1, s\,\alpha_i, s^2U, s^2I_\pm, t/s) \tag{11}$$

with a scalar $s > 0$.)

The two cases $\mathcal{E}_{SP}^{\chi=0} < \mathcal{E}_{thresh}$ and $\mathcal{E}_{SP}^{\chi=0} > \mathcal{E}_{thresh}$ are qualitatively quite different:

A. Regime of quenched selfpulsing

For $r < G$ we have $0 < \mathcal{E}_{thresh} < \mathcal{E}_{SP}^{\chi=0}$. Inspection of the eigenvalues of the matrix A^0 shows that in this case solution Eq.(6) is stable where it exists with positive intensities, i.e. for $\mathcal{E}_1 > \mathcal{E}_{thresh}$, regardless of the value of $\mathcal{E}_{SP}^{\chi=0}$. The full analytical expressions for the eigenvalues are too complicated to be given explicitly. Instead, the typical behaviour that all real parts of the eigenvalues become negative, when $\mathcal{E}_1$ is increased above the NDOPO threshold, is illustrated in Fig.2.

More interesting is the case that $\mathcal{E}_{SP}^{\chi=0}$ is only slightly larger than $\mathcal{E}_{thresh}$; then two eigenvalues may develop imaginary parts for increasing $\mathcal{E}_1$, but retain a negative real part, as shown in Fig.3.

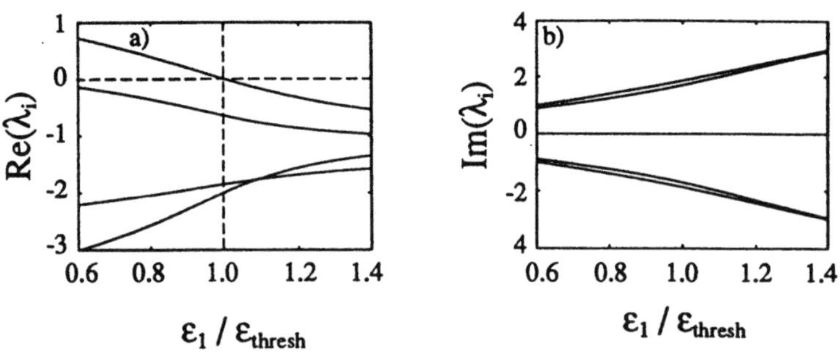

FIG. 2. The real and imaginary parts of the eigenvalues of the linearized semiclassical equations of motion as a function of the pump field $\mathcal{E}_1$ (arbitrary units). The parameters are $G = 2.25$ and $r = 1.2$. All real parts turn negative, i.e. the solution under investigation becomes stable, when crossing $\mathcal{E}_{thresh}$ from below.

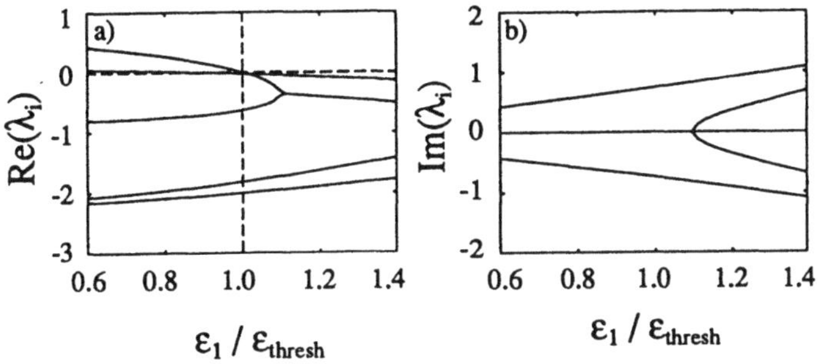

FIG. 3. The real and imaginary parts of the eigenvalues as in Fig.2, except for $G = 1.25$. Note the 'bifurcation' point, where two real parts merge and a pair of complex conjugate eigenvalues appears. This feature gives rise to damped selfpulsing, cf. Fig.4.

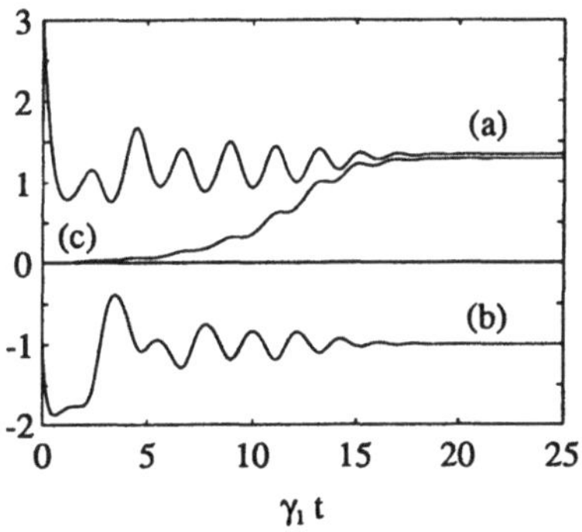

FIG. 4. Numerical integration of the equations of motion for the same parameters as in Fig.3 and $\mathcal{E}_1 = 2\mathcal{E}_{SP}^{X=0} = 2.09\mathcal{E}_{thresh}$. Curves (a), (b), and (c) correspond to a_1, a_2, and $\sqrt{I}$, respectively. Transient selfpulsing occurs in the amplitudes (and the intensities, not shown). As initial condition $a_1(0) = 3, b_1(0) = 1$, $a_2(0) = -1, b_2(0) = 0$, and a very small, but finite, intensity in the NDOPO has been chosen to prevent it from staying in the unstable trivial solution)

This means that selfpulsing in the fundamental and the second harmonic can only occur in the *transient* behaviour. Such a case with damped selfpulsing is demonstrated in Fig.4, where the result of direct numerical integration of the equations of motion (with suitably chosen initial values) is depicted. There it is clearly seen that the system tends towards the stationary solution of Eq.(6), even for $\mathcal{E}_1 > \mathcal{E}_{SP}^{X=0}$. This effect may account for the observed suppression of selfpulsing in a MOTIRR [11].

B. Multistability regime

In the other case $r > G$ we get $0 < \mathcal{E}_{SP}^{\chi=0} < \mathcal{E}_{thresh}$, i.e. for increasing $\mathcal{E}_1$ the selfpulsing threshold is reached before the NDOPO can oscillate.

Well below the NDOPO threshold, i.e. $\mathcal{E}_1, \mathcal{E}_{SP}^{\chi=0} \ll \mathcal{E}_{thresh}$, the NDOPO degrees of freedom of the stable stationary solution have zero mean values, $u^0 = v^0 = 0$ and $I^0 = 0$ and the system resembles pure SHG (at least in the semiclassical limit); therefore the selfpulsing threshold $\mathcal{E}_{SP}^{\chi=0}$ plays a similar role as for $\chi = 0$ [2]. If the two thresholds are close to each other, a complicated stability exchange involving stationary solutions as well as stationary limit cycles can occur. Numerical integration of the equation has shown that multistability and hysteresis effects are present. However, there is some indication that for a large range of initial values, the solution (6) is stable for $\mathcal{E}_{SP}^{\chi=0} < \mathcal{E}_{thresh} < \mathcal{E}_1$.

In conclusion, a semiclassical theory has been presented which describes the competition of two resonant nonlinear processes in an intracavity $\chi^{(2)}$ medium. It has been demonstrated that under certain conditions the selfpulsing instability is suppressed and a stable solution exists with combined oscillations of the second harmonic modes and the parametric oscillator modes.

ACKNOWLEDGEMENTS

This work was supported by the Österreichischer Fonds zur Förderung der wissenschaftlichen Forschung Projekt Nr. H012-PHY. Part of this work was carried out during a stay with J. Mlynek's group at the University of Konstanz.

References:

[1] See H.J. Kimble, *Quantum fluctuations in quantum optics, squeezing and related phenomena*, and C. Fabre and S. Reynaud, *Quantum noise in optical systems: a semiclassical approach*, Les Houches Lecture Notes, Session LIII, eds. J. Dalibard, J.-M. Raimond, and J. Zinn-Justin, North Holland (Amsterdam, 1992)

[2] P.D. Drummond, K.J. McNeil, and D.F. Walls, Optica Acta **27**, 321 (1980); P.D. Drummond, K.J. McNeil, and D.F. Walls, Optica Acta **28**, 211 (1981)

[3] P. Mandel and T. Erneux, Optica Acta **29**, 7 (1982); P. Mandel and T. Erneux, *Laser Physics*, eds. J.D. Harvey and D.F. Walls, Springer (Heidelberg, 1983)

[4] M.J. Collett and D.F. Walls, Phys. Rev. A **32**, 2887 (1985)

[5] S.F. Pereira, M. Xiao, H.J. Kimble, and J. Hall, Phys. Rev. A **38**, 4931 (1988)

[6] A. Sizmann, R.J. Horowicz, G. Wagner, and G. Leuchs, Opt. Commun. **80**, 138 (1990)

[7] P.D. Drummond and M.D. Reid, Phys. Rev. A **37**, 1806 (1988)

[8] A. Heidmann, R.J. Horowicz, S. Reynaud, E. Giacobino, C. Fabre, and G. Camy, Phys. Rev. Lett **59**, 2555 (1987)

[9] S. Schiller, L. Yu, M. Fejer and R. Byer, Opt. Lett. **17**, 378 (1992); S. Schiller and R.L. Byer, J. Opt. Soc. Am. B **10**, 1696 (1993)
[10] P. Kürz, R. Paschotta, K. Fiedler, A. Sizmann, G. Leuchs, and J. Mlynek, Appl. Phys. B **55**, 216 (1992); K. Fiedler, S. Schiller, R. Paschotta, P. Kürz, and J. Mlynek, Optics Letters **18**, 1786 (1993)
[11] K. Fiedler, PhD Thesis, University of Konstanz (1994)
[12] K.M. Gheri, D.F. Walls, and M.A. Marte, Physical Review A **46**, 6002 (1992)

Measurement Effects on Quantum Optical Systems

M. Orszag and R. Ramírez

Facultad de Física, Pontificia Universidad Católica de Chile,
Casilla 306, Santiago, Chile

We study a model of 3-level atoms passing through a micromaser with a classical externally applied field. We find that the measurement of the outcoming atoms accelerates the trapping of the quantum fields. We also explore the three-level atom crossing a double cavity. By performing a field measurement, one can accurately and uniquely locate the atom.

1 Introduction

In this work, we will describe optical phenomena that is strongly influenced by measurement. This is a way of preparing the state of a system via a measurement. In the first case presented here, we will deal with three-level atoms going through a micromaser. The atom will interact with both a quantum and a classical field and our subject of study will be the trapping state of the quantum field. We will discover, that if we measure the atomic state of the outgoing atom, this will influence the quantum field inside the cavity, accelerating the trapping of the field.

The second example deals again with three-level atoms, but this time they cross a double cavity (Fig. 1) with a quantum field and a classical field. Assuming a large detuning in such a way that we can neglect spontaneous emission, the atomic deflection is due to stimulated emission. A photon of one of the travelling waves is absorbed and re-emitted into the other wave travelling in the opposite direction, thus creating a momentum difference that produces the atomic deflection. Now, by making a phase-sensitive field measurement, a definite sharpening of the position distribution [1][2] takes place. This has been proven in a system with two-level atoms and one mode. It has the disavantage, however, that the field measurement determines very accurately how close the atom is to a node or an antinode of the field, but it does not determine the node itself, because of the multipeak structure of the position distribution. We will show, however, that with an appropiate choice of the parameters, we can suppress some of the peaks and, if the

Springer Proceedings in Physics, Vol. 77 Quantum Optics VI
Editors: D.F. Walls · J.D. Harvey

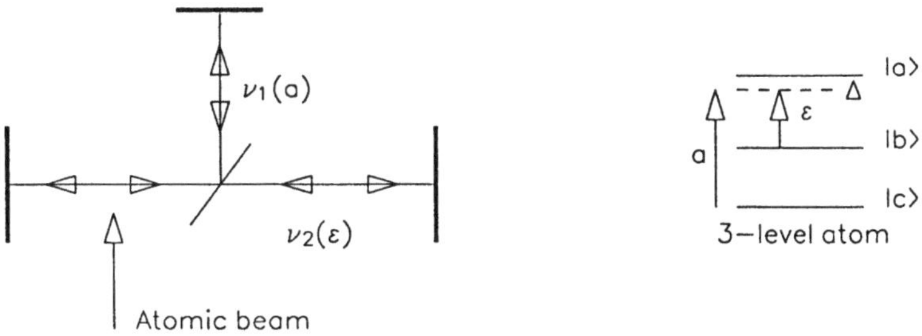

Figure 1 An atomic beam of three-level atoms crosses a double cavity with two near resonant fields. Δ is the common detuning between these waves and the corresponding atomic transitions.

initial atomic beam is approximately 1.8 λ we can *uniquely* locate the atom.

2 The Model

The two systems considered above can be described by a common Hamiltonian and time evolution operator. The main difference is that in the micromaser case, the two coupling coefficients (corresponding to two pairs of atomic levels, Fig. 1) are taken as constants, while in the second example, they are modulated by periodic functions of the atomic positions. Also, for the micromaser we take $\Delta = 0$, i.e. both modes are resonant with the corresponding atomic transition, while, in the second case we take a large detuning.

The Hamiltonian, in the interaction picture, can be written as:[3]

$$\begin{aligned} V &= \hbar g_1 \left(a e^{i\Delta t} |a><c| + a^{\dagger} e^{-i\Delta t} |c><a| \right) \\ &+ \hbar g_2 \left(\varepsilon e^{i\Delta t} |a><b| + \varepsilon^* e^{-i\Delta t} |b><a| \right) \end{aligned} \tag{1}$$

where Δ is the detuning, taken as the same for both fields and $g_1 = g\cos(kx)$, $g_2 = g'\cos(k'x)$ are the position dependent coupling constants. It is not difficult to find the time evolution operator for this system.

The result is:

$$U = \begin{pmatrix} U_{aa} & U_{ab} & U_{ac} \\ U_{ba} & U_{bb} & U_{bc} \\ U_{ca} & U_{cb} & U_{cc} \end{pmatrix}, \tag{2}$$

where

$$U_{aa} = e^{i\Delta t/2}(C - \frac{i\Delta}{2}S), \quad U_{ab} = -ig_2\varepsilon e^{i\Delta t/2}S, \tag{3}$$

$$U_{ac} = -ig_1 e^{i\Delta t/2}Sa, \quad U_{ba} = -ig_2\varepsilon^* e^{-i\Delta t/2}S, \tag{4}$$

$$U_{bc} = g_1 g_2 \varepsilon^* \frac{[e^{-i\Delta t/2}(C + \frac{i\Delta}{2}S) - 1]a}{\lambda}, \quad U_{ca} = -ig_1 a^\dagger e^{-i\Delta t/2}S, \tag{5}$$

$$\begin{aligned} U_{bb} &= 1 + \frac{g_2^2|\varepsilon|^2[e^{-i\Delta t/2}(C + \frac{i\Delta}{2}S) - 1]}{\lambda}, \\ U_{cb} &= g_1 g_2 \varepsilon a^\dagger \frac{[e^{-i\Delta t/2}(C + \frac{i\Delta}{2}S) - 1]}{\lambda}, \\ U_{cc} &= 1 + \frac{g_1^2 a^\dagger a[e^{-i\Delta t/2}(\bar{C} + \frac{i\Delta}{2}\bar{S}) - 1]}{\bar{\lambda}}, \end{aligned} \tag{6}$$

$$\mu \equiv g_2^2|\varepsilon|^2 + g_1^2 a a^\dagger + \Delta^2/4, \quad \lambda \equiv g_2^2|\varepsilon|^2 + g_1^2 a a^\dagger, \tag{7}$$

$$\bar{\mu} \equiv g_2^2|\varepsilon|^2 + g_1^2 a^\dagger a + \Delta^2/4, \quad \bar{\lambda} \equiv g_2^2|\varepsilon|^2 + g_1^2 a^\dagger a, \tag{8}$$

$$S \equiv \frac{\sin\sqrt{\mu}t}{\sqrt{\mu}}, \quad \bar{S} \equiv \frac{\sin\sqrt{\bar{\mu}}t}{\sqrt{\bar{\mu}}}, \quad C \equiv \cos\sqrt{\mu}t, \quad \bar{C} \equiv \cos\sqrt{\bar{\mu}}t, \tag{9}$$

3 The micromaser and the trapping states

As mentioned earlier, we take here $\Delta = 0$ and $g_1 = g$ and $g_2 = g'$ are considered constants. The time t is the flight time through the cavity.[4]

The density operator, after the $(k+1)$ atom has gone through, is:

$$\rho^{(k+1)} = U\rho^{(k)}U^{-1} \tag{10}$$

which is a quantum map for the atom radiation system.

Making use of (2), (6) and (9), it is easy to compute the diagonal matrix elements of the density operator, that is:

$$
\begin{aligned}
(\rho_{aa})^{(k+1)}_{nn'} &= \aleph_a \frac{\sin\sqrt{\mu(n)}\tau \sin\sqrt{\mu(n')}\tau}{\sqrt{\mu(n)}\sqrt{\mu(n')}} [\rho_{bb} g'^2 |\varepsilon|^2 \rho^{(k)}_{nn'} \\
&+ g^2\sqrt{(n+1)(n'+1)}\rho_{cc}\rho^{(k)}_{n+1,n'+1} + gg'|\varepsilon|\sqrt{(n+1)}e^{i\theta}|\rho_{cb}|\rho^{(k)}_{n+1,n'} \\
&+ gg'|\varepsilon|\sqrt{(n'+1)}e^{-i\theta}|\rho_{cb}|\rho^{(k)}_{n,n'+1}],
\end{aligned} \tag{11}
$$

$$
\begin{aligned}
(\rho_{bb+cc})^{(k+1)}_{nn'} &= \aleph_{bc}[A(n,n')\rho^{(k)}_{nn'} + B(n,n')e^{i\theta}\rho^{(k)}_{n+1,n'} \\
&+ B(n',n)e^{-i\theta}\rho^{(k)}_{n,n'+1} + C(n,n')e^{i\theta}\rho^{(k)}_{n,n'-1} + C(n',n)e^{-i\theta}\rho^{(k)}_{n-1,n'} \\
&+ D(n,n')\rho^{(k)}_{n+1,n'+1} D(n-1,n'-1)\rho^{(k)}_{n-1,n'-1}\frac{\rho_{bb}}{\rho_{cc}}]
\end{aligned} \tag{12}
$$

where

$$
\begin{aligned}
A(n,n') &= 1 - g'^2|\varepsilon|^2\rho_{bb}(\frac{1-C(n)}{\mu(n)} + \frac{1-C(n')}{\mu(n')}) - g^2\rho_{cc}(n\frac{1-\bar{C}(n)}{\bar{\mu}(n)} \\
&+ n'\frac{1-\bar{C}(n')}{\bar{\mu}(n')}) + g'^4|\varepsilon|^4\rho_{bb}(\frac{1-C(n)}{\mu(n)})(\frac{1-C(n')}{\mu(n')}) \\
&+ g^4\rho_{cc}(\frac{1-\bar{C}(n)}{\bar{\mu}(n)})(\frac{1-\bar{C}(n')}{\bar{\mu}(n')})nn'
\end{aligned} \tag{13}
$$

$$
\begin{aligned}
B(n,n') &= |\rho_{cb}|[-gg'|\varepsilon|\frac{1-C(n)}{\mu(n)}\sqrt{n+1} \\
&+ |\varepsilon|^3 gg'^3(\frac{1-C(n')}{\mu(n')})(\frac{1-C(n)}{\mu(n)})\sqrt{n+1}]
\end{aligned} \tag{14}
$$

$$
\begin{aligned}
C(n,n') &= -\sqrt{n'}|\rho_{cb}|(\frac{1-C(n'-1)}{\mu(n'-1)})gg'|\varepsilon| \\
&+ |\varepsilon|g'g^3|\rho_{cb}|(\frac{1-\bar{C}(n)}{\bar{\mu}(n)})(\frac{1-C(n'-1)}{\mu(n'-1)})n\sqrt{n'}
\end{aligned} \tag{15}
$$

$$
D(n,n') = |\varepsilon|^2(gg')^2\rho_{cc}\sqrt{(n+1)(n'+1)}(\frac{1-C(n)}{\mu(n)})(\frac{1-C(n')}{\mu(n')}) \tag{16}
$$

$$\rho_{cb} \equiv |\rho_{cb}| e^{i\theta_{cb}}, \varepsilon \equiv |\varepsilon| e^{i\theta_\varepsilon}, \theta \equiv \theta_{cb} - \theta_\varepsilon. \quad (17)$$

$(\rho_{aa})^{(k+1)}_{nn'}$ is the matrix element of the diagonal element of, corresponding to the a-a position. Similarly $(\rho_{bb+cc})^{(k+1)}_{nn'}$ is the (nn') matrix element of the bb and cc elements.

If we perform no measurement on the outcoming atoms, then we choose $\aleph_a = \aleph_{b+c} = 1$ and $\rho^{k+1}_{nn'} = (\rho_{aa})^{k+1}_{nn'} + (\rho_{bb+cc})^{k+1}_{nn'}$ is just the reduced field density matrix, after tracing over the atom. On the other hand, if we measure each atom and determine whether the atom is in the a-level or in the b or c level, then we proceed as follows: We generate a random number between 0 and 1 and calculate in each step the probability for the atom being in the a-level, P_a, or in the b or c levels P_{b+c} satisfying the relation $P_a + P_{b+c} = 1$. If the random number comes out between 0 and P_a , we decide that the atom was in the a-state. Obviously if that's not the case, the atom was in the b or c state. Next, we renormalize the corresponding density matrix, the normalization constants being $\aleph_a$ and $\aleph_{b+c}$ respectively.[5] In this way, as more atoms go through the cavity, we calculate numerically the field density matrix, as a function of time both with or without atomic measurement. Now, we look at the trapping states of the field. If, after k-atoms crossed the cavity, the field is given by $|\psi>_{\text{field}} = \sum_{n=n_d}^{n_u} C_n|n\rangle$, where n_d and n_u are the downwards and upwards boundaries of the trap, then, when a $(k+1)$-th atom goes through the cavity, the wave function is given by

$$|\psi\rangle^{(k+1)} = U|\psi\rangle^{(k)} = U \sum_{n_d}^{n_u} C_n|n\rangle \begin{pmatrix} 0 \\ \beta \\ \gamma \end{pmatrix}, \quad (18)$$

where we assume that the atom is in the "ground state", that is either in the b or c state with their respective probabilities being $|\beta|^2$ and $|\gamma|^2$. The field is in a trapping state if the coefficients of $|n_d\rangle$ and $|n_u\rangle$ in (8) vanish, which leads to the conditions:

$$(g\tau)\sqrt{n_u + 1 + |\frac{g'}{g}\varepsilon|^2} = 2\pi p, \qquad (g\tau)\sqrt{n_d + |\frac{g'}{g}\varepsilon|^2} = 2\pi q, \quad (19)$$

p and q being integer numbers.

The precise state of the field is given by the condition:

$$U \sum_{n=n_d}^{n_u} C_n|n\rangle \begin{pmatrix} 0 \\ \beta \\ \gamma \end{pmatrix} = \sum_{n=n_d}^{n_u} C_n|n\rangle \begin{pmatrix} \alpha' \\ \beta' \\ \gamma' \end{pmatrix} \quad (20)$$

that is, once the trapping state for the field is reached, the next

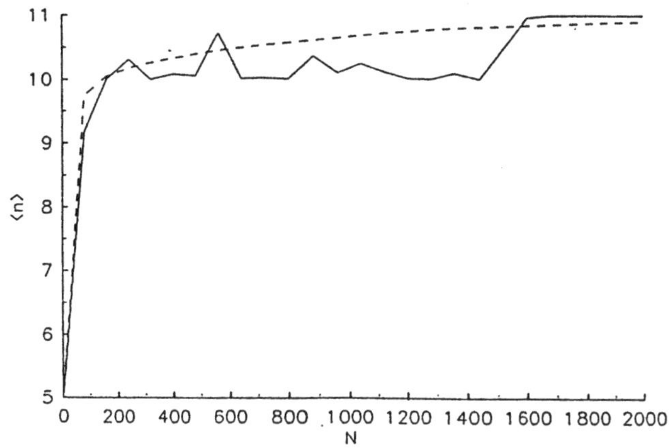

Figure 2 Average photon number versus N=number of atoms, for $g\tau = \pi$, $\varepsilon = 2$. The initial field was taken $\rho_{nn'}(0) = \delta_{n,5}\,\delta_{n',5}$. The solid line corresponds to the case with measurement and the dotted line to the freely evolving system.

atom will not modify the expansion coefficients C_n.The equation (10) gives us the relation between the various coefficients $\beta, \gamma, \beta', \gamma'$ as well as a recursion relation for the C_n. It is simple to prove that eq (10) is satisfied for the following sets of conditions:

$$\alpha' = 0, \quad \beta' = \beta, \quad \gamma' = \gamma, \quad C_{n+1} = -C_n(\frac{g'\beta}{g\gamma})\frac{\varepsilon}{\sqrt{n+1}} \tag{21}$$

A particularly simple example is when the atoms are injected in the b-state.

In this case $\rho_{bb} = 1$, $\rho_{cc} = 0$, $|\rho_{cb}| = 0$, $|\beta|^2 = 1$, $|\gamma|^2 = 0$, and the field will go to the pure Fock state $|n_u\rangle$. If $g = g'$, $g\tau = \pi$, $p = 2$, $|\varepsilon| = 2$, then $n_u = 11$.

In figure 2, we see a plot of the average photon number $\langle n\rangle$ versus N=number of atoms crossing the cavity, for the conditiones specified above. The full curve corresponds to the case with measurement and the dotted curve is the free evolution of the system. We notice that, in the former case, after the usual oscillations caused by the measurement $\langle n\rangle$ locks abruptly to $n_u = 11$ at $N \approx 1700$, while the free evolution curve tends to converge slowly to the same limit. Figure 2 corresponds to a particular random sequence. We performed several runs for different sequences. The N values, at which $\langle n\rangle$ locked, changed at each of these sequences, but the qualitative effect is the same. The net effect observed here is that the atomic measurements accelerate the trapping of the field. This effect is further confirmed in figure 3, where we plotted

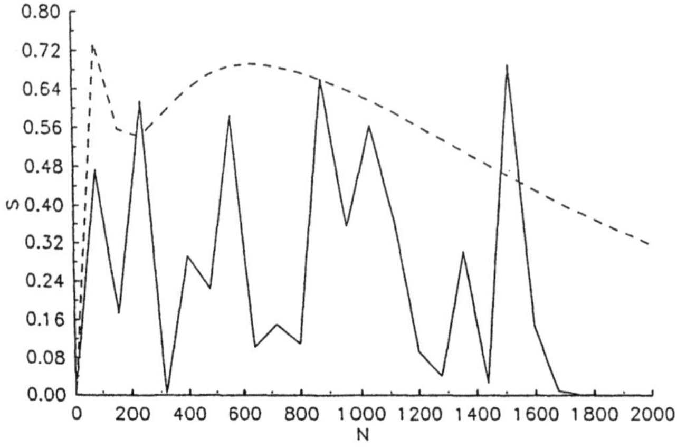

Figure 3 Entropy versus N, with the same parameters as in figure 2. The curve with solid line corresponds to the case with measurement and it falls sharply to zero at $N \approx 1700$. The dotted line corresponds to the freely evolving case.

the entropy S versus N. While the entropy for the free evolution curve smoothly goes down to zero, when we measure the atoms coming out from the cavity the entropy abruptly goes down to zero at $N \approx 1700$.[6] By changing the seed of the random number generator, one get's a different transient behavior or for $\langle n \rangle$and S but, again, $\langle n \rangle$ locks abruptly to $n_u = 11$ (at a different time) and S drops sharply to zero at the same point.

As a conclusion, when the parameters of the system correspond to a trapping state, the fact that we measure the state of the outcoming atoms affects the field inside the cavity producing a faster arrival or locking to the trapped state. This locking happens at random times and it can be seen either from the $\langle n \rangle$ curve or the sharp decrease of the entropy at one particular time. A possible interpretation of this results is the following: a measurement introduces a stochastic component in the system, producing random oscillations in the relevant variables. On the other hand, a trapping state corresponds to an infinite potential well in the n-variables[7] ($\rho_{nn} = e^{-V(n)}$), so that if, in this wandering caused by the measurement the system falls into this well, it will never come out, again.

4 The atomic microscope

Atomic interference and beam deflection are, at the present time, very active research areas, both theoretically and experimentally.

Early experiments show that neutral atoms can be diffracted by standing light waves.[8] This is the so called Kapitza-Dirac effect applied to neutral atoms rather than electrons. Also, atomic interferometry has been recently demonstrated experimentally[9][10][11].

As previously mentioned in the introduction, there has been theoretical work on the position sharpening of neutral atoms crossing standing waves. In the present model, the dynamics of our system is described by equations (2), (6) and (9). By making use of the results above, one can calculate the time evolution of the system's wave function $|\psi(t)>$:

$$|\psi(t)>=\int dx K(x)|x>[(U_{bb}\beta+U_{bc}\gamma)|b>+(U_{cb}\beta+U_{cc}\gamma)|c>]|\alpha> \tag{22}$$

where the atom is initially in a coherent superposition of levels b and c, with probabilities $|\beta|^2$ and $|\gamma|^2$ respectively. Also the standing wave is in a coherent state $|\alpha>$ and $K(x)$ is the initial probability amplitude for the atomic position x.

Now, suppose that the quadrature

$$X_\theta = ae^{-i\theta} + a^\dagger e^{i\theta} \tag{23}$$

is measured, giving an eigenvalue x_θ and a corresponding eigenstate $|x_\theta>$. One can easily show that $|x_\theta>$ is an infinitely squeezed state, given by [2]:

$$|x_\theta>=\frac{1}{(2\pi)^{1/4}}\exp\left[-\frac{1}{2}(a^\dagger e^{i\theta}-x_\theta)^2+\frac{1}{4}x_\theta^2\right]|0> \tag{24}$$

which can be expanded in terms of $|n>$ states as:

$$|x_\theta>=\sum_n b_n|n>, \qquad b_n=\frac{\aleph}{\sqrt{n!}}(\frac{1}{2}e^{i\theta})^{n/2}H(z) \tag{25}$$

where $z=(\alpha e^{-i\theta/2}+\alpha^* e^{i\theta/2})/2$ and $\aleph$ is a normalization constant. After the measurement, the wave function of the system is the projected version of equation (22), that is:

$$
\begin{aligned}
|\psi'> &= <x_\theta|\psi(t)> \\
&= \aleph' \sum_{n,m} b_n^* <n| \int dx K(x)|x> [M|b> + N|c>]\, c_m|m>
\end{aligned} \tag{26}
$$

where $M = (U_{bb}\beta + U_{bc}\gamma)$, $N = (U_{cb}\beta + U_{cc}\gamma)$, $|\alpha> \equiv \sum_m c_m|m>$ and $\aleph'$ is a normalization constant. Finally we can write the probability distribution for the atomic position:

$$
P(x) = < x|(\mathrm{Tr}_{\mathbf{atom}}|\psi' >< \psi'|)|x> \tag{27}
$$

which can be easily computed using Equations (26) and (6). The result is:

$$
P(x) = \aleph'|K(x)|^2 \left[|\sum \alpha_n|^2 + |\sum \beta_n|^2 \right] \tag{28}
$$

where:

$$
\begin{aligned}
\alpha_n &\equiv C_n b_n^* (1 + g_2^2|\varepsilon|^2 \Lambda_n)\beta + C_{n+1} b_n^* g_1 g_2 \varepsilon^* \Lambda_n \sqrt{n+1}\gamma, &(29)\\
\beta_n &\equiv C_{n-1} b_n^* g_1 g_2 \varepsilon \Lambda_{n-1} \sqrt{n}\beta + C_n b_n^* (1 + g_1^2 n \Lambda_{n-1})\gamma, &(30)
\end{aligned}
$$

and:

$$
\Lambda_n \equiv \frac{e^{i\lambda_n t} - 1}{\lambda_n}, \qquad \lambda_n = \lambda(a^\dagger a \to n), \qquad C_n \equiv \frac{\alpha^n}{\sqrt{n!}} e^{-|\alpha|^2/2}. \tag{31}
$$

The approximation $\Delta >> \sqrt{g^2 n + g'^2|\varepsilon|^2}$ was used in the derivation of equation (31). We have numerically calculated $P(x)$ for various regions of the paramater space. The optimized results are shown in Fig. 4. The initial position distribution is a step function between -0.9 and 0.9 (in units of λ). The parameters used were $g = 1$, $\Omega \equiv g'|\varepsilon| = 0.1$, $k = 1$, $k' = .5$, $\alpha = \sqrt{8}$, $\beta = 1$, $\gamma = 0$, $x_\theta = 0$. For the interaction times we took two values:

(i) $t/\Delta = \pi$ (ii) $t/\Delta = 1$

In the case (i), the two central peaks in the (0,1) region are strongly suppressed to about 15 % of the two main peaks. This compares favourably with the two-level atom, one mode case, where the four peaks have a uniform height. Perhaps the most interesting result corresponds to case (ii). In this case, again, with an initial beam 1.8λ wide and choosing the right parameters, we can *uniquely* determine the atomic position with a resolution better than 0.1λ. This is the main result of this section. We believe that this behaviour can be explained by a combination of two effects. On one

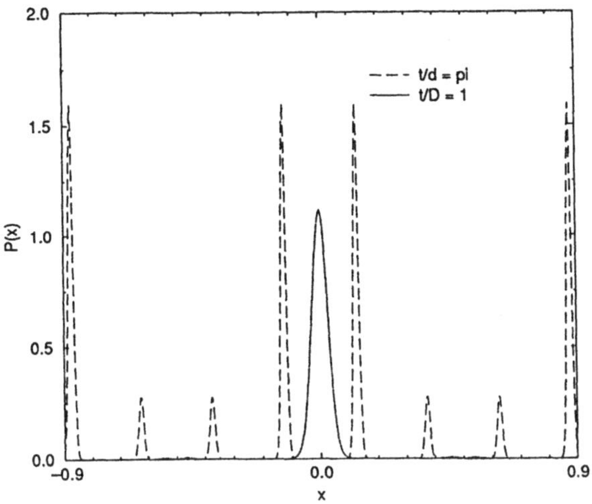

Figure 4 The position probability distribution $P(x)$ versus x (in units of λ), for the following set of parameters: $g = 1$, $\Omega \equiv g'|\varepsilon| = 0.1$, $k = 1$, $k' = .5$, $\alpha = \sqrt{8}$, $\beta = 1$, $\gamma = 0$, $x_\theta = 0$ and $t/\Delta = 1$ (solid), $t/\Delta = \pi$ (dashed)

hand, the longer classical wave modulates the quantum field. This combined with the quantum interference between the various n-dependent terms will finally produce the peak suppression for some particular values of the interaction times and $x_\theta = 0$

Acknowledgement

The authors would like to acknowledge the partial support by Fondecyt grants 1930568 and 1930563. One of us (M.O.) would also like to thanks Fundacion Andes for additional support.

References

[1] Pipa Storey, M. Collett, D. Walls, Phys. Rev. Letters, **68**, 472(1992), see also M. Marte, P. Zoller, App. Phys. **54**, 477(1992)

[2] Pipa Storey, M. Collett, D. Walls, Phys. Rev., **A47**, 405(1993)

[3] M. Orszag, R. Ramírez, J.C. Retamal, C. Saavedra, submitted Optical Communications.

[4] M. Orszag, R. Ramírez, Optical Comm. **101**, 377(1993)

[5] The idea of continuous measurement consists in having a continuous string of atoms interacting with the cavity for a very short time and being monitored at the output. It was developed by: M. Ueda, N. Iomoto, H. Nagaska, T. Ogawa, Phys. Rev. A **46**,2859(1992).

[6] Although figure 3 shows the behaviour of the entropy for 2000 atoms only, the actual calculation was performed up to 10000 atoms. The entropy stayed constant at zero above 2000 atoms.

[7] In the presence of finite losses, this potential well is finite and has a width, see for example: L. Roa, C. Saavedra, J.C.Retamal, Phys. Rev. A, **46**, (1992).

[8] P.L. Gould, G.A. Ruff, D. Pritchard, Phys. Rev. Letters, **56**, 827(1986),

[9] O. Carnal and J. Mlynek, Phys. Rev. Letters, **66**, 2689(1991),

[10] D.W. Keith, Q.A. Turchette and D.A. Pritchard, Phys. Rev. Letters,**66**, 2693(1991)

[11] M. Kasevich and S. Chu, Phys. Rev. Letters, **67**, 181(1991)

Quantum-Optical Phase and Its Measurement

D.T. Pegg[1], *J.A. Vaccaro*[2], and *S.M. Barnett*[3]

[1]School of Science, Griffith University, Nathan, Australia Q 4111
[2]Faculty of CAD, Griffith University, Nathan, Australia Q 4111
[3]Department of Physics and Applied Physics, University of Strathclyde, Glasgow G4 0NG, UK

Abstract. We discuss a simple and readily acceptable operational approach to defining quantum optical phase, and examine possible ways of measuring the resulting phase probability density.

1. Introduction

Until relatively recently the quantum mechanical phase properties of light have not received much attention. This is attributable to the difficulties in finding a unitary or hermitian operator to represent phase as an observable [1-3]. The best approximation to a representation of phase was given by the Susskind-Glogower operators $\hat{e}$ and $\hat{e}^\dagger$ [2]. These are non-unitary and therefore, according to quantum mechanics, do not represent the exponentials of any observable. On the other hand the phase states, which comprise equally weighted superpositions of all the number states $|n\rangle$, are right-side eigenstates of $\hat{e}$. It is therefore tempting to infer that $\hat{e}$, although not unitary, represents phase in some sense. This line of reasoning has dangers, however. The vacuum state $|0\rangle$ is *also* a right eigenstate of $\hat{e}$, but $|0\rangle$ is a number state and is therefore as far removed from a phase state as it possibly can be. At best $\hat{e}$ yields sensible phase properties for some states but often does not, particularly for states with a large vacuum component.

Some time ago we proposed two alternative approaches: the operational, or measured-phase, approach [3] and the Ψ-space formalism [4-7]. Any quantum theory of phase must have the correct classical limit but, as this condition alone allows an infinity of theories, some quantum mechanical criterion is also necessary. In the Ψ-space approach we chose number-phase complementarity as this criterion and proposed a mathematical limiting procedure which allowed this concept to be incorporated consistently. This approach is mathematically consistent with all previous results of quantum electrodynamics [7] and also allows the incorporation of a phase conjugate to photon number. The results obtained are very reasonable physically but, because the approach is a slight departure from the conventional paradigm, even though we believe it to be correct it is yet to find universal acceptance. In our earlier, operational, approach a measured-phase operator was defined in terms of a particular type of experiment [3]. The operational approach was later revived by Noh *et al.* [8] who showed the important result that different experiments yield different measured-phase operators. This means that, although defining phase by a measurement process has the advantage of yielding a directly measurable quantity, the procedure is not particularly useful because it does not lead to a unique, or even generally agreed, operator. Such experiments merely measure a quantity called "phase" which is defined to be the result of the measurement. They do not confirm, invalidate or even measure phase defined by other approaches.

Here we look again at the fundamental questions of what phase is and how it can be measured. We adopt an operational approach, but different from those above. This does not involve Ψ-space and all the hermitian operators involved act on the usual infinite-dimensional Hilbert space, but we find that the results of this operational approach are perfectly consistent with those obtained by the use of Ψ-space.

2. Operational Phase

Phase is well understood classically. What quantum properties should it have? Classically it is a relative quantity, with only phase differences or phase shifts having a physical meaning. It is possible, however, to speak of the phase of a field by arbitrarily selecting an origin of time or by comparison with the phase of a reference field whose phase is defined to be zero. There would be general agreement that this property should also apply in the quantum regime. After all, quantum mechanics does not introduce an absolute origin of time. Another property which would also meet with overall agreement is that an intense coherent state should have similar properties to those of a classical field, which includes a well-defined phase. In quantum mechanical terms this means that the phase probability density $P(\theta)$ for such a state should approach a delta-function, or actually a 2π-periodic combination of delta-functions, that is a series of sharp spikes separated by 2π showing the periodic nature of phase. Quantum mechanical probability for a pure state is obtained from the square of an amplitude or, in the case of a continuous distribution of possible outcomes of measurements, as expected for phase, the probability density $P(\theta)$ should be the square of the modulus of a wave function $|\psi(\theta)|^2$.

Unfortunately, although the above properties would be generally accepted, they are not sufficient to define phase uniquely. At least one more property is needed, which we find from our operational approach. Well-known operations in quantum mechanics include measurement and preparation or generation of a physical quantity. Our previous remarks indicate that the former is not particularly useful, so we now examine the latter. The physical quantities involved are phase differences and phase shifts. We choose the latter of these while noting that a phase shift operation on one of a pair of identical states will also generate a phase difference. To shift the phase of a classical travelling field we simply insert an extra optical path length. This is equivalent to introducing a time shift Δt in the field at the point of detection, with the resulting phase shift being $\omega\Delta t$. This is the property which we now use to define phase in the quantum mechanical regime. In essence we are saying that in both classical and quantum mechanics phase is that which is shifted by a phase-shifter. This operational definition has the advantage of potential universal agreement, indeed we know of no experiments involving phase shifts which assume anything inconsistent with this definition. Another advantage is that we know how to apply a time shift to a quantum state. This is generated by the field hamiltonian $\hat{N}\omega$, with $\hbar = 1$, so a phase shift $\Delta\theta = \omega\Delta t$ is generated by $\hat{N}$. The phase shifter changes the field wave function from $\psi(\theta)$ to $\psi(\theta + \Delta\theta)$. If the field is in a number state $|n\rangle$, however, the action of $\exp(-i\hat{N}\Delta\theta)$ is merely to multiply the wave function, which we label as $\psi_n(\theta)$, by $\exp(-in\Delta\theta)$. This leads to the functional equation [9]

$$\exp(-in\Delta\theta)\,\psi_n(\theta) = \psi_n(\theta + \Delta\theta) \quad . \tag{1}$$

We can show [9] that this functional equation has the unique solution

$$\psi_n(\theta) = g_n \exp(-in\theta) \ , \tag{2}$$

where g_n is independent of θ. Normalization of the periodic wave function (2) over any 2π range yields $(2\pi)^{-1/2}\exp(i\beta_n)$ for the value of g_n where β_n is real. The orthonormality relation for $\psi_n(\theta)$ for any 2π interval can then be verified, allowing us to express a general wave function $\psi(\theta)$ as a linear expansion in $\psi_n(\theta)$ with coefficients c_n. From this we obtain the probability density

$$P(\theta) = (2\pi)^{-1} \left| \sum_n c_n \exp[-i(n\theta - \beta_n)] \right|^2 \ . \tag{3}$$

For a photon number state all the coefficients are zero except for one, so $P(\theta)$ is uniform and thus all number states have a random phase, which is consistent with number-phase complementarity.

The requirement discussed earlier for the phase distribution of an intense coherent state removes some of the flexibility in the choice of β_n. Choosing β_n as a constant satisfies this condition [9]. Further, choosing this constant to be zero sets the phase of the strong coherent state $|\alpha\rangle$ equal to the argument of the corresponding eigenvalue of the annihilation operator. Choosing other values for the constant merely shifts the phase. For convenience, we choose the value zero and obtain

$$P(\theta) = (2\pi)^{-1} \left| \sum_n c_n \exp(-in\theta) \right|^2 \ . \tag{4}$$

This operationally derived expression gives sensible results, indeed it is identical to that derived from the Ψ-space approach and from the phase probability operator measure [10]. We have seen in this section that our operational approach, based on well-accepted quantum mechanical concepts and on the action of a phase-shifter, is consistent with number-phase complementarity and gives the same results as two other approaches. As the operation involved was not one of measurement, the question now arises as to how $P(\theta)$ can be measured for a particular state of the field.

3. Measurement of Phase

A question which is sometimes asked is whether or not it is possible, even in principle, to measure the phase probability distribution. To measure a probability distribution we need reproducible states, which we shall assume to be available. Further we shall assume that the states are pure as we are interested in quantum mechanical probability. Clearly we can find the phase probability distribution (4) if we can find the c_n. Measuring the photon number distribution gives only $|c_n|$, which is not sufficient to determine the state of the field. Very recently, however, measurements have been made of the state of the field by Smithey *et al.*[11,12] using optical homodyne tomography [13], which allow the phase probability distribution to be determined. This important work provides a clear affirmative answer to the question at the beginning of this paragraph. It leads, however, to a another question. We do not need to obtain sufficient information to determine the state of the field in order to measure, say, the number state distribution, so can we measure the phase probability distribution, at least to a reasonably good approximation, without acquiring so much information? To address this question we discuss methods for measuring the probability density for a particular states of the field.

3.1 Homodyne Detection

Although other phase-dependent measurements exist, such as those involving prepared atoms [3], we concentrate here on homodyne experiments, the most promising of which is the balanced homodyne detector involving a 50%:50% beam splitter with the signal obtained by subtraction of the outputs of the two photodetectors [8]. Classically the signal is proportional to $\sqrt{I(t)}\,\sqrt{I_L}\,\sin[\theta(t) - \theta_L]$, where the subscript refers to the local oscillator field which we have assumed is stable in intensity and phase. Quantum mechanically this will be the case when the local oscillator field is in a very intense coherent state. $I(t)$ and $\theta(t)$ are the intensity and phase of the unknown field before it enters the beam splitter. Defining the local oscillator phase before the beam splitter to be zero makes the phase difference equal to the phase of the unknown field. (Note that in previous discussions [3,14] we have worked in terms of the phase difference between the unknown and local oscillator fields at the detector, which will differ by $\pi/2$ from that used here.) The signal involves both intensity and phase fluctuations of the unknown field so, as we wish to measure only the phase fluctuations, we need somehow to separate these. Classically we can stabilize the intensity without affecting the phase, but this is not true for weak quantum fields. In the latter case, removing the intensity fluctuations completely randomizes the phase. The only option appears to be to adjust the phase so as to at least minimize the effect of the intensity fluctuations on the signal and if possible to maximize the effect of the phase fluctuations. This can be done by adjusting the phase to obtain zero mean signal, that is, so that $\langle\sin\theta(t)\rangle = 0$ [14]. Then, provided the fluctuations $\Delta\sqrt{I}$ are much smaller than $\langle\sqrt{I}\rangle$, we obtain a signal which is proportional to $\sin\theta(t)$ which is now equal to the fluctuations $\Delta\sin\theta$ from the mean. The corresponding signal in the quantum mechanical case is just proportional to the original measured-phase operator [3]:

$$\widehat{\sin}_M\phi = (\hat{a} - \hat{a}^\dagger)/[2i(\langle n\rangle + 1/2)^{1/2}] \,, \tag{5}$$

where $\langle n\rangle$ is the mean photon number. We note that (5) is proportional to the electric field component operator. The idea of using the electric field strength as a measure of phase was revived in Ref. [15]. Now the annihilation operator $\hat{a}$ contains both intensity and phase information [4], so we again need to adjust the phase to minimize the effect of the intensity fluctuations if we want (5) to be an approximate measure of the sine of the phase angle. Guided by the classical case we take as the appropriate condition the vanishing of the mean signal, $\langle \sin_M\theta \rangle = 0$.

The size of the phase shift needed to give a zero mean signal would give an indication of the mean phase of the unknown field in state $|f\rangle$ and the final probability distribution of the signal as a function of the signal size, which we call the measured-phase probability density $P_M(\sin\theta)$, would be an approximation to the actual sine phase distribution [14]

$$P_{\sin\phi}(\sin\theta) = (2\pi\cos\theta)^{-1} \sum_{n,n'}^{\infty} \{\exp[i(n'-n)\theta]$$

$$+ (-1)^{n'-n}\exp[-i(n'-n)\theta]\}\langle f|n'\rangle\langle n|f\rangle , \tag{6}$$

with $P_{\sin\phi}(\sin\theta)d(\sin\theta)$ being the probability for the sine of the phase angle to have a value between $\sin\theta$ and $\sin\theta + d(\sin\theta)$. For states with narrow phase distributions around the mean phase of zero, the sine can be replaced by the angle itself, and the sine distribution approximates the phase distribution. The value of $\langle n \rangle$ for use in (5) could be found by a separate measurement, or alternatively for narrow distributions the scaling factor relating signal size to phase angle could be found by applying a small phase shift and finding the resultant shift in the mean of the distribution [14].

To set the mean to zero we would have to monitor the change in measured distribution as the phase is shifted. This suggests a modified way of obtaining phase information, that is, to plot the values Pr(0) of the probability of the signal being in a small bin at zero as a function of the phase shift [14,15]. These values are then scaled to give a suitable area under the curve. We call the scaled values $Q_M(\theta)$ [14]. $Q_M(\theta)$ will be proportional to $P_M(0)$ for the phase-shifted state, which we call $|f(-\theta)\rangle$ where $|f(0)\rangle$ is the state with zero mean phase. Pr(0) is easily calculated from the projection of the state $|f(-\theta)\rangle$ onto the zero eigenvalue eigenstate $|y=0\rangle$ of $\hat{a} - \hat{a}^\dagger$ as given, for example, by Merzbacher [16]. To find what actual phase distribution $Q(\theta)$ corresponds to $Q_M(\theta)$ we can, in as much as $\hat{\sin}_M\phi$ corresponds to $\sin\phi$, simply replace $P_M(0)$ by $P_{\sin\phi}(0)$ given by (6) with $\theta = 0$ and with $|f(-\theta)\rangle$ substituted for $|f\rangle$. This gives

$$Q(\theta) = (2\pi)^{-1} \sum_{n,n'}^{\infty} \{1 + (-1)^{n'-n}\}\langle f(-\theta)|n'\rangle\langle n|f(-\theta)\rangle$$

$$= P(\theta) + P(\theta + \pi) , \tag{7}$$

which follows from (4) after substituting $\exp(-i\hat{N}\theta)|f(0)\rangle$ for $|f(-\theta)\rangle$ and using c_n for $\langle n|f(0)\rangle$. We now examine some states to which these measurable distributions might apply with reasonable accuracy.

3.2 Partial Phase and Other States

Partial phase states [6] have number state coefficients c_n of the form $|c_n|\exp(in\mu)$. Examples include coherent states, squeezed states, phase-optimized states [17] and number states. The phase distribution is symmetric about μ and thus μ is equal to the mean phase in the appropriate 2π window [6] and characterizes the phase of the state. Any state can be written as a superposition of partial phase states. This superposition may have a number of characteristic phases which might be associated with different peaks of the phase probability density. It is natural,therefore, to look first at a state which we know, for example by virtue of its preparation, is a partial phase state with a single peaked phase distribution. We note that determining the phase probability distribution for such a state is particularly easy. We need only measure $|c_n|$ from a photon number distribution and then measure the mean phase μ. This is sufficient to determine the coefficients c_n and thus the phase probability density from (4). The important question for this case is therefore - how accurately can we measure the mean phase of a partial phase state? Perhaps surprisingly the answer is that it can in principle be measured precisely. It is not difficult to show for such a state that $\langle \sin_M \phi \rangle$ is proportional to $\sin\mu$, so if we shift the phase until the mean signal first becomes zero the size of the shift is either μ or $\pi - \mu$. Which one it is can be determined from the direction in which the mean shifts as a phase shift is applied. As this measurement relies on knowing the field is in a partial phase state, it is useful, as a check, to verify experimentally that the signal is symmetric when its mean is zero, and has only one peak.

Partial phase states provide an accuracy test for the measurable distributions. Surprisingly, both distributions are reasonably close to the corresponding actual phase distributions for many states, even for states with extremely small phase fluctuations [14] where the necessary associated largeness of the intensity fluctuations might be expected to invalidate them [18]. Such states include phase-optimized states or Airy states [17]. On the other hand, for the state $(|7\rangle + |8\rangle + |9\rangle)/\sqrt{3}$ with a variance in $\sin\theta$ of 1/3, the variance of the $P_M(\sin\theta)$ is 0.34 but $Q_M(\theta)$ is constant, as for a random phase. This is due to the projection onto $|y=0\rangle$ being sensitive only to even number states, so only $|8\rangle$ contributes to $Q_M(\theta)$. For coherent or Airy states, however, with smooth changes in number state coefficients, this lack of sensitivity to odd number states is not so important. For an Airy state the greatest discrepancy occurs for $\langle n \rangle$ near unity, and even this case is not too bad, as shown on the left side of Fig. 1.

It is clear from (7) that we can approximate $P(\theta)$ by $Q(\theta)$ for states for which $P(\theta + \pi)$ is negligible, and thus measure it approximately by using $Q_M(\theta)$. Such a state must have a sufficiently narrow phase distribution to be virtually fully contained in a range of width π. Consider a state $|f(0)\rangle$, not necessarily a partial phase state, with a mean phase of zero which has such a distribution. This means that we can phase shift the state to $|f(-\theta)\rangle$ where $-\pi/2 < \theta < \pi/2$ while maintaining zero probability of being found with a phase of π. Since $|f(-\theta)\rangle$ is $\exp(-i\hat{N}\theta)|f(0)\rangle$, the number state coefficients for $|f(-\theta)\rangle$ are $c_n\exp(-in\theta)$ where $|f(0)\rangle$ has coefficients c_n. It follows from (4) that $P(\theta)$ for $|f(0)\rangle$ is the same as $P(0)$ for $|f(-\theta)\rangle$, which we can write as

$$P(0) = (2\pi)^{-1} \Big| \sum_{n\ \mathrm{even}} c_n \exp(-in\theta) + \sum_{n\ \mathrm{odd}} c_n \exp(-in\theta) \Big|^2 . \tag{8}$$

To find $P(\pi)$ for $|f(-\theta)\rangle$ we add π to θ in (8), which does not alter the first term inside the modulus but changes the sign of the second term. Thus for a state $|f(-\theta)\rangle$

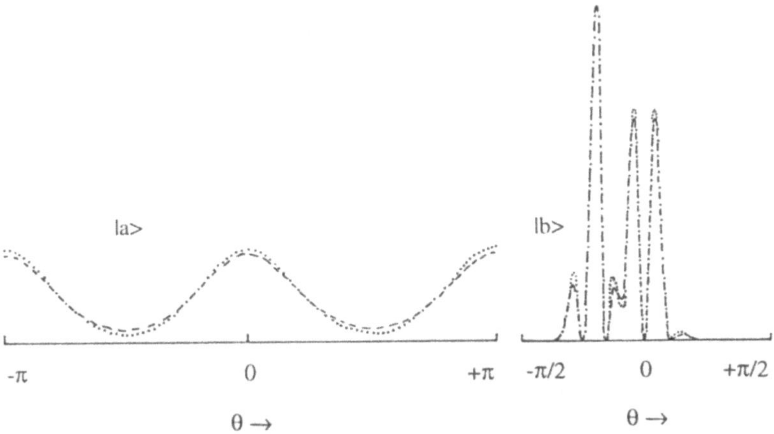

Fig. 1. Comparison of the measurable probability distribution $Q_M(\theta)$ (dots) and the actual phase probability distribution $Q(\theta)$ (dashes) for an Airy state $|a\rangle$ with $\langle n\rangle = 1$, and a comparison of $Q_M(\theta)$ (dots) and $P(\theta)$ (dashes) for the phase randomized state $|b\rangle$.

for which $P(\pi)$ vanishes, both terms in the modulus in (8) become equal. We then have

$$P(0) = (2/\pi) \left| \sum_{n \text{ even}} c_n \exp(-in\theta) \right|^2 , \tag{9}$$

which is also equal to $P(\theta)$ for $|f(0)\rangle$. This can be compared with

$$Q_M(\theta) \propto |\langle f(-\theta)|y=0\rangle|^2$$

$$= \left|\sum_n b_n c_n \exp(-in\theta) \right|^2 , \tag{10}$$

where b_n are the number state coefficients of $|y=0\rangle$. Thus for states with a phase probability distribution contained in a range of π, the fact that b_n are zero for n odd does not matter. The only difference between (9) and (10) is that b_n are not constant but reduce gradually with n [16]. For many states this variation over the range for which c_n are significantly non-zero will not lead to huge discrepancies. As a numerical test we changed the phases of the number states in an Airy state randomly to make a combination of five or six partial phase states. The result was a state with a reasonably broad phase distribution, but which was contained in the π range centred on zero. The normalized distributions obtained from (9) (dashes) and from (10) (dots) are shown on the right side of Fig. 1. In this case at least, the measurable distribution is quite a good representation of the actual phase distribution. For an unknown state, a possible method of establishing whether the probability density vanishes over half of the possible range is to examine the signal distribution corresponding to the measured-phase operator (5) as the phase is being changed. If it is impossible to find a phase for which this distribution vanishes for either all negative or all positive values of signal, then this method of measurement will not be suitable.

4. Conclusion

We have discussed the problem of the nature of phase from an operational point of view. Loosely, our definition of phase is that which is shifted by a phase shifter, with which most would agree. This leads to a description of phase which is perfectly consistent with photon number-phase complementarity. In order to ensure that the phase density of a strong coherent state tends to a delta-function type of distribution, we have set an arbitrary parameter to a constant. Choosing this constant to be zero leads us to the same probability distribution as we have obtained previously, but this time without the need to use any hermitian operators or states of the $(s + 1)$-dimensional Ψ-space. This may be viewed both as a verification of our previous approach, while perhaps being at the same time more acceptable to some than our previous approach. Since our operational approach was based on the operation of generation rather than measurement, we also discussed ways of measuring the phase probability distribution. These included exact determination by measurement of the state and some more direct, but less accurate, methods that are suitable for particular types of states. In conclusion, we feel that the correct phase probability density for a state of light is now reasonably well established and the question as to its measurement can be answered both in principle and in practice.

References

1. P.A.M. Dirac: Proc. R. Soc. London, Ser.A **114**, 243 (1927)
2. L. Susskind and J. Glogower: Physics **1**, 49 (1964)
3. S.M. Barnett and D.T. Pegg: J. Phys. A **19**, 3849 (1986)
4. D.T. Pegg and S.M. Barnett: Europhys. Lett. **6**, 483 (1988)
5. S.M. Barnett and D.T. Pegg: J. Mod. Opt. **36**, 7 (1989)
6. D.T. Pegg and S.M. Barnett: Phys. Rev. A **39**, 1665 (1989)
7. J.A. Vaccaro and D.T. Pegg: Physica Scripta **T48**, 22 (1993)
8. J.W. Noh, A. Fougéres and L. Mandel: Phys. Rev. Lett. **67**, 1426 (1991); Phys. Rev. A **45**,424 (1992)
9. D.T. Pegg, J.A. Vaccaro and S.M. Barnett: in preparation
10. C.W. Helstrom: *Quantum Detection and Estimation Theory* (Academic Press, New York, 1976) p. 53-57; J.H. Shapiro and S.R. Shepard: Phys. Rev. A **43**, 3795 (1991)
11. D.T. Smithey, M. Beck, M.G. Raymer and A. Faridani: Phys. Rev. Lett. **70**, 1244 (1993)

12. D.T. Smithey, M. Beck, J. Cooper and M.G. Raymer: Phys. Rev.A **48**, 3159 (1993)
13. K. Vogel and H. Riskin: Phys. Rev. A **40**, 2847 (1989)
14. J.A. Vaccaro and D.T. Pegg: Opt. Commun. (in press)
15. W. Vogel and W. Schleich: Phys. Rev. A **44**, 7642 (1991)
16. E. Merzbacher: *Quantum Mechanics,* 2nd ed. (Wiley, New York, 1970), p. 362
17. G.S. Summy and D.T. Pegg: Opt. Commun. **77**, 75 (1990)
18. H.-H. Ritze: Opt. Commun. **92**, 127 (1992)

Nonlinear Quantum Optics with Few Atoms

U. Martini, C. Wagner, and A. Schenzle

Sektion Physik der Universität München, Theresienstr. 37,
D-80333 München, Germany

Abstract. Under appropriate conditions already a single atom is sufficient to induce nonlinear optical effects like laser or maser action or optical bistability. In this extreme microscopic limit of individual atoms and small photon numbers, signatures of nonclassical effects must be visible. Manifestations of quantum behavior are observed either in the statistical properties of the emitted light or in the population statistics of the atomic beam which passes through the interaction zone. While the tools for a quantitative description of the field statistics are well developed, the level specific counting probabilities for atomic beams have not been described in a quantitative way yet.

1 Introduction

Nonlinear phenomena in quantum optics usually involve a macroscopic number of atoms in the gaseous, liquid or solid state. Consequently, the fields involved are intense and contain a dramatic number of photons. Under these conditions the semiclassical approach is the adequate description. Examples are the wellknown models of maser or laser action or of optical bistability. In order to handle the vast number of degrees of freedom of a macroscopic sample, the semiclassical approximation is essential. Starting from here, a first step toward a consistent inclusion of quantum fluctuations is based on the expansion in inverse powers of the atom number - since macroscopic systems are expected to behave classically.

In contrast, for a system consisting of a small number of atoms only such approximations become invalid and a complete quantum picture must be developed. Quantum optical systems are open systems which exchange energy with their environment in an irreversible fashion. In the present case this is due to spontaneous emission or absorptive losses of photons by imperfect resonator walls. The dynamic evolution of the coupled system of atoms and fields is governed by the master equation for the density operator ρ. The dimension of its matrix representation and more so of the generator of time evolution becomes forbiddingly large - too large for any numerical treatment, as soon as the number of atoms typically exceeds 10 and the photon number exceeds 100 substantially. However, in recent years quantum optical experiments became possible in exactly this domain, where the presence of one or two atoms already resulted in nonlinear optical phenomena. [1, 2].

Springer Proceedings in Physics, Vol. 77 Quantum Optics VI
Editors: D.F. Walls · J.D. Harvey

While most quantum optical devices do not require quantization of the field modes of interest, fields with 3-30 photons have no classical correspondence which has subtle but experimentally observable consequences. We distinguish two different situations. In the optical domain information about the interacting system is most directly obtained from measurements on the emitted field. Correlation functions or the photon counting distributions are well established tools. In the micro wave domain only closed superconducting cavities provide the necessary quality with the immediate consequence that outside the resonator no radiation can be detected. The only way for obtaining information on the state of the field inside, is to study the level population statistics of the atomic beam which interacts with the cavity mode. For this purpose we derived the concept of atom-counting and waiting statistics in close analogy to the photon counting statistics developed by [3] in the classical and by [4] and [5] in the quantum regime. We show that there is a close connection between waiting-time densities and two-atom-correlations.

2 Atomic counting statistics

The photon counting statistics is governed by the probability $W(n,T,\eta)$ for observing n counting events with a detector of efficiency η, exposed for given time interval T [4]. The mathematical structure of W suggests that all that is needed is the no-count probability $W(n=0,T,\eta)$. The general expression for an arbitrary count rate $n \neq 0$ is obtained by mere differentiation with respect to the quantum efficiency η [6]. Of special experimental interest are two associated probabilities, which are closely related to the no-event probability:

- the probability $P_1(T)dT$ for observing no event during an observation period T following the last recorded photo-electron and
- the probability $P_2(T)dT$ for a dark time of length T sandwiched between two recorded events.

The concept of counting probabilities is also well suited for describing the level-population statistics in an atomic beam. The particle distribution of a well collimated atomic beam follows the Poissonian law. This is the inevitable classical randomness of a thermal particle source. A detector exposed for a certain interval T will count a random number of events. The probability for detecting n particles during this interval is:

$$F(n,T) = \frac{(rT)^n}{n!} e^{-rT} \,. \tag{1}$$

r is the average rate of arrival. This is the pure thermal part of the statistics. Atoms that enter the interaction region with the field in a definite atomic state are redistributed over the resonant levels. A detector sensitive only to atoms occupying a specific level will sense this population change by counting fewer events with different statistical properties. Let us assume that the resonance condition only selects two atomic levels: an excited state $|a\rangle$ and and a lower state $|b\rangle$. The actual sequence of atoms that emerges from the interaction region at certain times $t_1, t_2, t_3...$ in state $|a\rangle$ or $|b\rangle$ is irrelevant. Relevant is only the relative frequency of atoms recorded in these states. This information is contained in the joint probability:

$$W(n, |a\rangle; m, |b\rangle; T), \tag{2}$$

i.e. the probability of observing n atoms in state $|a\rangle$ and m in $|b\rangle$ during a selected observation time T. The counting probability for a detector sensitive to lower state atoms only is obtained by summing over all non-identified atoms, i.e. the atoms in the excited state $|a\rangle$:

$$W(m, |b\rangle; T) = \sum_{n=0}^{\infty} W(n, |a\rangle; m, |b\rangle; T). \tag{3}$$

An equivalent expression results for detectors which count atoms in the excited state only.

In analogy to the photon case, we may restrict ourselves to the no-count probability $m = 0$ i.e. the probability of observing no atom in the lower state $|b\rangle$ while the detector is exposed to the beam for a time interval T:

$$W(m = 0, |b\rangle; T) = \sum_{n=0}^{\infty} W(n, |a\rangle;\ m = 0, |b\rangle;\ T). \tag{4}$$

A concept directly suggested by experiment is the waiting or dark time statistics for atoms. The probability density $P_1(|b\rangle, T)$ is the statistical measure for observing no atom in the lower state $|b\rangle$ during a time interval T after the observation of the last lower state atom. The probability density $P_2(|b\rangle; T)$, the waiting time density, characterizes the time between two successive atoms in the lower state. P_2 is normalized to ensure that an arbitrary waiting time occurs with unit probability:

$$\int_0^{\infty} P_2(|b\rangle; T) dT = 1\,. \tag{5}$$

The functions P_1 and P_2 are exclusive probability densities for lower state atoms registered at the beginning and/or at the end of the chosen time interval. In between only upper state atoms may occur. If we now relax this restriction and ask for the probability of observing two lower state atoms separated by a time T, irrespective of all the events that occur in between, we are lead to the two-atom-correlation function $g_2(|b\rangle; T)$. All these statistical measures, while motivated by their optical counter part, are also well suited for describing atomic counting experiments.

2.1 An illustrative example: The micromaser

In order to illustrate the method of atom counting and waiting time probabilities we choose the example of the one-atom maser, where a weak beam of excited two-level Rydberg atoms passes through a resonant high Q cavity. While inside the cavity, the atoms interact with the field and the joint state evolves according to the reversible Jaynes-Cummings dynamics. Between successive atoms, the cavity suffers dissipation. With this simplifying assumption the time evolution can be split into a pumping and a damping cycle of statistical duration. The time dependence in these individual cycles is easily found by simple integration. While during the pumping cycle, the state of the system evolves according to the elementary Jaynes-Cummings Hamiltonian H, for the notation see [7]:

$$\dot{\rho} = -i/\hbar[H, \rho], \tag{6}$$

between two atomic transits only dissipation remains:

$$\dot{\rho} = L\rho = \frac{\kappa}{2}([a, \rho a^\dagger] + [a\rho, a^\dagger]). \tag{7}$$

Starting at time t_1 with the field in an arbitrary initial state $\rho_f(t_1)$ and the atom entering the cavity in the excited state $|a\rangle$, the density matrix of the combined system at the exit time $t_1 + \tau$ of the atom is given by:

$$S(t_1 + \tau) = e^{-iH\tau/\hbar}(\rho_f(t_1) \otimes |a\rangle\langle a|)e^{iH\tau/\hbar}. \tag{8}$$

Here we assume that the atoms arrive at random times, but we neglect the variation of the transit time τ through the cavity. Now the state of the atom is registered by the detector and the future of the system depends on the outcome of the meter reading. If the atom is recorded in the excited state $|a\rangle$, the field inside the cavity is projected into the state:

$$\rho_f(t_1 + \tau) = \frac{1}{N_{|a\rangle}}\langle a|S(t_1 + \tau)|a\rangle = \frac{1}{N_{|a\rangle}}\hat{O}_{|a\rangle}\rho_f(t_1), \tag{9}$$

while registration of an atom in the lower state $|b\rangle$ projects the cavity field into:

$$\rho_f(t_1 + \tau) = \frac{1}{N_{|b\rangle}}\langle b|S(t_1 + \tau)|b\rangle = \frac{1}{N_{|b\rangle}}\hat{O}_{|b\rangle}\rho_f(t_1). \tag{10}$$

The normalization constants $N_{|a\rangle}$ and $N_{|b\rangle}$ guarantee that the trace of $\rho_f(t_1+\tau)$ over the states of the field after the measurement remains unity. The operators $\hat{O}_{|a\rangle}$ and $\hat{O}_{|b\rangle}$ act as the field-propagators with the additional condition that the atom is observed afterward in the corresponding state. In the time between the transit of two successive atoms the field freely relaxes with a time constant γ^{-1} according to the Liouvillian L:

$$\rho_f(t) = e^{L(t-t_1-\tau)}\rho_f(t_1 + \tau). \tag{11}$$

The present simplified treatment neglects dissipation while the atom is inside the cavity. In order to find the state of the field for arbitrary times t it is necessary to iterate the alternating sequence of pumping and damping cycles for an arbitrary number of times, weighted by the probability of their occurrence in a Poissonian beam and to sum over all possible time sequences. In a Poissonian beam the probability of finding no atom in an interval T is $F(n=0,T) = \exp(-rT)$. The probability of observing a single atom in the infinitesimal interval dT is $F(n=1,dT) = rdT$. It is now our aim to construct the propagator for the density matrix of the cavity field under the condition that atoms are leaving the cavity in a given state. This is the propagator for a statistical sub-ensemble the trace of which is not preserved. By $G_0(T)$ we denote the propagator for free evolution i.e. that no atom enters the cavity during the time T. It is given by the time evolution of free decay $\exp(LT)$ multiplied by the probability that during the entire interval no atom has passed through the cavity $F(0,T)$:

$$G_0(T) = e^{LT} F(0,T). \tag{12}$$

In order to simplify the expressions later, we use the abbreviation $F(T) = F(0,T)$. F is a c-number and therefore commutes with L, and only e^{LT} acts on the states of the field. When averaged over the stationary state of the field $\rho_f = \rho_{stst}$ we obtain the zero count probability for atoms in either of its two levels:

$$\mathrm{tr}\rho_{\mathrm{st\,st}} G_0(T) = W(0,|a\rangle;0,|b\rangle;T) = F(T). \tag{13}$$

This simple result could have been guessed easily: The probability decays exponentially, indicating that it becomes more and more improbable to detect not a single atom the longer the observation time. The propagator for one and only one atom leaving the resonator in the upper state at time t_1 during the interval of length T, while the field was in a given state at $t=0$ is:

$$G_1(T) = r\int_0^T dt_1 e^{L(T-t_1)} F(T-t_1)\hat{O}_{|a\rangle} e^{L(t_1)} F(t_1)\,. \tag{14}$$

This result is evident as well:

- $F(t_1)$ is the probability that no atom enters during the initial period t_1;
- rdt_1 is the probability that an atom is present at t_1 during the interval of length dt_1;
- $F(T-t_1)$ is the probability that no further atom enters in the remaining period;
- $\hat{O}_{|a\rangle}$ collapses the present state according to the meter reading ;
- e^{Lt_1} and $e^{L(T-t_1)}$ takes into accout the relaxation of the field between atomic transits.

Since the upper state atom can occur at any time during the interval, we have to sum over all possibilities.

The generalization of this concept to an arbitrary number of excited atoms is straightforward. These atoms are recorded in the excited state at times $t_1, t_2, t_3 ...$, which calls for a factor $\hat{O}_{|a\rangle}$ each time, while in between two recordings t_i, t_j no atom is observed, this happens with probability $F(t_i - t_j)$. Since the times t_j are arbitrary, one must sum over all realizations.

Finally, the probability for observing no lower state atom but an arbitrary number of atoms in the excited state during the time interval T is obtained by summing over n:

$$W(0,|b\rangle) = \langle \sum_{n=0}^{\infty} G_n(T) \rangle \tag{15}$$

$$= \mathrm{tr}\rho_{\mathrm{st}}\ {}_{\mathrm{st}}e^{(L+r(\hat{O}_{|a\rangle}-1))T}. \tag{16}$$

The infinite sequence of iterated convolution integrals has been summed up leading to the exponential form.

In order to arrive at the result above, we did not only assume that no atom in state $|b\rangle$ was detected, we assumed that no atom in this state had passed through the cavity. This is not the same as soon as we consider detectors with a finite quantum efficiency η. We can include an imperfect detector by repeating the previous arguments for an ideal detector and assume in addition that at random instances t_j a lower state atom had occurred, but went undetected with probability $(1-\eta)$. This leads to the final, experimentally relevant expression:

$$W(0,|b\rangle;T) = \mathrm{tr}\rho_{\mathrm{st}}\ {}_{\mathrm{st}}e^{(L+r(\hat{O}_{|a\rangle}+(1-\eta)\hat{O}_{|b\rangle}-1))T}. \tag{17}$$

From here it is obvious how the waiting time probabilities $P_2(T)$ can be derived directly form the above result, since we only have to ensure that at the initial and the final instant of the interval T, a lower state atom was detected:

$$P_2(|b\rangle;T) = \frac{\langle \hat{O}_{|b\rangle} e^{(L+r(\hat{O}_{|a\rangle}+(1-\eta)\hat{O}_{|b\rangle}-1))T} \hat{O}_{|b\rangle} \rangle}{\langle \hat{O}_{|b\rangle} \rangle^2}. \tag{18}$$

For a detector with an ever smaller quantum efficiency η atoms in state $|b\rangle$ can freely pass the recording zone undetected. In the limit of vanishing η and with the appropriate normalization $P_2(|b\rangle;T)$ turns into the two-atom-correlation function $g_2(T)$:

$$g_2(|b\rangle;T) = \frac{\langle \hat{O}_{|b\rangle} e^{(L+r(\hat{O}_{|a\rangle}+\hat{O}_{|b\rangle}-1))T} \hat{O}_{|b\rangle} \rangle}{\langle \hat{O}_{|b\rangle} \rangle^2}. \tag{19}$$

In the definition above, $P_2(|b\rangle;T)$ has not yet been normalized properly in order to show the close connection g_2. The waiting time probability density $P_2(|b\rangle;T)$ must vanish asymptotically for large times T due to normalizability

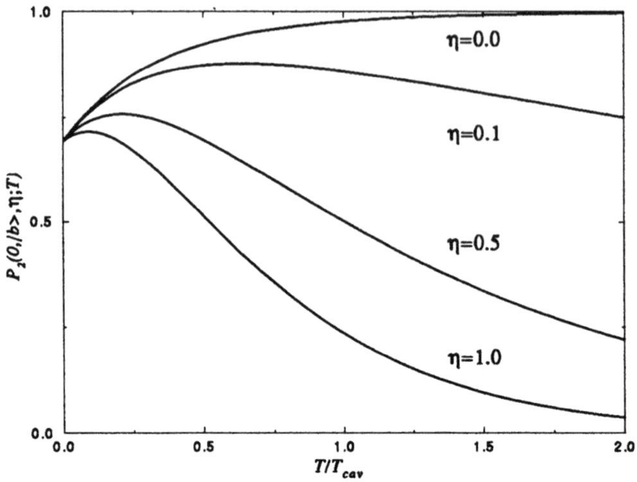

Figure 1: Waiting-time distributions $P_2(0, |b\rangle, \eta; T)$ for different detection efficiencies η. The time is given in units of the cavity decay time T_{cav}. The ratio of the atomic injection rate to the cavity decay rate is $N_{ex} = 4.0$. The number of thermal photons was chosen to be $n_{th} = 0.15$ and the product of coupling strength and interaction time is $g\tau = 1.54$. The curves belong to $\eta =$ 0%, 10%, 50% and 100% detection efficiency (top to bottom). The uppermost line is equivalent to the "two-atom correlation" function $g_2(|b\rangle, T)$. The anti-bunching of atoms, indicated by a value below 1 at time zero, is related to the sub-Poissonian photon statistics of the cavity field.

and physically because arbitrary large waiting times are extremely improbable for beams with finite intensity. In contrast, the two-atom correlation function approaches unity in this limit, since the two detection events become statistically independent.

In Fig.(1) we have plotted the atomic waiting time probability $P_2(|b\rangle; T)$ for varying quantum efficiencies η. $\eta = 1$ is the waiting time density for a perfect detector, $\eta = 0$ represents the two-atom correlation function. The different tendency of the asymptotic behavior is clearly visible. $g_2(|b\rangle; T)$ is smaller than one for all times indicating nonclassical features in the arrival statistics of the atoms in the lower state: they arrive more regularly than predicted by the Poissonian law. The nonclassical features of the field inside is reflected in the nonclassical behavior of the emerging atomic beam.

This brings us back to the question that arose in the beginning: In which way can one infer from the statistical properties of the atomic beam those of the field inside? Certainly, the average number of photons inside the cavity is simply related to the number of atoms that leave in the ground state - by mere energetic arguments. Not that direct is the relation between the noise in the field and the number fluctuations of the beam. A signature of trapping states, where the atoms undergo multiple 2π rotations leaving the cavity predominantly in the excited state $|a\rangle$ are indicated by exceptionally long waiting times for atoms in the ground state. These periods are interrupted by bursts of atoms in state $|b\rangle$ lasting typically for a cavity decay time κ^{-1}. These bursts occur when fluctuations upset the condition for trapping.

3 Non-classical photon statistics

The one-atom maser is a realization of the basic concept of atom-field interaction in the micro wave regime. Due to recent breakthrough in mirror technology, optical resonators are available with unprecedented quality. This allows one to carry out few atom experiments in the optical domain: A weak atomic beam now crosses an *optical* resonator which is driven itself by a coherent external laser beam. At a given time only a few atoms are simultaneously present inside the interaction zone [2]. The field transmitted through the cavity carries the information about the quantum mechanical energy exchange between the beams. The most prominent non-classical features are expected when either few atoms or few photons are involved in the interaction process.

We will model this situation by one or two atoms at rest in a cavity interacting with the resonant cavity mode and, if necessary, with the thermal background radiation. The cavity field itself suffers losses from the imperfections of the mirrors. The number of atoms in semiclassical description is accounted for by a scaling parameter for the coupling constant g, while the complete quantum picture must add the additional degrees of freedom for any additional atom. For a large number of atoms, the semi-classical description is is valid - for 1,2,3.. atoms we expect noticeable differences. In order to study the turn over from the global description by a scaled dipole coupling constant g to the individual behavior of a few atoms we compare the microscopic two-atom case with the case of a single atom but with twice the interaction strength.

The evolution equation to consider is the *standard master equation* for atom field interaction including the atom and field reservoirs:

$$\dot{\rho} = L\rho = -i/\hbar[H,\rho] + L_a\rho + L_f\rho\,, \tag{20}$$

where ρ is the statistical operator for atoms and field, $H = H(g)$ the resonant coherent interaction in rotating wave approximation and $L_a(\gamma)$ and $L_f(\kappa)$ are the contributions of thermal reservoirs to atom and field dissipation.γ is the spontaneous emission rate and κ the loss rate of the resonator. The two essential parameters that characterize the quantum behavior are:

- $C = g^2/\kappa\gamma$ is the one-atom cooperativity,
- $n_s = \frac{1}{8}\gamma/g^2$ is the saturation photon number.

The scaled amplitude of the external field is $y = \alpha/\sqrt{n_s}$. The formal integral of the master equation in the product Hilbert space of one and two atoms and a single mode of the field is:

$$\rho(t) = e^{Lt}\rho(t=0) \,. \tag{21}$$

Due to spontaneous emission and cavity losses the system evolves toward a unique steady state:

$$\rho_{stst} = \lim_{t\to\infty} e^{Lt}\rho(t=0) \,, \tag{22}$$

where $\rho(t=0)$ is any initial density operator.

Already for the elementary problem of a single atom interacting with a single mode of the field in the presence of external driving and damping, there is no analytic treatment and the solution must be found numerically. If we are interested in the stationary state only, we can considerably save computing time by a reasonable guess of $\rho(0)$. For calculating the dynamic properties, however, like one- and two-photon correlation functions we cannot choose the initial conditions freely. Given a numerical algorithm for calculating the time evolution of ρ starting from an arbitrary initial condition, we can also evaluate any two time correlation function using the same procedure:

$$\begin{aligned} g_1(t) &= \langle a^\dagger(t)a(0)\rangle & (23)\\ &= \mathrm{tr}(a^\dagger e^{Lt} a\rho_{\mathrm{st\ st}}) \,, & (24) \end{aligned}$$

and

$$\begin{aligned} g_2(t) &= \langle a^\dagger a^\dagger(t)a(t)a\rangle & (25)\\ &= \mathrm{tr}(a^\dagger a e^{Lt} a\rho_{\mathrm{st\ st}} a^\dagger) , & (26) \end{aligned}$$

where $a = a(t=0)$ and $a^\dagger = a^\dagger(t=0)$. We only have to use somewhat unusual initial conditions: $a\rho_{stst}$ or $a\rho_{stst}a^\dagger$. The decoration of ρ_{stst} with a's and $a^\dagger$'s is essential, otherwise e^{Lt} would act directly on the steady state and would simply reproduce it $e^{Lt}\rho_{stst} = \rho_{stst}$. An alternative method for calculating correlation functions was proposed by Haken [8], where small perturbations are applied to upset the steady state conditions inducing transient fluctuations about the steady state. We have used both methods for the numerical solution of the problem [9]. Comparing the results of the independent procedures allowed us to estimate the accuracy of the approach which turned out to be excellent.

In Fig.(2-3) we have summarized some of the essential results of the numerical solution. In Fig.(2) we have plotted the two-photon correlationfunction for different amplitudes of the externally applied field for a single atom. As expected intuitively, the nonclassical feature of anti-bunching is rapidly lost when the amplitude of the driving field is increased. This is also the case when more than one atom is present. One is tempted to conclude that all non-classical features are lost as soon as the field intensity inside the cavity exceeds a certain - rather small - value. That this is not correct can be seen in Fig.(3). Here we compare one- and two-atom calculations for the two-photon correlation function. For small times g_2 assumes values large compared to 1 consistent with classical statistics. However, for intervals of the order of 1.5 times the cavity decay time the two atom model leads to rather small values of g_2. One should

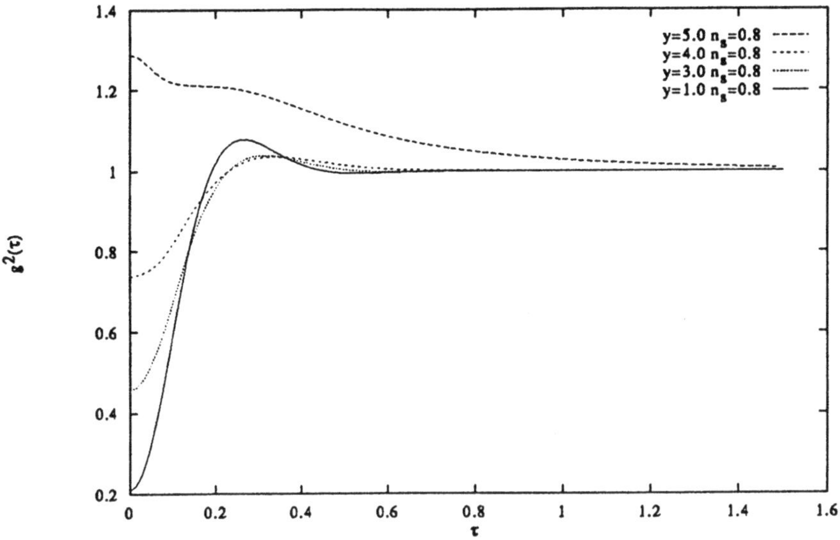

Figure 2: Increasing the intensity of the external field decreases the antibunching effect and turns it eventually into bunching. In this plot, only one atom is present in the cavity. The cooperativity is $C = 6$.

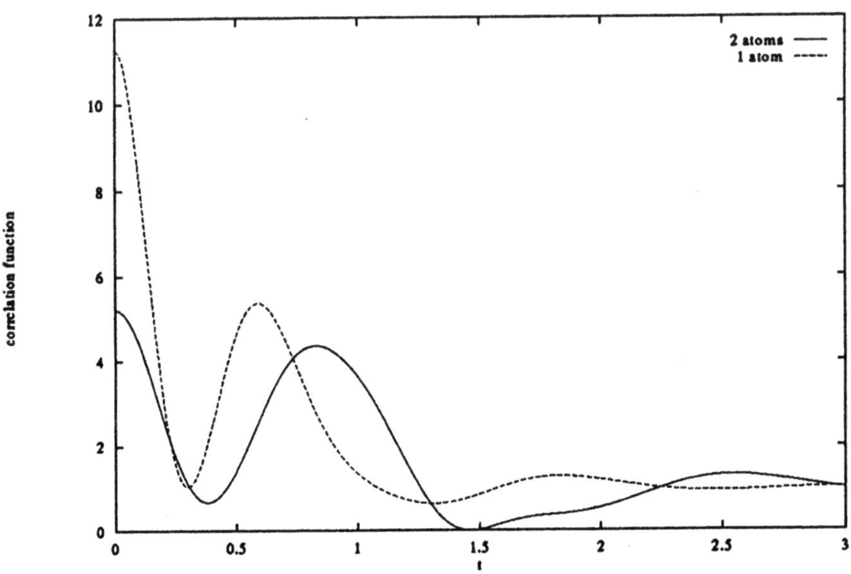

Figure 3: Even bunched light may show nonclassical behavior. The zero of the intensity correlation function for two atoms (solid line) at $t \approx 1.5$ cavity decay times is a quantum feature and classically forbidden. The one atom case (dotted line) shows more drastic bunching and no violation of the classical inequality

remember that anti-bunching is not the only non-classical feature of a two-photon correlation function. Using the Cauchy-Schwarz inequality one easily derives the following inequality based on classical statistics:

$$g_2(t) \geq \frac{g_1(t)^2}{g_1(0)^2} \ . \tag{27}$$

This inequality is obviously violated in a certain region around $t = 1.5\kappa^{-1}$ as seen in Fig.(3). The violation is only obtained for the two atom model, while the rescaled single atom model satisfies almost the classical limit. The reason for the difference of one and two atom results is found in the atom-atom correlation which is absent in the rescaled single atom approach.

References

[1] D.Meschede, H.Walther, and G.Mueller. *Phys.Rev.Lett* **54**, 551 (1985).

[2] G.Rempe, R.J.Thompson, R.J.Brecha, W.D.Lee, and J.Kimble. *Phys.Rev.Lett.* **67**, 1727 (1991).

[3] L.Mandel. *Proc.Phys.Soc.London* **72**, 1037 (1958).

[4] R.Glauber. *Quantum Optics and Electronics, Les Houches* (1965).

[5] P.W.Kelley and W.H.Kleiner. *Phys.Rev.* **136**, 316 (1964).

[6] A.Schenzle. *J.Stat.Phys.* **54**, 1243 (1989).

[7] E.T.Jaynes and F.W.Cummings. *Proc.IEEE* **51**, 89 (1963).

[8] H.Haken. "Light and Matter", vol. XXIV2c. Springer Verlag Heidelberg (1970).

[9] U.Martini, Ch.Ginzel, and A.Schenzle. *Opt.Commun.* **102**, 379 (1993).

Dissipative Environments: Quantum Jumps or Phase-Space Localization?

B.M. Garraway and P.L. Knight

Optics Section, The Blackett Laboratory, Imperial College, Prince Consort Road, London SW7 2BZ, UK

Abstract. We discuss the evolution of individual realizations of quantum superpositions conditioned by two distinct measurement schemes: amplitude (heterodyne) measurements and excitation number (counting) measurements. These two schemes result in either localization or quantum jumps in phase space, which we illustrate for some non-classical initial superpositions.

1 Introduction

The decay of coherence when a quantum system interacts with a much larger environment is usually described by a master equation for the system reduced density matrix, and emphasises the evolution of an entire ensemble [1]. Coherences are destroyed at a rate which is proportional to the degree of excitation of the system (much faster than energy decay) as the system relaxes to a statistical mixture of relevant "pointer bases" [2,3]. Such an ensemble-averaged approach does not easily address the question of how an individual member of an ensemble evolves in a dissipative environment, even though this issue is accessible to experiment [4]. A number of methods have been developed recently to simulate the evolution of single realizations. One centres on quantum jumps, reflecting the information gained from the detection of decay quanta, and involves a continuous evolution which is randomly interrupted by "instantaneous" jumps as the system state vector is conditioned by the information gained from the register of counts in detection of decay [5,6]. The other method involves both a quantum state *diffusion* where the individual quantum trajectory fluctuates through a Wiener process deriving from the environment, and *localization* to a coherent state, an eigenstate of the relevant Lindblad operator describing the coupling of the system to the environment [7]. Wiseman and Milburn [8] have related these two simulation methods to specific measurement schemes: the quantum jump method to continuous direct counting of decay quanta, and the quantum state diffusion method to a heterodyne measurement of oscillation amplitudes. Both methods, of course, lead to the *same* ensemble behaviour when averaged over many realizations [5–9]. The new simulation schemes have been widely adopted for their utility in computing averages in a numerically efficient way on small computers. A quite separate issue is what, if any, meaning can be ascribed to individual trajectories, for these do differ in the different simulation schemes, reflecting the distictly different conditionings applied, and reflecting the acquisition of different information in each simulation. Individual quantum trajectories "unravelled" from the master equation reflect the particular measurement scheme [6,8].

Here, we examine single realizations of a superposition of coherent states of an oscillator (for example, a single electromagnetic field mode) interacting with a dissipative environment. We show that such realizations evolve very differently in the two simulation methods described above. The quantum jump, or state vector Monte-Carlo, method can lead to what we have called [10] a "jumping cat" where the conditioned state vector evolves through a series of jumps from one pure state "cat" (e.g. an even coherent state) to another (an odd coherent state) as individual quanta are removed at random by the dissipation. The quantum state diffusion model leads to a rapid localization to one or other of the coherent state components of the initial superposition, reflecting the approximate projection onto one potential outcome of an amplitude measurement. The coherent states are an overcomplete set and form an interesting pointer basis for an amplitude (heterodyne) measurement on any state: we exploit this in a study of individual trajectories of a chosen state (a Fock state) which localizes to one coherent state component (picked at random) of that state expressed in a coherent state basis. We employ phase-space descriptions of the evolution, and conclude with a comment on the "tiling" of phase-space and localization.

In order to compare the simulation methods with the density matrix result we need to solve the master equation for that density matrix. In a typical dissipative process the master equation for ρ takes the form (with $\hbar = 1$)

$$\frac{\mathrm{d}}{\mathrm{d}t}\rho = -i\,[H,\rho] + \left(\hat{R}\rho\hat{R}^\dagger - \tfrac{1}{2}\hat{R}^\dagger\hat{R}\rho - \tfrac{1}{2}\rho\hat{R}^\dagger\hat{R}\right), \tag{1}$$

where the Lindblad reservoir operator $\hat{R}$ represents the influence of the environment. We have discussed elsewhere [9] the case of a two-photon absorber but $\hat{R}$ would typically be $\sqrt{\kappa}\hat{a}$ for ordinary single photon dissipation. In the case of a dissipative two-level system we would have $\hat{R} = \sqrt{\kappa}\sigma_-$ where σ_- is one of the Pauli spin matrices. In both of these cases the parameter κ signifies the loss rate of energy into the reservoir and the Hamiltonian H would represent the coherent evolution of the lossless system. The methods used to solve (1) depend specifically on the choice of operators $\hat{R}$ and the Hamiltonian H. Sometimes there is more than one reservoir operator, e.g. for a cavity in a thermal heatbath (characterised by a photon occupation $\bar{n}$) we have both $\sqrt{(\bar{n}+1)\kappa}\hat{a}$ and $\sqrt{\bar{n}\kappa}\hat{a}^\dagger$ as reservoir operators. However, all the methods we describe below can be extended to cope with this kind of situation.

2 Quantum Jumps

When we perform a simulation of a density matrix problem (by quantum jumps or quantum state diffusion) we consider not the density matrix ρ, but a state vector $|\psi\rangle$ which is one member of an ensemble of state vectors. Each member, or sample, undergoes a different evolution under the influence of a different sequence of random events. To obtain results that do not depend on a specific history of events and which can be compared with the density matrix results we have to average the observables over the ensemble (as would be done in an experiment). Thus given a set of state vectors at a time t and an observable $\hat{O}$, the averaged value may be denoted

$$\mathsf{M}\left\{ \langle\psi(t)|\hat{O}|\psi(t)\rangle \right\}. \tag{2}$$

A sufficient number of samples are taken to ensure an acceptable average.

We now focus on single photon dissipation as simulated by quantum jumps, or the state vector Monte-Carlo, and we start by considering the two things that can happen to a single state vector over a very short time interval δt: there may be a quantum jump, or there may be no quantum jump. Our approach here most closely follows that of Ref. [5]. If we observe a quantum jump we project the operator $\hat{a}$ onto the state vector because of the acquisition of information through the decay measurement:

$$|\psi(t)\rangle \longrightarrow \hat{a}|\psi(t)\rangle . \tag{3}$$

The probability of this quantum jump within a short time δt is

$$\Delta P = \langle\psi(t)|\hat{R}^\dagger\hat{R}|\psi(t)\rangle\delta t = \kappa\langle\hat{n}\rangle\delta t . \tag{4}$$

The state (3) must be normalized to give the result

$$\begin{aligned} |\psi(t)\rangle \longrightarrow \quad & \frac{\hat{a}|\psi(t)\rangle}{\sqrt{\langle\psi(t)|\hat{a}^\dagger\hat{a}|\psi(t)\rangle}} \\ & = \frac{\hat{a}|\psi(t)\rangle}{\sqrt{\Delta P}}\sqrt{\delta t} , \end{aligned} \tag{5}$$

which is conditional on a quantum jump having taken place within δt.

Now we consider what happens to the wave function if we observe no quantum jump in the time interval δt. We propagate $|\psi(t)\rangle$ with the effective Hamiltonian

$$H_{eff} = -\tfrac{1}{2}i\kappa\hat{a}^\dagger\hat{a} . \tag{6}$$

Because H_{eff} is not Hermitian the normalization of the state vector will shrink; this can be attributed to the possibility of quantum jumps and the state vector must be renormalized so that in a time step δt:

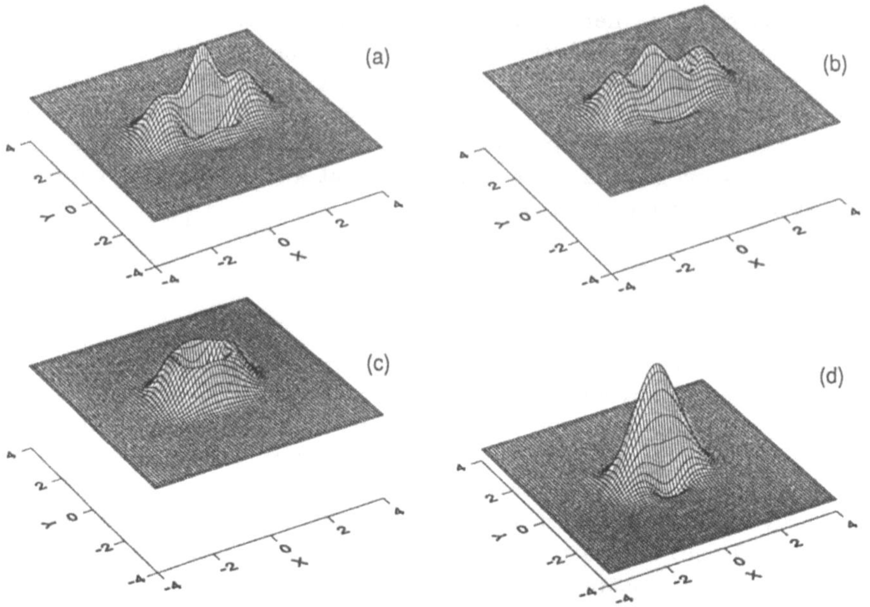

Figure 1. A "jumping cat" which was initially comprised of the coherent states $|\alpha\rangle$ and $|-\alpha\rangle$ with $\alpha = 2$. We show the Wigner functions before and after the first jump (at $\kappa t = 0.668$) in (a) and (b). In (c) and (d) we show the Wigner functions before and after the second jump (at $\kappa t = 1.884$).

$$
\begin{aligned}
|\psi(t)\rangle \longrightarrow \quad & \frac{\mathrm{e}^{-i\delta t H_{eff}}|\psi(t)\rangle}{\sqrt{\langle\psi(t)|\,\mathrm{e}^{i\delta t H_{eff}^{\dagger}}\,\mathrm{e}^{-i\delta t H_{eff}}|\psi(t)\rangle}} \\
\sim & \frac{\mathrm{e}^{-i\delta t H_{eff}}|\psi(t)\rangle}{\sqrt{\langle\psi(t)|\,\mathrm{e}^{-\delta t \hat{a}^{\dagger}\hat{a}}|\psi(t)\rangle}} \\
\sim & \frac{\mathrm{e}^{-i\delta t H_{eff}}|\psi(t)\rangle}{\sqrt{1-\Delta P}}\,,
\end{aligned}
\tag{7}
$$

where the approximations are correct to order δt.

The above forms a procedure for the discretised evolution of a state vector which describes a single member of the ensemble. Each time step δt proceeds in the following way. We calculate the small, but finite probability of a jump ΔP, Eq. (4). Then with a random number we decide with probability ΔP whether there will be a quantum jump or not. If there is a jump, the state vector is projected and renormalised according to Eq. (5). However, if there is to be no jump, the state vector evolves according to Eq. (7). Because the probability of jumping is small the usual decision at each time step is the non-linear evolution (7). When there is a jump a discontinuity in the state vector takes place and as a result the typical behaviour of observables for a single state vector is smooth evolution interrupted by jumps.

In Fig. 1 we illustrate the time evolution with some snapshots of the Wigner function of a single ensemble member as it undergoes two quantum jumps. The initial state was a Schrödinger "cat" state of the form $(|\alpha\rangle + |-\alpha\rangle)/N_+(\alpha)$, where $N_+(\alpha)$ is the normalization and $|-\alpha\rangle$ and $|\alpha\rangle$ are coherent states. During the time interval when there are no jumps the Schrödinger "cat" shrinks as $\alpha(t) = \alpha(0)\,e^{-\kappa t/2}$ when the evolution (7) is considered. This process has taken place from the initial state up to Fig. 1(a). When there is a jump, then according to Eq. (5) we find that the Schrödinger "cat" jumps to a different type of "cat" with the form $(|\alpha(t)\rangle - |-\alpha(t)\rangle)/N_-(\alpha(t))$. The jump is seen as we go from Fig. 1(a) to Fig. 1(b); the jump from an "even" to an "odd" Schrödinger "cat" state is particularly visible in the Wigner functions as the interference fringes change sign. Subsequently the Schrödinger "cat" shrinks again as we evolve according to Eq. (7) in the absence of any jumps. This takes us to Fig. 1(c) where the two principal components of the 'cat" are seen to have moved in towards the phase space origin. Then there is another quantum jump as we go from Fig. 1(c) to Fig. 1(d); the Wigner function fringes again change their sign.

3 Quantum State Diffusion

The second simulation method we consider is quantum state diffusion in which the evolution of $|\psi(t)\rangle$, representing one member of the ensemble, is determined by a non-linear stochastic differential equation [11,7]. We write the equation (for the specific case of one-photon damping of a field state) in the Itô form involving drift and diffusion processes so that

$$\begin{aligned}|\psi(t+\delta t)\rangle = {} & (1 - i\delta t H)\,|\psi(t)\rangle + \kappa\left[\langle \hat{a}^\dagger\rangle \hat{a} - \tfrac{1}{2}\hat{a}^\dagger \hat{a} - \tfrac{1}{2}\langle \hat{a}^\dagger\rangle\langle \hat{a}\rangle\right]|\psi(t)\rangle\,\delta t \\ & + \sqrt{\kappa}\,[\hat{a} - \langle \hat{a}\rangle]\,|\psi(t)\rangle\,d\xi_t\,,\end{aligned} \tag{8}$$

where $d\xi_t$ is a random complex Wiener variable which perturbs the system. It varies randomly between each time step and each sample run so that when averaged:

$$\overline{d\xi_t} = 0 \quad , \quad \overline{d\xi_t d\xi_{t'}} = 0 \quad , \quad \overline{d\xi_t^* d\xi_{t'}} = \delta_{t,t'}\,\delta t\,. \tag{9}$$

In consequence the time evolution is not smooth on a very small scale because of the discretised fluctuations of the state vector in its Hilbert space. This distinguishes the quantum state diffusion calculations from the state vector Monte-Carlo approach in a characteristic way; in the latter case the state vector (and hence the observables) jump only at certain places rather than at every time point. This manifestation of the constant jiggling of the state vector is clearly seen in the single trajectory examples of Fig. 2 (solid and dotted curves). However, as in the state vector Monte-Carlo the results for an ensemble average smooth out as we increase the number of samples.

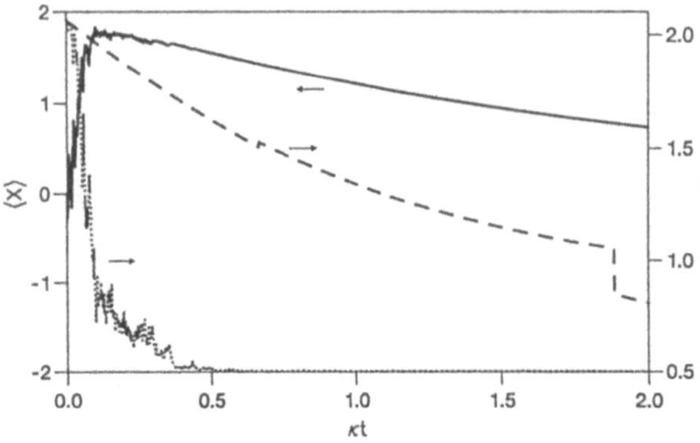

Figure 2. Localization of a Schrödinger "cat" state which is initially comprised of the coherent states $|\alpha\rangle$ and $|-\alpha\rangle$ with $\alpha = 2$. We show $\langle\hat{X}\rangle$ (solid curve—left-hand scale) and ΔX (dotted curve—right-hand scale). The localization is seen in the development of the non-zero value of $\langle\hat{X}\rangle$ and in the rapid reduction of the uncertainty. The dashed curve shows ΔX for the quantum jump trajectory of Fig. 1. The value of $\langle\hat{X}\rangle$ is zero in this case because there is no localization.

It is interesting to note that if it happens that $|\psi(t)\rangle$ is an eigenstate of the Lindblad operator $\hat{a}$ the random fluctuations do not contribute to the system evolution in (8). This must result in a vanishing of the random perturbations and can be seen in Fig. 2 as equilibrium is established. That $|\psi(t)\rangle$ is an eigenstate of $\hat{a}$ is not a sufficient condition for the steady state; an inspection of Eq. (8) shows that we also require $|\psi(t)\rangle$ to be an eigenstate of $H - \frac{1}{2}i\hat{a}^\dagger\hat{a}$ as is found in the state vector Monte-Carlo case.

The stochastic equation (8) amounts to a Langevin equation for the state vector (rather than the usual case of observables). As in the case of state vector Monte-Carlo simulations we can recover the master equation (1) when we average over the ensemble. Equation (8), originally constructed by Gisin and Percival, has recently been interpreted by Wiseman and Milburn as the state vector evolution in a balanced heterodyne detection scheme [8]. A related equation is found for the case of balanced homodyne detection by Carmichael [6]. In both cases a quantum jump theory is developed for the measurement process in the presence of a strong classical local oscillator. This quantum jump theory, unlike the single mode case presented above, contains two jump operators, i.e. one for each of the detectors. Using the fact that the local oscillator is strong and creates many jumps in the detector, the number of jumps in a short time interval can be determined by a stochastic function $d\xi$ as found in Eq. (8) above. Wiseman and Milburn then treat the heterodyne case by smoothing out the fastest oscillations to obtain:

$$|d\tilde{\psi}\rangle = \left[-\frac{\kappa}{2}\hat{a}^\dagger\hat{a}dt + \kappa\hat{a}\langle\psi(t)|\hat{a}^\dagger|\psi(t)\rangle dt + \sqrt{\kappa}\hat{a}d\xi\right] |\tilde{\psi}(t)\rangle , \qquad (10)$$

where $|\tilde{\psi}(t)\rangle$ is an un-normalized state vector and $|\psi(t)\rangle$ is the normalized state vector. To obtain the equation of motion for the normalized state vector one has to be careful about differentiation because of the normalization of the noise term (Eq. (9)). The Itô calculus may be used under which, for example,

$$d(\langle\tilde{\psi}(t)|\tilde{\psi}(t)\rangle) = \langle\tilde{\psi}(t)|d\tilde{\psi}\rangle + \langle d\tilde{\psi}|\tilde{\psi}(t)\rangle + \langle d\tilde{\psi}|d\tilde{\psi}\rangle\,. \tag{11}$$

Furthermore, to obtain a connection with the Gisin-Percival equation (8) it is necessary to include a phase factor as well as renormalizing the state vector so that

$$|\psi(t)\rangle = \frac{e^{i\zeta(t)}|\tilde{\psi}(t)\rangle}{\sqrt{\langle\tilde{\psi}(t)|\tilde{\psi}(t)\rangle}}\,. \tag{12}$$

The phase factor is stochastic because

$$i d\zeta = \frac{\langle\tilde{\psi}(t)|\hat{a}^\dagger|\tilde{\psi}(t)\rangle d\xi^* - \langle\tilde{\psi}(t)|\hat{a}|\tilde{\psi}(t)\rangle d\xi}{2\langle\tilde{\psi}(t)|\tilde{\psi}(t)\rangle} \tag{13}$$

resulting in the equation for the normalized $|\psi(t)\rangle$:

$$\begin{aligned}|d\psi\rangle = \kappa\,&\left[-\tfrac{1}{2}\hat{a}^\dagger\hat{a} + \hat{a}\langle\psi(t)|\hat{a}^\dagger|\psi(t)\rangle + \tfrac{1}{2}\langle\psi(t)|\hat{a}^\dagger\hat{a}|\psi(t)\rangle\right.\\ &\left.- \langle\psi(t)|\hat{a}^\dagger|\psi(t)\rangle\langle\psi(t)|\hat{a}|\psi(t)\rangle\right]|\psi(t)\rangle dt\\ &+ \tfrac{1}{2}\kappa\left[-\langle\psi(t)|\hat{a}^\dagger\hat{a}|\psi(t)\rangle\right.\\ &\left.+\langle\psi(t)|\hat{a}^\dagger|\psi(t)\rangle\langle\psi(t)|\hat{a}|\psi(t)\rangle\right]|\psi(t)\rangle d\xi^* d\xi\\ &+ \sqrt{\kappa}\left[\hat{a} - \langle\psi(t)|\hat{a}|\psi(t)\rangle\right]|\psi(t)\rangle d\xi\,.\end{aligned} \tag{14}$$

This equation yields the Gisin-Percival equation (8) if $d\xi^* d\xi \longrightarrow dt$. The only difference between these equations is that the state vector in Eq. (8) is only normalized for the ensemble average whilst in Eq. (14) it is always normalized. In practice this seems to result in no noticeable differences because whilst our computer program integrates Eq. (8) it renormalizes the state vector at every time step.

We should emphasize that the non-linear equations we consider here in no way violate the principles of quantum mechanics. Each kind of simulation is non-linear because it is conditional on a specific unravelling of the master equation (1) and when we ensemble average over a large number of state vectors we always recover the master equation (1) with the appropriate reservoir operators [5–7,9].

In Fig. 3 we demonstrate the localization process which takes place in quantum state diffusion from an initial superposition. We plot the Wigner quasi-probability function (which reproduces the correct marginal distributions and also indicates coherence through the existence of interference fringes). Such fringes can be seen in the initial state, Fig. 3(a), which consists of a Schrödinger "cat" coherent superposition of two coherent states $|\alpha\rangle$ and $|-\alpha\rangle$ (with $\alpha = 2$ in this case), with the two components representing the live "cat" and the

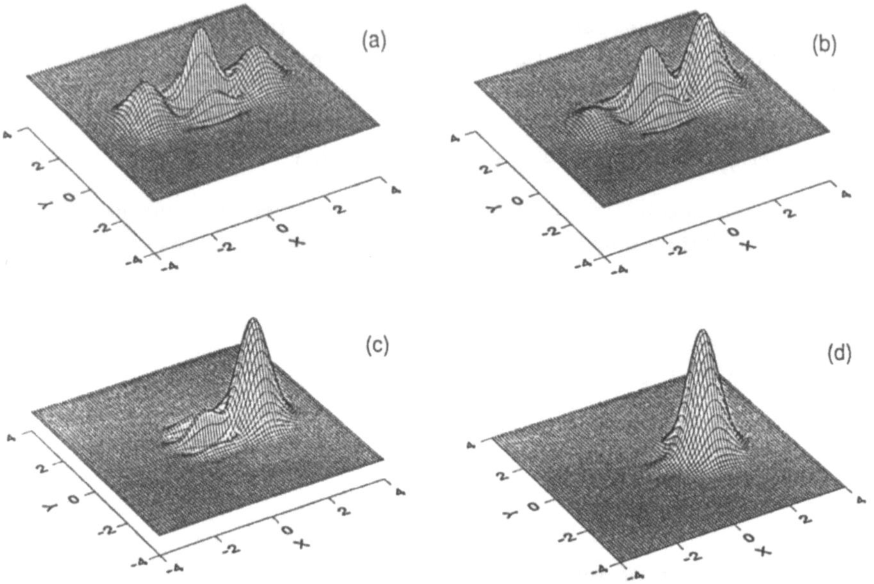

Figure 3. Wigner functions from the quantum state diffusion of a single sample We see $\kappa t = 0, 0.05, 0.2$ and 0.8 in (a-d). Localization is seen as the increasing dominance of one component of the "cat".

dead "cat". If we now examine the time sequence in Figs.3(a-d) we observe that one of the components of the "cat" diminishes as the other component grows. This is accompanied by a rapid loss of the fringes. The Wigner function clearly shows localization to one of the coherent state components of the original "cat", the choice of component being determined randomly according to the stochastic input. In Fig. 2 we show the values of the quadrature $\langle \hat{X} \rangle = \langle \psi(t)|(\hat{a} + \hat{a}^\dagger)|\psi(t)\rangle/2$ and uncertainty $\Delta X = (\langle \hat{X}^2 \rangle - \langle \hat{X} \rangle^2)^{1/2}$ as a function of time and including the times at which the Wigner functions of Fig. 3 are appropriate. The localization manifests itself in $\langle \hat{X} \rangle$ through the rapid development of a non-zero value of the displacement which takes place as one of the component coherent states is chosen. Subsequently, the value of $\langle \hat{X} \rangle$ decays as the coherent state itself approaches the ground state. The localization is seen in ΔX as a rapid reduction to a value of one half, the uncertainty value for a coherent state. The value of one half is maintained as the coherent state dissipates to the origin. The picture would be quite different if we were to consider the density matrix approach. In that case the fringes in the Wigner function would also rapidly disappear, but we would be left with both of the components reflecting a statistical mixture of outcomes [12].

We have so far presented the heterodyne experiment with a simple choice of two components (outcomes) of a Schrödinger "cat". Of course *any* state of the oscillator can be represented as a superposition of coherent states, and the

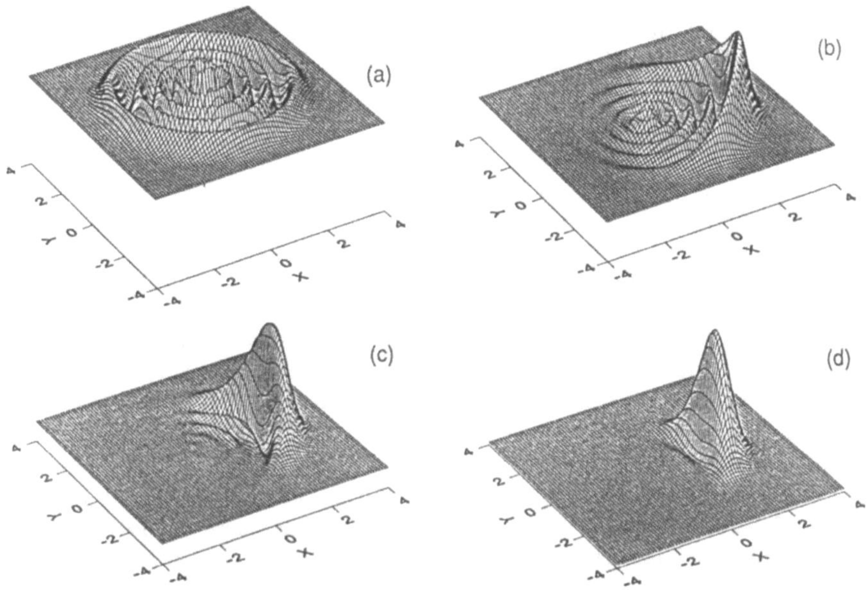

Figure 4. Localization of a Fock state $|n\rangle$ with $n = 9$. These results are for a single sample undergoing quantum state diffusion. The Wigner functions are shown for $\kappa t = 0, 0.1, 0.5$ and 1.0 in (a-d).

quantum state diffusion simulation will describe localization to a single coherent state component. We next consider a complex superposition of coherent states: a Fock state. It is well known that the Fock state $|n\rangle$ has an integral representation in terms of a phase-space ring of coherent states $|\alpha\rangle \equiv |r\,\mathrm{e}^{i\theta}\rangle$ [13]:

$$|n\rangle = \frac{\mathrm{e}^{r^2/2}\sqrt{n!}}{2\pi r^n} \int_0^{2\pi} \mathrm{d}\theta\; \mathrm{e}^{-in\theta}\, |r\,\mathrm{e}^{i\theta}\rangle\,. \tag{15}$$

Thus a ring of coherent states with any radius r can form a Fock state provided the correct phase factors $\mathrm{e}^{-in\theta}$ are present. However, when localization takes place a coherent state is selected from that circle for which each component coherent state has nearly the same energy as the original Fock state. The process is illustrated in Fig. 4. The initial state in (a) contains nine quanta and its Wigner function, basically a Laguerre polynomial, is made up of an outer rim with inner concentric fringes. As time progresses, (b,c), we observe that localization starts with the preferential orientation of the rim and fringes in one direction. This chosen direction is different for each realization. As the outer rim localizes, the inner fringes wash out with the vanishing of quantum interference between coherent state components. In (d) we see that the fringes have almost disappeared and the remaining localized state approaches a coherent state. During the subsequent time evolution the Wigner function becomes quite rounded and moves in towards the phase space origin. We have also investigated the localization from initially squeezed states, and again we see localization (and under some conditions, the formation of *Schrödinger "cats"*).

4 Conclusions

One can regard the evolution of single quantum trajectories as generating potential outcomes contingent on the acquisition of information in specific measurement schemes. The conditioning of the initial state vector reflects this time-dependent gain in information. The potential outcomes depend upon the measurement process: the pointer basis reflects this, and has to be chosen appropriately. If the level of excitation of the quantum system is monitored by quantum counting, then the appropriate basis reflects the successive loss of quanta observed in the decay process. But the quantum system may instead be studied by homodyne or heterodyne measurements where the local oscillator is very intense: then the effect of a single quantum jump is so small and the frequency of jumps is so high that the evolution appears to be a stochastic excursion.

Any initial state can be described in terms of its relevant phase space quasiprobability (for example, the Wigner function). The phase space can be built up in terms of a superposition of pointer state bases. Fock states tile the phase space by a set of annulae centred on the origin, reflecting (at least in the semi-classical approximation) the lack of phase information in energy eigenstates. However, it may be prefereble to employ a basis with amplitude *and* phase information, and this may be accomplished by tiling the phase space with a set of coherent states. The acquisition of information through observation conditions the wave function in a manner dependent on the specific measurement scheme, and the conditioning emphasises a particular tiling. Localization in the quantum state diffusion model reflects the coherent state tiling in a remarkable way. These issues will be addressed elsewhere.

Acknowledgements

This work was supported in part by the United Kingdom Science and Engineering Research Council, and by the European Community.

References

[1] See, e.g., W. Louisell, *Quantum Statistical Properties of Radiation*, John Wiley and Sons, Chichester, 1973.

[2] J.P. Paz, S. Habib and W.H. Zurek, Phys. Rev. **D 47**, 488 (1993); W.H. Zurek, Phys. Today **44**, 36 (1991); *ibid.* **46**, 84 (1993); W.H. Zurek, Prog. Theor. Phys. **89**, 281 (1993); W.H. Zurek, S. Habib and J.P. Paz, Phys. Rev. Lett. **70**, 1187 (1993) and references therein.

[3] E. Joos and H.D.Zeh, Z. Phys. **B 59**, 223 (1985); D.F. Walls and G.J. Milburn, Phys. Rev. **A 31**, 2403 (1985); S.J.D. Phoenix, Phys. Rev. **A 41**, 5132 (1990); G.J. Milburn, Phys. Rev. **A 36**, 5271 (1987).

[4] See, for example: N. Gisin, P.L. Knight, I.C. Percival, R.C. Thompson and D.C. Wilson, J. Mod. Opt. **40**, 1663 (1993) and references therein; M. Brune, S. Haroche, J.M. Raimond, L. Davidovich and N. Zagury, Phys. Rev. **A 45**, 5193 (1992).

[5] J. Dalibard, Y. Castin and K. Mølmer, Phys. Rev. Lett. **68**, 580 (1992); K. Mølmer, Y. Castin, and J. Dalibard, J. Opt. Soc. Am. **B 10**, 524 (1993).

[6] H.J. Carmichael, *An Open Systems Approach to Quantum Optics*, Lecture Notes in Physics (Springer-Verlag, Berlin, 1993).

[7] N. Gisin and I.C. Percival, J. Phys. **A 25**, 5677 (1992); *ibid.* **26**, 2233 (1993); *ibid.* **26**, 2245 (1993).

[8] H.M. Wiseman and G.J. Milburn, Phys. Rev. **A 47**, 1652 (1993).

[9] B.M. Garraway and P.L. Knight, to appear in Phys. Rev. **A 49** (1994).

[10] B.M. Garraway and P.L. Knight, submitted to Phys. Rev. Lett.

[11] N. Gisin and I.C. Percival, Phys. Lett. **A 167**, 315 (1992).

[12] V. Bužek, A. Vidiella-Barranco and P.L. Knight, Phys. Rev. **A 45**, 6570 (1992).

[13] P. Adam, J. Janszky and A.V. Vinogradov, Opt. Commun. **80**, 155 (1990); see also C.W. Gardiner, *Quantum Noise*, (Springer-Verlag, Berlin, 1991); V. Bužek and P.L. Knight, Opt. Commun. **81**, 331 (1991); V. Bužek and P.L. Knight, Progress in Optics, (ed. E. Wolf, North Holland, Amsterdam, 1994) to appear.

Ponderomotive Control of Quantum Noise

P. Tombesi

Dipartimento di Matematica e Fisica Università di Camerino,
I-62032 Camerino, Italy

Abstract. It is shown that a linear Fabry-Perot cavity with an oscillating end mirror can be used as a quantum nondemolition measurement device or as a quantum noise eater. For high quality factor of the mechanical oscillator and high mechanical frequency, the output quantum fluctuations of a monocromatic light beam can be significantly squeezed at a frequency very close to that of the impinging light. At lower mechanical frequency, by measuring the statistics of the mechanical momentum, the measurement of the photon number is obtained without the demolition of the quantum state of the light inside the cavity. The analysis is performed by taking into account the coupling of the system with the external world.

1. Introduction

The effects of the ponderomotive force (i.e. the elettromagnetic pressure force) were known well before the confirmation of Einstein's theory of the photoelectric effect. Lebedev [1], indeed, measured the purely classical effect of the pressure of the light concluding that any portion of energy carries a mechanical momentum. Thus, the method of measuring electromagnetic energy avoiding photon absorption is well known. Only in the last decade, however, the development of the measurements technology has reached such a level of accuracy that the realization of a measurement, which uses the ponderomotive force, can be no more considered as a thought experiment but a real measurement and a good challenge for an experimenter. Furthermore, in recent years the electromagnetic pressure force was considered by Meystre *et al.* [2] showing that its coupling to a freely suspended end-mirror of a Fabry-Perot cavity, introduces a nonlinearity which gives rise to the bistability. Experimental demostrations of this effect have been given in both the optical [3] and microwave [4] regimes. The ponderomotive force induces an intensity dependent phase shift, then its effect can be considered equivalent to an optical Kerr effect. Stenholm [5] proposed the use of a cavity with a mouvable mirror, considered as a harmonic oscillator, as a device to generate squeezed light. Its analysis, however, did not take into account the thermal fluctuations on the oscillating mirror which can completely destroy the obtainable squeezing; as was shown by Hilico *et al.* [6] with a semiclassical analysis. Recently, Heidmann and Reynaud [7], within a

Springer Proceedings in Physics, Vol. 77
Editors: D.F. Walls · J.D. Harvey
Quantum Optics VI

corpuscolar model, proposed a regulation mechanism of the photon flux by using the optomechanical coupling of a laser beam with a single movable mirror. They show that the reflected light may exhibit sub-Poissonian statistics.

In this paper the Hamiltonian model of the ponderomotive coupling to a mouvable mirror, recently introduced by Pace *et al.*[8] in the context of optical interferometers for gravitational wave detection, will be reviewed. It will be shown that two opposit limits exist, for the mechanical oscillator frequancy ν_m, where a linear Fabry-Perot cavity may be used either as a quantum nondemolition (QND) measurement device [9] or as a quantum noise eater [10]. At low frequecy the momentum fluctuations of the freely suspended mirror may be used as a meter which provides information on the photon number. At high ν_m, and for high quality factors of the mechanical oscillator, the device may be used as a quantum noise eater: because the output quantum fluctuations can be significantly squeezed at a frequency very close to that of an impinging monocromatic coherent beam. Since the mechanical motion is wery slow, with respect to the electromagnetic frequency, it can be considered adiabatic. It is well known that the number of photons in the cavity is left unaffected when the cavity is adiabatically deformed. To evaluate the quality of the scheme as a QND device, the criteria introduced by Holland *et al.* [11] will be used. The cavity loss and the backaction of the noise on the meter will be considered in the model. In the second case the analogies with an optical Kerr medium will be enphasized to show that the device at higher mechanical frequency, but still much lower than the inverse round trip time of the cavity, gives high quality squeezed light in the output port of the cavity. It will be shown that the squeezing is not spoiled by the thermal fluctuations on the oscillating mirror at a frequency very close to that of the input field even at room temperature. This result is in agreement with that of a recent paper by Fabre *et al.* [12] where the former pessimistic conclusion of Hilico *et al.* [6] was revised.

2. The Hamiltonian of the model

We consider a linear Fabry-Perot empty cavity with one fixed mirror with transmissivity T_r and one perfectly reflecting end miror. The completely reflecting mirror can move undergoing harmonic oscillations damped by the coupling to a thermal bath in equilibrium at temperature T. We assume that the light impinging the fixed mirror has a small detuning with respect to one cavity mode. The cavity resonances are calculated in absence of the impinging field. If L is the equilibrium cavity lenght the resonant angular frequency of the cavity will be $\omega_c = n\pi c/L$ where n is an arbitrary integer number and c the speed of light. Furthermore, we assume that at the frequency of the impinging field $\omega_0/2\pi$ the fixed mirror with transmissivity T_r does not introduce any excess noise but the input field one. We also assume that the retardation effects, due to the oscillating mirror, in the intracavity field are negligible. We will use a field intensity such that the correction to the radiation pressure force, due to the Doppler frequency shift of the photons [13], on the moving mirror is completely negligible. This account in considering the damping coefficient of the oscillating mirror only due to the coupling with the thermal bath. We are thus able to write the Hamiltonian [8] as

$$
\begin{aligned}
H = \hbar\omega_c a^+ a + \frac{p^2}{2m} + \frac{1}{2} m\omega_m^2 x^2 - \hbar g a^+ a x + \frac{1}{2}\int_0^{\infty} d\Omega\{[P(\Omega) - \kappa(\Omega)x]^2 \\
+ \Omega^2 Q^2(\Omega)\} + i\hbar \int_{-\infty}^{+\infty} d\omega\, s(\omega)\, [b^+(\omega)a - b(\omega)a^+] \,,
\end{aligned}
\tag{1}
$$

where a, a^+ are the boson operators of the resonant cavity mode at frequency $\omega_c/2\pi$; p and x respectively the momentum and the displacement, from the equilibrium position, of the oscillating mirror with mass m and oscillation frequency $\nu_m = \omega_m/2\pi$; $b(\omega)$, $b^+(\omega)$ are the bose operators for the infinite modes of the external radiation field at frequency $\omega/2\pi$, $s(\omega)$ the coupling costant and we have neglected the vacuum terms. The interaction between the cavity mode and the oscillating mirror is governed by the coupling constant $g = \omega_c/L$ [8]. We have also taken into account the damping of the oscillating mirror due to the coupling with the thermal bath modeled with a big number of harmonic oscillators [14,15] and the cavity loss in the rotating wave approximation [16].

3. The dynamics of the system

With the standard procedure, well described in the Gardiner's book [15], by solving their dynamical equations we can eliminate the thermal bath variables and the variables of the radiation bath. In a frame rotating at the frequency $\omega_0/2\pi$ of the impinging field we can write the equations of motion for the cavity mode and the mechanical oscillator

$$
\begin{aligned}
\frac{da}{dt} &= -i\,(\omega_c - \omega_0 - gx)a - \frac{\gamma_c}{2}\, a + \sqrt{\gamma_c}\, a_{in}(t) \,, \\
\frac{da^+}{dt} &= i\,(\omega_c - \omega_0 - gx)a^+ - \frac{\gamma_c}{2} a^+ + \sqrt{\gamma_c}\, a^+_{in}(t) \,, \\
\frac{dx}{dt} &= \frac{p}{m} \\
\frac{dp}{dt} &= -m\omega_m^2 x + \hbar g a^+ a - \frac{\gamma_m}{2m} p - \sqrt{\gamma_m}\epsilon_{in}(t) .
\end{aligned}
\tag{2}
$$

These equations are obtained in the Markovian approximation [15] and have to be considered as coupled stochastic differential equations in the Stratonovich sense and are valid in the low damping regime where $\gamma_m/m << \omega_m$. The constants $\gamma_c = 2\pi s^2(\omega)$ and $\gamma_m/m = \pi\kappa^2(\Omega)/m$ respectively represent the cavity linewidth and the damping constant of the mechanical oscillator, due to the interaction with the external baths. They are constant in the so-called first Markovian limit [15]. Of course, γ_c accounts for the loss of photons in the cavity mode due to the interaction with the radiation field i.e. $\gamma_c = cT_r/2L$ and $a_{in}^+(t)\, a_{in}(t)$ is the flux of photons hitting the first cavity mirror in one second. In the rotating frame for an impinging monochromatic coherent field we have

$$a_{in}(t) = \frac{-1}{\sqrt{2\pi}} \int_0^\infty d\omega\, e^{-i\omega t}\, b(\omega) = \alpha_{in} + \delta a_{in}(t) \;, \tag{3}$$

where $\delta a_{in}(t)$ represents the quantum fluctuation with respect to the mean amplitude $< a_{in} >= \alpha_{in}$ of the impinging mode. $\delta a_{in}(t)$ can be considered as a "white quantum noise" [15] at least in a narrow bandwidth at ω_0 larger than γ_c, with expectations

$$\begin{aligned} &< \delta a_{in}(t) >= 0 \quad , \qquad < \delta a_{in}^+(t)\delta a_{in}(t') >= \bar{n}\delta\,(t-t') \;, \\ &< \delta a_{in}(t)\delta a_{in}^+(t') >= (1+\bar{n})\,\delta\,(t-t') \;, \end{aligned} \tag{4}$$

where $\bar{n}$ represents the number of photons at the frequency of the impinging field which is negligibly small at optical or infrared frequencies. All the contribution of the thermal bath variables is included in $\sqrt{\gamma_m}\,\epsilon_{in}(t)$ which can be considered as a stochastic noise force depending only on the state of the thermal bath at time t_o, where t_o is the initial time of the interaction. It can be shown [15] that this noise term at high temperature behaves as a classical noise and has the following expectation values with respect to the equilibrium distribution of the thermal bath at time t_0;

$$< \epsilon_{in}\,(t) >= 0\,; < \epsilon_{in}\,(t)\,\epsilon_{in}(t') >= K_B T \delta\,(t-t'). \tag{5}$$

The high temperature condition in the present case reads $\hbar\omega_m \ll K_B T$, where K_B is the Boltzman's constant, which is always satisfied for non extremely low temperature. It means that we can disregard the quantum nature of the mirror's motion and we can consider it as a classical oscillator. The equations (2) are valid after an initial transient and must then be considered for times $t > t_o$ [15].

4. The steady state

We are actually interested in the steady state regime and in small fluctuations with respect to the steady state. Then, as is usual in the semiclassical approximation [17], we take the expectation with respect to the steady state of Eq.(2). Neglecting all the fluctuations we put $\frac{d<a>_s}{dt} = \frac{d<a^+>_s}{dt} = \frac{d<x>_s}{dt} = \frac{d<p>_s}{dt} = 0$. Thus we get

$$
\begin{aligned}
&i\left(\omega_c-\omega_0-g x_s\right) \alpha_s+\frac{\gamma_c}{2} \alpha_s-\sqrt{\gamma_c} \alpha_{in}=0 , \\
&i\left(\omega_c-\omega_0-g x_s\right) \alpha_s^*-\frac{\gamma_c}{2} \alpha_s^*+\sqrt{\gamma_c} \alpha_{in}^*=0 , \\
&p_s=0 , \\
&m \omega_m^2 x_s-\hbar g a_s^* \alpha_s=0 ,
\end{aligned} \tag{6}
$$

where with the subscript s we intend the steady state value i.e. ($\alpha_s =< a >_s$, $\alpha_s^* =< a^+ >_s$, $x_s =< x >_s$, $p_s =< p >_s$). The steady state solutions are

$$
p_s=0 ; \quad x_s=\frac{\hbar g}{m \omega_m^2}\left|\alpha_s\right|^2 , \qquad \alpha_s=\frac{\sqrt{\gamma_c} \alpha_{in}}{\frac{\gamma_c}{2}+i\left(\omega_c-\omega_0-g x_s\right)} \quad . \tag{7}
$$

5. QND device

It is easily seen that, due to the presence of the oscillating mirror, the mode inside the cavity is detuned. The detuning is responsible of the bistable behaviour [10] and the system behaves as an effective Kerr medium. For the poupose of obtaining a good QND device the Kerr medium analogy is not essential. We can get rid of this detuning, however, by introducing *ab initio* a cavity detuning δ which can be adjusted either by varying the length of the cavity or by varying the frequency of the external field. Thus the frequency ω_0 in Eq.(1) is modified to $\omega_0+\delta$ and by choosing the detuning δ so that $\delta=\omega_c(1-y)-\omega_0$ (with $y=x_s/L$) the steady state equations (6) simplify and we get the solutions

$$
x_s=\frac{8 \hbar \omega_0}{m c \omega_m^2 T_r}\left|\alpha_{in}\right|^2 , \; p_s=0 ; \; \alpha_s^* \alpha_s=\frac{8 L}{c T_r}\left|\alpha_{in}\right|^2 , \tag{8}
$$

where we used the definitions given above for g and γ_c.

We now consider the evolution equations of small fluctuations with respect to the steady state values given in Eq.(8). To this end we define $\delta a=a-\alpha_s$; $\delta x=x-x_s$; $\delta p=p-p_s$ and the fluctuations with respect the mean value of the input field $\delta a_{in}=a_{in}-\alpha_{in}$. To the lowest order in the fluctuations we get

$$
\frac{d}{dt} \vec{\xi}=\mathbf{A} \vec{\xi}+\vec{\eta} \quad . \tag{9}
$$

The vectors $\vec{\xi}$ and $\vec{\eta}$ are defined by

$$
\vec{\xi}^T=\left(\delta a, \delta a^+, \delta x, \delta p\right) \quad ; \vec{\eta}^T=\left(\sqrt{\gamma_c}\, \delta a_{in}, \sqrt{\gamma_c}\, \delta a_{in}^+, 0, -\sqrt{\gamma_m}\, \epsilon_{in}\right) , \tag{10}
$$

where for simplicity we showed the transpose of the column vectors $\vec{\xi}$ and $\vec{\eta}$. The matrix $\mathbf{A}$ is given by

$$\mathbf{A} = \begin{pmatrix} -\gamma_c/2 & 0 & ig\alpha_s & 0 \\ 0 & -\gamma_c/2 & -ig\alpha_s^* & 0 \\ 0 & 0 & 0 & 1/m \\ \hbar g\alpha_s^* & \hbar g\alpha_s & -m\omega_m^2 & -\gamma_m/2m \end{pmatrix}. \tag{11}$$

By using the Fourier Transform $f(t) = (2\pi)^{-1/2} \int d\omega \tilde{f}(\omega) exp(i\omega t)$ we get

$$\tilde{\vec{\xi}}(\omega) = \mathbf{B}^{-1}\tilde{\vec{\eta}}(\omega) \tag{12}$$

with $\mathbf{B} = i\omega\mathbf{I} - \mathbf{A}$ where $\mathbf{I}$ is the identity matrix. We are interested in studying the two-time correlation functions of fluctuations with respect to the steady state which are given, in terms of Fourier transforms, by

$$\langle\tilde{\vec{\xi}}(\omega)\tilde{\vec{\xi}}^T(\omega')\rangle = \mathbf{B}^{-1}(\omega)\langle\tilde{\vec{\eta}}(\omega)\tilde{\vec{\eta}}^T(-\omega)\rangle(\mathbf{B}^{-1}(-\omega))^T\delta(\omega+\omega'). \tag{13}$$

The correlations of the noise terms are given by the Fourier transforms of Eqs. (4) and (5) [18]. and we already used the δ-function. We are now able to calculate the various correlation functions of interest.

The measurement will be performed by considering the statistics of the momentum of the oscillating mirror. It represents the probe which gives all the information we need on the signal which is the electromagnetic field inside the cavity. The correlation function of fluctuations of the momentum of the oscillating mirror, after lengthy but straigthforward algebra, is given by

$$\langle\delta\tilde{p}(\omega)\delta\tilde{p}(\omega')\rangle = \delta(\omega+\omega')\frac{\frac{8|\alpha_{in}|^2 g^2\hbar^2(\bar{n}+1/2)}{\omega^2+\gamma_c^2/4} + \gamma_m K_B T}{[(\omega_m^2-\omega^2)^2 + \omega^2\gamma_m^2/4m^2]}\,\omega^2. \tag{14}$$

In order to assert that this system is a good QND measurement scheeme, the criteria introduced by Holland *et al.* [11] should be fulfilled. Indeed, the quality of the measurement is determined by the level of correlation between the probe field and the signal field, specified by the coefficient $C^2_{X^{in}Y^{out}}$. The scheme should not degrade the signal field; this fact is specified by the coefficient $C^2_{X^{in}X^{out}}$. Finally, the scheme should be a good state-preparation device; this can be evaluated by the coefficient $C^2_{X^{out}Y^{out}}$. Where

$$C^2_{AB} = \frac{|\langle AB\rangle - \langle A\rangle\langle B\rangle|^2}{V_A V_B} \tag{15}$$

with V_C the variance of C. For a good QND measurement device any coefficient C^2_{AB} have to be equal to one. After lenghty but straightforward algebra, by using the definition for X^{in}, i.e. the input quadrature of the signal incident on the apparatus, $X^{in} = (a_{in} + a^+_{in})$, and taking as a probe the momentum of the oscillating mirror, i.e. $Y^{out} = \epsilon_{in} + \frac{\sqrt{\gamma_m}}{m}p$, we get

$$C^2_{X^{in}Y^{out}}(\omega) = \left[1 + \frac{K_B T m^2(\omega^2+\frac{\gamma_c^2}{4})[(\omega_m^2-\omega^2)^2 + \frac{1}{4}\frac{\gamma_m^2\omega^2}{m^2}]}{4\gamma_m\omega^2|\alpha_{in}|^2 g^2\hbar^2(1+2\bar{n})}\right]^{-1}. \tag{16}$$

Further, being $X^{out} = \sqrt{\gamma_c}(a + a^+) - X^{in}$ the output signal, we get

$$C^2_{X^{in}X^{out}}(\omega) = 1. \tag{17}$$

Finally

$$C^2_{X^{out}Y^{out}}(\omega) = C^2_{X^{in}Y^{out}}(\omega). \tag{18}$$

The maximum of Eqs.(16) is obtained for $\omega = \pm\omega_m$, then we have

$$(C^2_{X^{in}Y^{out}})_{max} = [1 + \frac{1}{16} K_B T \gamma_m \frac{(\omega_m^2 + \frac{\gamma_c^2}{4})}{|\alpha_{in}|^2 g^2 \hbar^2 (1 + 2\bar{n})}]^{-1}. \tag{19}$$

We see that it is possible to get $(C^2_{X^{in}Y^{out}})_{max} = (C^2_{X^{out}Y^{out}})_{max} \simeq 1$ for $\omega_m \ll \gamma_c$ depending on the values of the various external parameters. We are thus lead to conclude that for low ω_m the scheme is a good QND device at least in a well defined temperatures range. The maximum correlation between the momentum fluctuations of the mirror and the amplitude quadrature fluctuations occurs at $\omega = \pm\omega_m$. Thus, one should observe the amplitude of the variance of the output probe fluctuations at this maximum. For $\omega = \pm\omega_m$ we get

$$V_{Y^{out}}(\omega_m; |\alpha_{in}|^2) = K_B T(1 + \frac{64}{K_B T}(\frac{g\hbar}{\gamma_c})^2 \frac{|\alpha_{in}|^2}{\gamma_m}). \tag{20}$$

Thus, the number of photons inpinguing the cavity at frequency ω_0 is $|\alpha_{in}|^2 = \frac{1}{\Lambda}(V_{Y^{out}}(\omega_m; |\alpha_{in}|^2) - V_{Y^{out}}(\omega_m; 0))$ with $\Lambda = (8g\hbar/\gamma_c)^2/\gamma_m$ fixed by the experimental set up.

6. Quantum noise eater

The search for high squeezing generated by materials with $\chi^{(3)}$ nonlinearities is motivated by the possibility of building-up a quantum-noise-eater. In these materials, indeed, the squeezing is produced without frequency conversion, in contrast to the $\chi^{(2)}$ case where it is usually generated in proximity of the first sub-harmonic frequency of the incident beam. Media with $\chi^{(3)}$ nonlinearities are the so-called Kerr media in nonlinear optics [19]. We have already observed that our system might mimic a Kerr medium when we stay in the bistable situation. Thus, if we do not introduce the detunung δ we are able to write the evolution equations for the fluctuations up to first order with respect to the stable steady state values in Eq.(7) as

$$\frac{d}{dt}\delta a = -i\left(\omega_c - \omega_0 - \frac{\hbar g^2}{m\omega_m^2}|\alpha_s|^2\right)\delta a - \frac{\gamma_c}{2}\delta a + ig\alpha_s\delta x + \sqrt{\gamma_c}\,\delta a_{in}\,, \tag{21a}$$

$$\frac{d^2}{d^2t}\delta x = -\omega_m^2\delta x + \frac{\hbar g}{m}\alpha_s^*\delta a + \frac{\hbar g}{m}\alpha_s\delta a^+ - \frac{\gamma_m}{2m}\frac{d}{dt}\delta x - \frac{\sqrt{\gamma_m}}{m}\epsilon_{in}\,, \tag{21b}$$

and, obviously, the Hermitean conjugate of Eq.(21a). From Eqs.(21) we see the effect of the coupling between the cavity mode and the oscillating mirror. As the intracavity intensity field is $|\alpha_s|^2$, there will be an intensity dependent phase shift. It is obvious that δx depends linearly on δa^+ thus introducing a coupling between the fluctuation δa and its conjugate δa^+. Thus, as a consequence of the dependence of δa on δx, the fluctuations of the internal field could be squeezed. However, δx depends on δa as well so a further dynamical phase shift and damping is introduced by the coupling of the cavity mode with that mirror. In the frequency domain, by eliminating $\delta\tilde{x}$ through the Fourier transform of Eq.(21b) and introducing the dimensionless dynamical response factor of the mirror:

$$\chi(\omega) = \frac{\omega_m^2}{(\omega_m^2 - \omega^2) - i\,\dfrac{\gamma_m \omega}{2m}} = \chi_1(\omega) + i\,\chi_2(\omega), \tag{22}$$

we can write

$$\frac{\gamma_c}{2} + i(\phi - \omega - \chi(\omega)|K|)\delta\tilde{a} - i\chi(\omega)K\delta\tilde{a}^+ = \sqrt{\gamma_c}\delta\tilde{a}_{in} - \rho\chi(\omega)\frac{\sqrt{\gamma_m}}{m\omega_m^2}\tilde{\epsilon}_{in}, \tag{23}$$

and its Hermitean conjugate, where $K = \frac{\hbar g^2}{m\omega_m^2}\alpha_s^2$, $\rho = ig\alpha_s$ and $\phi = \omega_c - \omega_0 - \frac{\hbar g^2}{m\omega_m^2}|\alpha_s|^2 = \omega_c(1-y) - \omega_0$. We see that besides an additional frequency independent detuning the interaction with the moving mirror introduces a frequency dependent dissipative term in Eq.(23); *i.e.* $\chi_2(\omega)|K|$ which is maximum at $\omega = \omega_m$, while the frequency dependent phase shift $\chi_1(\omega)|K|$ is zero at the same frequency. We want to avoid such an additional dissipation which might destroy the squeezing at any frequency and in what follows we will require $\chi_2(\omega)|K| \ll \gamma_c/2$. Furthermore, we see from Eq.(23) that a thermal noise contribution is added to the usual radiation noise of the input field. The input-output theory [18] gives the following relation among the incoming field, the internal field and the output field as a consequence of boundary conditions at the fixed mirror surface, $\delta\tilde{a}_{out} + \delta\tilde{a}_{in} = \sqrt{\gamma_c}\delta\tilde{a}$. Finally, by defining the Fourier transform of the output quadrature as $X_\theta^{out}(\omega) = \frac{1}{2}\left(e^{-i\theta}a_{out}(\omega) + e^{i\theta}a_{out}^+(\omega)\right)$, where θ is controlled by the experimenter in a homodyne detection scheme, we get for the optimum quadrature squeezing $S_{opt}(\omega)$

$$\begin{aligned}
S_{opt}(\omega) = 1 + &\frac{2\gamma_c K}{(1+2\bar{n})|\Delta(\omega)|^2}\Big\{(1+2\bar{n})\Big(\gamma_c K|\chi(\omega)|^2 - 2\omega\phi\chi_2(\omega)\Big)\\
&+ (\gamma_c^2/4 + \omega^2 + \phi^2)\Big(\frac{|\chi(\omega)|^2 K_B T}{Q_m \hbar\omega_m} - \chi_2(\omega)\Big)\\
&- \{[(1+2\bar{n})\gamma_c\Big(K|\chi(\omega)|^2 - \phi\chi_1(\omega)\Big)\\
&+ (\gamma_c^2/4 + \omega^2 - \phi^2)\Big(\frac{|\chi(\omega)|^2 K_B T}{Q_m \hbar\omega_m} - \chi_2(\omega))]^2\\
&+ [(1+2\bar{n})\Big(2\phi K|\chi(\omega)|^2 + \chi_1(\omega)(\gamma_c^2/4 + \omega^2 - \phi^2)\Big)\\
&+ \gamma_c\phi\Big(\frac{|\chi(\omega)|^2 K_B T}{Q_m \hbar\omega_m} - \chi_2(\omega))]^2\}^{\frac{1}{2}}\Big\},
\end{aligned} \tag{24}$$

where $\Delta(\omega)$ is the determinant of the Fourier transform of Eqs.(21) and the mechanical quality factor $Q_m = m\omega_m/\gamma_m$ was introduced. The optimum quadrature squeezing is obtained by choosing $\theta(\omega)$ in such a way that $dS_\theta(\omega)/d\theta = 0$, where the normalized quadrature spectrum is $S_\theta(\omega) = \int d\omega' < \delta X_\theta^{out}(\omega)\delta X_\theta^{out}(\omega') > /[\frac{1}{4}(1+2\bar{n})]$. In order to have squeezing the second term in the right hand size must be negative and greater or equal to -1.

It is worth to consider also the normalized output intensity spectrum $S_I(\omega)$ which is defined by $S_I(\omega) = \int d\omega' < \delta I_{out}(\omega)\delta I_{out}(\omega') > /|\alpha_{out}|^2$ where $\delta I_{out}(\omega) = \alpha^*_{out}\delta a_{out}(\omega) + \alpha_{out}\delta a^+_{out}(\omega)$ and $\alpha_{out} = \sqrt{\gamma_c}\alpha_s - \alpha_{in}$ for the boundary conditions at the fixed mirror surface. After some algebra, by considering only the local minimum for $\omega \ll \omega_m$ and setting $\bar{n} = 0$ for simplicity we get

$$S_I(\omega) \simeq 1 - \frac{4K\phi\gamma_c^2\omega^2}{(\gamma_c^2/4 + \phi^2)|\bar{\Delta}(\omega)|^2}\left(\frac{1}{2} - \frac{\phi n_{th}}{\gamma_c Q_m}\right), \tag{25}$$

where $n_{th} = K_B T/\hbar\omega_m$ represents the number of thermal excitations due to the interaction with the thermal bath and $\bar{\Delta}(\omega)$ is the value of the determinant where we considered $\chi_2(\omega \ll \omega_m) = 0$ and $\chi_1(\omega \ll \omega_m) = 1$. This function has a minimum at $\omega = \pm\left(\frac{\gamma_c^2}{4} + \phi^2 - 2\phi K\right)^{1/2}$. The minimum of Eq.(25) is then

$$S_I(\omega) = 1 - 2K\phi(\frac{\gamma_c^2}{4} + \phi^2)^{-1}(1 - 2\frac{\phi n_{th}}{\gamma_c Q_m}) \,. \tag{26}$$

We see that for $\phi \ll \gamma_c Q_m/n_{th}$ the intensity squeezing becomes independent of the termal noise and can be high for the right choice of the external parameters.

This work has been partially supported by the European Economic Community under the Uman Capital and Mobility programme.

REFERENCES

1. P.N. Lebedev, *Journal of Russian Physical-Chemical Society*, **33**, 53 (1901).
2. P. Meystre, E.M. Wright, J.D. McCallen and E. Vignes, *JOSA B*, **2**, 1830 (1985).
3. A. Dorsel, J.D. McCullen, P. Meystre, E. Vignes and H. Walther, *Phys. Rev. Lett.*, **51**, 1550 (1983).
4. A. Gozzini, F. Maccarone, F. Mango, I. Longo and S. Barbarino, *JOSA B*, **2**, 1841 (1985).
5. S. Stenholm, in *Second International Workshop on Squeezed States and Uncertainty Relations* D. Han, Y.S. Kim and V.I. Man'ko Eds. (NASA Conference Publications, Maryland 1993).
6. L. Hilico, J.M. Courty, C. Fabre, E. Giacobino, I. Abrahm and J.L. Oudar, *Appl. Phys.B.* ,**55**, 202 (1992).

7. A. Heidmann and S. Reynaud, *Photon Noise Reduction by Reflection from a Mouvable Mirror* , (preprint).
8. A.F. Pace, M.J. Collett and D.F. Walls, *Phys. Rev. A.* **47**,3173(1993).
9. K. Jacobs, P. Tombesi, M.J. Collett and D.F. Walls, *to appear in Phys. Rev. A,*
10. S. Mancini and P. Tombesi, *Quantum Noise Reduction by Radiation Pressure* , (preprint).
11. M.J.Holland, M.J.Collett, D.F.Walls, M.D.Levenson, *Phys.Rev.* **A42**, 2995 (1990).
12. C. Fabre, M. Pinard, S. Bourzeix, A.Heidmann, E. Giacobino and S. Reynaud, *Quantum Noise Reduction Using a Cavity with a Mouvable Mirror,* preprint
13. W. Unruh, in *Quantum Optics, Experimental Gravitation and Measurements Theory,* P. Meystre and M.O. Scully Eds.(Plenum, New York 1983).
14. A. O. Caldeira and A.J. Leggett, *Phys. Rev. A* **31**, 1059(1985).
15. C. W. Gardiner, *Quantum Noise,* (Springer, Heidelberg, 1991).
16. W. Louiselle, *Quantum Statistical Properties of Radiation,* (Wiley, New York, 1974).
17. L. Lugiato, in *Progress in Optics,* Vol. XII, E. Wolf Ed.(North Holland, 1984).
18. C.W. Gardiner and M.J. Collett, *Phys. Rev. A* **31**, 3761(1985).
19. X.R.Shen *The Principles of Nonlinear Optics* , (Wiley, New York, 1984)

Quantum Soliton Collisions: Logic and QND

P.D. Drummond, J. Breslin, W. Man, and R.M. Shelby

Physics Department, University of Queensland,
St. Lucia, Australia Q4072 and
IBM Almaden Research Laboratory, San Jose, CA 95120, USA

The Wigner function technique for the quantum theory of nonlinear Schroedinger equation soliton collisions is developed. This is applied to experiments on reversible pulse-position switching or logic gates, and to quantum non-demolition (QND) measurements of photon number. A strong limit to cascaded logic gate performance is found, due to rapidly increasing error rates as the number of sequential logic operations is increased. We also obtain bounds on the achievable QND signal-to-noise ratio for quantum soliton collisional measurements of photon number. In this case, a modified strategy of combined phase/amplitude quadrature measurement can give essentially unlimited QND performance, provided thermal noise effects are small.

Experiments on the collisions of quantum solitons combine the interest of a fundamental test of quantum field theory, with technologically important applications. It is not often that new quantum scattering experiments can be accomplished with table-top apparatus, and yet uncover an aspect of physics that has not previously been tested.

In order to treat the theory of these novel experiments, it is necessary to develop techniques that are capable of treating highly nonlinear operator equations for the propagating quantum field, with an initially coherent superposition of large particle number. In this paper, an approximate technique using a combination of the Wigner representation with a singular perturbation theory for time-evolution is described. This technique is valid in the limit of large mean soliton photon number. In the experiments analysed here, the mean soliton quantum number is of order 10^8, so the approximations should be relatively accurate. Numerical techniques can be used to improve this, if desired. The chief advantage of the theory is that it results in analytic expressions for correlation functions that can be used to gain an intuitive understanding of the physics.

We obtain predictions from the resulting quantum theory of soliton collisions for the position fluctuations of one pulse after interaction, and for the phase-amplitude correlations between different pulses. These are relevant to quantum pulse-position logic and to QND measurements of soliton photon number respectively. Both these types of experiment have significance in wider fields of physics and even possible technological applications. The idea of reversible quantum-limited logic is fundamental to recent investigations into the physical limits of computation [1]. It is important to have a physical model for this type of logic that is realisable experimentally, and soliton pulse-position logic appears to be the first reversible logic demonstrated in an experiment [2]. Similarly, QND measurements of soliton photon number [3] are a significant new experimental tool for low-noise, non-invasive measurement of total photon number.

In the case of soliton pulse-position logic, use is made of the position shift that occurs during a soliton collision. By giving different logical truth values to the unshifted and shifted positions, it is possible to logically detect the presence or absence of another

Springer Proceedings in Physics, Vol. 77 Quantum Optics VI
Editors: D.F. Walls · J.D. Harvey

pulse. As the interaction is energy-conserving, this is an example of a reversible logic gate. While this type of gate has been realised in experiment, any useful application of this type of switch would necessitate multiple, cascaded use of a long sequence of collisions. Realistically, there are many effects that could cause problems, including absorption due to the long interaction lengths required and Raman scattering mediated by molecular vibrations.

Our task here will be to ignore these material problems, and rather to ask what are the fundamental issues involved when quantum solitons interact many times. Surprisingly, the quantum uncertainties are large enough to completely randomise logic operations, after only a few hundred logic steps. This result is manifest even when using typical macroscopic laboratory soliton pulses with 10^8 or more photons. Even more surprisingly, the result is not especially due to the collisions, but just due to the initial Heisenberg uncertainties in the soliton pulse position and momentum, together with the long (scaled) times required for multiple interactions.

Measuring the photon number using collisions of coherent solitons in optical fibers [3], represents a new technique of performing a quantum non-demolition (QND) measurement [4]. It has an advantage over many previous approaches to QND, in that the measurement is performed on a propagating signal. This has a number of possible technological applications. Of most interest to the physicist is the fact that these recent experiments provide a fundamental test of the quantum field theory of interacting quantum solitons. In principle these tests have a wide applicability outside of fiber optics, due to the universal nature of soliton formation in nonlinear physical systems. Optical QND experiments are the first soliton collision measurements at the quantum limit.

For this reason, it is important to develop a quantitative theory that is amenable to direct experimental test. Previous papers that have treated QND soliton collisions [5] have not included a quantitative treatment of the initial coherent nature of the soliton. In practical experiments, the inputs are in a coherent state, with a Poissonian distribution of photon number. This is a vital component of a QND collision experiment, as it determines the quantum fluctuations in the probe. We will show that, even at zero temperature, the input coherent state fluctuations give bounds to the QND detection efficiency. If a strategy is followed in which the probe phase alone is used to infer the signal intensity, then this severely limits the QND correlations.

We also show a that new strategy, in which a linear combination of phase and amplitude is read out, can greatly improve the QND performance. The reason for this is straight forward. Quantum fluctuations in the probe amplitude cause additional probe phase noise due to self-phase-modulation. This additional noise is directly correlated to the probe amplitude, which is invariant in a collision. By measuring an appropriate combination of phase and amplitude, the unwanted noise can be cancelled.

To analyse collision experiments of the types described above, we suppose that in the nonlinear fibre medium, the photon (or, more technically polariton) field is subject to the quantum nonlinear Schrödinger equation. In a moving reference frame, the resulting Heisenberg equations of motion are[6–8]:

$$\frac{\partial \hat{\phi}}{\partial \tau} = i \left\{ \frac{1}{2} \frac{\partial^2 \hat{\phi}}{\partial \xi^2} + \hat{\phi}^\dagger \hat{\phi}^2 \right\}, \tag{1}$$

where $\tau = vt/x_o$, $\xi = (x/v-t)/t_o$ are scaled time and distance coordinates. The velocity

v is the electromagnetic group-velocity, $x_o = t_o^2/k"$ is a characteristic interaction length, $k"$ is the group velocity dispersion, and t_o is a typical pulse duration introduced for scaling purposes. The field $\hat{\phi}$ is scaled to give a dimensionless polariton density of $\bar{n}\left\langle\hat{\phi}^\dagger\hat{\phi}\right\rangle$. Here $\bar{n} = |k"|\, Ac/\left(\hbar n_2\omega_o^2 t_o\right)$ is a characteristic photon number for a fiber with core area A and nonlinear refractive index n_2 at the carrier frequency ω_o.

Typical experiments have $\bar{n} \sim 10^8$, which means that a functional Wigner [9] representation becomes a useful technique. The initial coherent state is then a Gaussian Wigner distribution. After dropping terms of $O\left(1/\bar{n}^3\right)$, the equation of motion for the characteristics of the Wigner representation corresponds to the classical equation with initial vacuum fluctuations $\Delta\phi$ correlated according to:

$$\begin{aligned} \left\langle\Delta\phi\left(0,\xi\right)\Delta\phi\left(0,\xi'\right)\right\rangle &= 0, \\ \left\langle\Delta\phi\left(0,\xi\right)\Delta\phi^*\left(0,\xi'\right)\right\rangle &= \delta\left(\xi-\xi'\right)/2\bar{n}. \end{aligned} \tag{2}$$

Here we have arbitrarily chosen $\tau = 0$ as the time when the pulses are injected into the fiber. Corrections due to non-simultaneous pulse injection are negligible if $x_o >> vt_o$.

We now assume that probe and signal were initially well separated in space. Each can be treated independently in this approximation, using classical soliton perturbation theory techniques [10], but with initial conditions that include vacuum fluctuations. Following techniques we have given elsewhere[11], the resulting solitons have a continuum (radiative) component, together with perturbed "emergent" soliton parameters. These include the local vacuum fluctuations in the vicinity of the injected probe and signal. Suppose the initial pulse amplitude is:

$$\begin{aligned} & \phi\left(0,\xi\right) = \bar{\phi}\left(\xi\right) + \Delta\phi\left(0,\xi\right), \\ where: \quad & \bar{\phi}\left(\xi\right) = \bar{A}sech\left(\bar{\theta}\left(\xi\right)\right)\exp\left(i\bar{\alpha}\left(\xi\right)\right), \\ & \bar{\theta}\left(\xi\right) = A\left(\xi-\bar{\xi}_o\right), \\ & \bar{\alpha}\left(\xi\right) = V\left(\xi-\bar{\xi}_o\right) + \bar{\alpha}_o. \end{aligned} \tag{3}$$

The coherent input therefore has an average amplitude $\bar{\phi}$, which would classically propagate as a sech-type [12, 13] solution to the nonlinear Schrödinger equation. The full Wigner-function solutions, including quantum noise, can now be expanded in eigenfunctions of the linearised classical Schrödinger equation, using Kaup's [10] perturbation theory approach. The new solution is an emergent soliton with modified parameters that depend on the quantum noise $\Delta\phi\left(0,\xi\right)$, together with a continuum background. This perturbation theory is an adaption of Poincare and Lindstedt's singular perturbation theory techniques, originally used in planetary dynamics. The result is that:

$$\begin{aligned} & \phi\left(\tau,\xi\right) = \tilde{\phi}\left(\tau,\xi\right) + \phi_c\left(\tau,\xi\right), \\ where: \quad & \tilde{\phi}\left(\tau,\xi\right) = \tilde{A}sech\left[\tilde{\theta}\left(\tau,\xi\right)\right]\exp\left[i\tilde{\alpha}\left(\tau,\xi\right)\right], \\ & \tilde{\theta}\left(\tau,\xi\right) = \tilde{A}\left(\xi-\tilde{\xi}_o-\tilde{V}\tau\right), \\ & \tilde{\alpha}\left(\tau,\xi\right) = \tilde{V}\left(\xi-\tilde{\xi}_o\right) + \tfrac{1}{2}\left(\tilde{A}^2-\tilde{V}^2\right)\tau + \tilde{\alpha}_o. \end{aligned} \tag{4}$$

The emergent parameters $\tilde{A}$, $\tilde{\xi}_o$, $\tilde{V}$, $\tilde{\alpha}_o$ describe modifications to the pulse amplitude, position, velocity and phase respectively which arise from the quantum fluctuations. These depend on $\Delta\phi\,(0,\xi)$, and are given to lowest order in $1/\bar{n}$, by:

$$\begin{aligned}
\Delta A &= \tilde{A} - A = Re \int d\theta \left(\Delta\phi e^{-i\alpha} sech\,(\theta)\right), \\
\Delta V &= \tilde{V} - V = Re \int d\theta \left(-i\Delta\phi e^{-i\alpha} sech\,(\theta) \tanh\,(\theta)\right), \\
\Delta\alpha_o &= \tilde{\alpha}_o - \alpha_o = Re \int d\theta \left(\Delta\phi e^{-i\alpha} \left(\tfrac{i}{A}\left(\theta \tanh\,(\theta) - 1\right) + \tfrac{V\theta}{A^2}\right) sech\,(\theta)\right), \\
\Delta\xi_o &= \tilde{\xi}_o - \xi_o = Re \int d\theta \left(\Delta\phi e^{-i\alpha} \theta sech\,(\theta)\right).
\end{aligned} \tag{5}$$

The nonvanishing quantum correlations are given by:

$$\begin{aligned}
\langle \Delta A^2 \rangle &= A/\left[2\bar{n}\right], \\
\langle \Delta V^2 \rangle &= A/\left[6\bar{n}\right], \\
\langle \Delta\xi_o^2 \rangle &= \pi^2/\left[24\bar{n}A^3\right], \\
\langle \Delta\alpha_o^2 \rangle &= \left[1 + \tfrac{\pi^2}{12} + \tfrac{V^2\pi^2}{4A^2}\right] / \left[6A\bar{n}\right], \\
\langle \Delta\xi_o \Delta\alpha_o \rangle &= V\pi^2/\left[24\bar{n}A^3\right].
\end{aligned} \tag{6}$$

The photon-number (proportional to A), and the momentum (proportional to V), are invariant under propagation. This is not true for the effective soliton phase and position, whose fluctuations increase with time according to:

$$\begin{aligned}
\tilde{\xi}_o(\tau) &= \tilde{\xi}_o + \tilde{V}\tau, \\
\tilde{\alpha}_o(\tau) &= \tilde{\alpha}_o + \tfrac{1}{2}\left(A^2 + V^2\right)\tau.
\end{aligned} \tag{7}$$

If two solitons that are well-separated compared to their width are injected, then their emergent soliton parameters will be approximately uncorrelated. This approximation corresponds to neglecting the (small) continuum components. The subsequent interaction of two emergent solitons can be simply calculated using classical inverse scattering theory. This states that in a collision, only the position ξ_o and phase α_o are changed. For the first pulse, if $V_2 > V_1$, the position shift is [14]:

$$\delta\xi_1 = \frac{1}{\tilde{A}_1} \ln \left[\frac{(\tilde{V}_2 - \tilde{V}_1)^2 + (\tilde{A}_2 - \tilde{A}_1)^2}{(\tilde{V}_2 - \tilde{V}_1)^2 + (\tilde{A}_2 + \tilde{A}_1)^2}\right] \tag{8}$$

and the phase shift is given, for either pulse, by:

$$\delta\alpha_i = 2\left[\arctan\left(\frac{\tilde{A}_i + \tilde{A}_j}{\left|\tilde{V}_i - \tilde{V}_j\right|}\right) - \arctan\left(\frac{\tilde{A}_i - \tilde{A}_j}{\left|\tilde{V}_i - \tilde{V}_j\right|}\right)\right]. \tag{9}$$

The phase is always increased in a collision, due to the positive sign of the nonlinear refractive index.

Here we have neglected the noise introduced by the collision between soliton (i) and the continuous spectrum of soliton (j).

With soliton (1) having zero average velocity in the frame of reference $\left(\overline{V}_1 = 0\right)$, the total variance of soliton position after one collision is given to the lowest order by

$$\left\langle \Delta\xi_1^2(\tau) \right\rangle = \frac{\pi^2}{24\overline{n}A^3} + \frac{A\tau^2}{6\overline{n}} + \frac{A}{3\overline{n}}\frac{\partial\delta\xi_1}{\partial V_1}\tau + \left\langle \Delta\delta\xi_1^2 \right\rangle . \tag{10}$$

Each logic operation will involve either one or zero collisions between the soliton pulses. On the average, approximately half the number of logic operations will involve collisions. Therefore, the typical position fluctuation $\Delta\xi_0$ of a particular logic soliton after a sequence of N pulse position logic operations can be given by the following expression:

$$\Delta\xi_0(\tau_N) = \Delta\xi_0(0) + \Delta V_0\tau + \sum_{n=1}^{\frac{N}{2}} \left(\frac{\partial\delta\xi_0}{\partial A_n}\Delta A_n + \frac{\partial\delta\xi_0}{\partial V_n}\Delta V_n \right)$$

$$+ \frac{N}{2}\left(\frac{\partial\delta\xi_0}{\partial A_0}\Delta A_0 + \frac{\partial\delta\xi_0}{\partial V_0}\Delta V_0 \right) . \tag{11}$$

The soliton position variance after N operations is given by:

$$\left\langle \Delta\xi_0(\tau_N)^2 \right\rangle = \frac{\pi^2}{24\overline{n}A^3} + \frac{A\tau^2}{6\overline{n}} + \sum_{n=1}^{\frac{N}{2}} \left(\left(\frac{\partial\delta\xi_0}{\partial A_n} \right)^2 \frac{A}{2\overline{n}} + \left(\frac{\partial\delta\xi_0}{\partial V_n} \right)^2 \frac{A}{6\overline{n}} \right)$$

$$+ \frac{N^2}{4}\left(\left(\frac{\partial\delta\xi_0}{\partial A_0} \right)^2 \frac{A}{2\overline{n}} + \left(\frac{\partial\delta\xi_0}{\partial V_0} \right)^2 \frac{A}{6\overline{n}} \right) + \frac{N}{6\overline{n}}\frac{\partial\delta\xi_0}{\partial\delta V_0}A\tau. \tag{12}$$

It is clear from this equation that as long as each logic operation takes a time duration $\tau >> 1$, the overall noise is dominated by the second term, proportional to τ^2. This is due to wave-packet spreading, or equivalently to initial frequency shot-noise in the logic pulse. Noise due to collisions is relatively small, except for the last term. This is caused by a cross-correlation between the velocity and collisional position-shift, and tends to reduce overall fluctuations slightly. Despite this, the overall trend is a quantum wave-packet spreading that gives a variance increasing quadratically with elapsed time. This type of logic therefore gives rapidly increasing error-rates as the number of logic operations increase, since reversible logic provides no mechanism for re-synchronizing the pulse positions.

We conservatively assume that the logic will fail if the soliton position fluctuation increases to the half of the pulse width. Using typical experimental data of $A_j = 1, V_2 - V_1 = 0.1, \overline{n} = 10^8$, and $\xi = 6$ pulse widths for each complete operation, we find the time taken for each logic operation is $\tau = 60$. The number of logic operations that can be performed before the logic fails can be obtained by solving the following quadratic equation:-

$$\frac{\overline{n}}{4} = \frac{\pi^2}{24} + 600N^2 + 33.6N. \tag{13}$$

With $\overline{n} = 10^8$, we find a result of $N \cong 200$ operations, which is a relatively small number compared with the number of operations used by a computer programme. Furthermore, the equations describe a perfect lossless fibre without any other source of noise, so that the performance of the logic gates would be even worse in practice. We conclude that using purely conservative pulse position logic in a real computer network is unlikely to be successful, unless these gates could be combined with other techniques that allow a dissipative re-synchronization of signals, before errors accumulate.

However, the use of pulses with sub-shot-noise velocity error (squeezed frequency pulses) could improve these estimates.

We now turn to the analysis of a QND measurement of soliton photon number using a soliton collision to transcribe information from a signal to a probe soliton. In this interaction, the photon number is unchanged, hence the use of the term QND, or quantum non-demolition. The real purpose of a QND measurement is to measure photon number with an accuracy better than the signal's own intrinsic quantum uncertainty, or shot-noise. This requires not just a complete preservation of the original photon number, but also a high degree of correlation between the input photon number and the output probe measurement. We therefore have to calculate the correlation between these two quantities.

There are several ways to normalise the resulting correlation functions, and we will use two different techniques for comparison here. The quantity of most interest in experiment is the error obtained when comparing a signal photon number fluctuation ΔA_s with an amplified version of the probe measurement ΔX. This error is then $\Delta A_s - g\Delta X$, where g is a gain coefficient, that can be chosen in the experiment to minimise the resulting variance.

It is necessary to normalise the error signal, which involves defining a reference noise level that is conventionally called the shot-noise level. One way to do this is to use the sum of the intrinsic variances in the two fluctuations being compared, i.e., $< [\Delta A_s]^2 + [g\Delta X]^2 >$. We call the variance normalised this way an "S" variance, since it is essentially normalised in a way similar to a squeezing measurement. However, when this normalisation is used, part of the reference shot-noise is actually due to the probe measurement X, which could be much larger than the signal variance. For this reason, it has become more usual to normalise by just the signal variance $< [\Delta A_s]^2 >$. We call this type of normalised variance the "C" variance, since it provides a true measure of the correlations of interest.

Including the effects of collision and free propagation, we find that the quantum phase fluctuations for a probe soliton depend on both it's own parameters, and on the parameters of the signal pulse. To leading order:

$$\Delta\alpha_p(\tau) = \Delta\alpha_p(0) + a\Delta A_s(0) + b\Delta A_p(0) + c\Delta V_s(0) + d\Delta V_p(0)\,. \tag{14}$$

In a coordinate frame in which $V_p = 0$, we find:

$$
\begin{aligned}
a = \left(\frac{\partial \delta\alpha_p}{\partial A_s}\right) &= 2\,|V_s| \left[\frac{1}{(A_p+A_s)^2+V_s^2} + \frac{1}{(A_p-A_s)^2+V_s^2}\right], \\
b = A_p\tau + \left(\frac{\partial \delta\alpha_p}{\partial A_p}\right) &= A_p\tau - 2\,|V_s| \left[\frac{1}{(A_p-A_s)^2+V_s^2} - \frac{1}{(A_p+A_s)^2+V_s^2}\right], \\
c = \left(\frac{\partial \delta\alpha_p}{\partial V_s}\right) &= \frac{2V_s}{|V_s|} \left[\frac{(A_p-A_s)}{(A_p-A_s)^2+V_s^2} - \frac{(A_p+A_s)}{(A_p+A_s)^2+V_s^2}\right], \\
d = \left(\frac{\partial \delta\alpha_p}{\partial V_p}\right) &= -c.
\end{aligned}
\tag{15}
$$

The problem of performing a QND measurement with a phase readout of the input amplitude of a signal is equivalent to optimising the readout coefficient a, while minimising the remaining coefficients b, c, d. These coefficients mix information on the *unwanted* parameters of probe amplitude (and the two velocities), with the *wanted* information on the signal photon number or amplitude. In no case is there any back-action on the measured quantity. However, the contamination of wanted information with unwanted information (noise) can be very large. This reduces the effectiveness of the QND measurement, which is chiefly intended for low-noise measurements beyond the usual quantum limits.

As a test of our procedure, we have calculated the optimum variance $C^{(1)}_{\min}$ in the difference signal between the probe phase and signal amplitude, normalised relative to the signal variance. This calculation has been optimized with respect to the relative electronic gain of the two signals being subtracted. It is sometimes useful to normalise by the sum of the signal and probe variances, as in the recent NTT experiments. We call this $S^{(1)}_{\min}$. We find:

$$
C^{(1)}_{\min} = 1 - \left[\frac{\langle \Delta A_s \Delta\alpha_p\rangle^2}{\langle \Delta A_s^2\rangle \langle \Delta\alpha_p^2\rangle}\right]
\tag{16}
$$

and:

$$
S^{(1)}_{\min} = 1 - \left[\frac{\langle \Delta A_s \Delta\alpha_p\rangle^2}{\langle \Delta A_s^2\rangle \langle \Delta\alpha_p^2\rangle}\right]^{\frac{1}{2}} .
\tag{17}
$$

Substituting relevant experimental parameters, which are: $V_s = -3.0, A_s = 1, A_p = 0.4, \tau = 2.5$, we obtain: $10 \log \left(S^{(1)}_{\min}\right) > -0.8\ dB$. This result includes detection losses, and the additional phase noise of the reference soliton used in the experimental phase measurement. Published non-optimised experimental measurement gave a relative variance of $-0.25\ dB$, which is within the expected bounds. Although recent measurements have improved on this, they are still compatible with the bound given above. With the alternate normalisation, we find $10 \log \left(C^{(1)}_{\min}\right) \approx -0.2\ dB$. We note that perfect QND correlations occur if $C\left[S\right] = 0$.

The readout errors are principally due to the self-phase-modulation noise in the above equation. This can be removed by taking a different experimental strategy, in which a linear combination of phase and amplitude is measured, as discussed by Levenson and Shelby [15]. Complete removal of self-phase-modulation noise is achieved by measuring the quadrature:

$$\hat{X}_{QND} = \Delta\hat{\alpha}_p(\tau) - b\Delta\hat{A}_p(\tau)\,. \tag{18}$$

This can be accomplished via optical homodyne detection of the probe pulse with local oscillator phase $\theta = -arccot\,(b)$. Substituting this new quadrature, and rederiving equations (16) and (17) it is easy to obtain a new expression for the variance, which we call $C^{(2)}_{\min}\left[S^{(2)}_{\min}\right]$. For the experimental parameters with detection losses and addition reference soliton phase noise included we find 10 log $\left(S^{(2)}_{\min}\right) \simeq -2.3\ dB$ and 10 log $\left(C^{(2)}_{\min}\right) \simeq -1.2\ dB$, a clear improvement on the values obtained above for the pure phase measurement. Velocity dependent noise, proportional to c and d, cannot be easily removed. However, this can be reduced to any desired level. The technique is to first set $A_p = A_s = 1$, then consider the limit of $|V_s| << 1$. In this limit of equal pulse amplitudes and long interaction times, we find the following limits:

$$\begin{aligned} a &\rightarrow -2/V_s\,, \\ b &\rightarrow \tau + 2/V_s, \\ c &\rightarrow 1\,; \\ d &\rightarrow -1. \end{aligned} \tag{19}$$

Since $a, b >> c, d$ in this limit, the velocity dependent correlations can be reduced arbitrarily relative to the amplitude dependent correlations, giving the limits:

$$\lim_{V_s \rightarrow 0} C^{(2)}_{\min} = \left(V_s^2/4\right)\left(1 + \pi^2/36\right) \tag{20}$$

and

$$\lim_{V_s \rightarrow 0} S^{(2)}_{\min} = \left(V_s^2/8\right)\left(1 + \pi^2/36\right)\,. \tag{21}$$

To lowest order, it is possible to perform an arbitrarily precise QND measurement with coherent soliton probes, provided a linear combination of probe phase and amplitude is measured. The calculations have ignored thermal noise effects however, and if thermal noise is large compared to self-phase modulation noise, it would provide a signal-to-noise floor which scales the same as the QND signal as the measured probe quadrature is varied. Thus, the advantage of measuring the quadrature defined by equation (18) would be reduced.

In experiments with quantum solitons, thermal noise effects can often be made negligible through the use of low temperatures, short pulses and/or by using measurement techniques which reduce their influence by probing the thermal noise with a reference pulse and then subtracting if from the measured QND signal. Future QND experiments using soliton collisions should therefore employ a combined phase-amplitude quadrature, rather than a pure phase quadrature readout. This allows the elimination of self-phase-modulation noise, which prevents reliable QND measurements in current experiments. Another possible strategy is to use a contractive state rather than a coherent state for the probe input:- this appears difficult, as it requires fibers with negative nonlinear refractive index, which are not currently available.

In summary, while reversible soliton logic has severe quantum noise limitations, the same is not true for QND measurements: here the use of sufficiently low temperatures and the freedom to choose an appropriate readout variable can eliminate almost all lowest order quantum noise effects.

References

[1] C. H. Bennet, IBM J. Res. Dev. **17**, 525 (1973).
[2] M. N. Islam and C. E. Soccolich, Optics Letters **16**, 1490 (1991).
[3] S.R. Friberg, S. Machida, Y. Yamamoto, Phys. Rev. Lett. **69**, 3165 (1992).
[4] V. B. Braginsky, Y. I. Vorontsov, K. S. Thorne, Science **209**, 547 (1980).
[5] H. A. Haus, K. Watanabe, Y. Yamamoto, J. Opt. Soc. Am. B **6**, 1138 (1989).
[6] S. J. Carter, P. D. Drummond, M. D. Reid & R. M. Shelby, Phys. Rev. Lett. **58**, 1841 (1987).
[7] P. D. Drummond, S. J. Carter, J. Opt. Soc. Am. B **4**, 1565 (1987).
[8] P. D. Drummond, R. M. Shelby, S. R. Friberg, Y. Yamamoto, Nature **365**, 307 (1993).
[9] P. D. Drummond & A. D. Hardman, Europhys. Lett. **21**, 270 (1993).
[10] D. J. Kaup, Phys. Rev. A **42**, 5689 (1990).
[11] P.D. Drummond, W. Man, (to appear in Opt. Commun).
[12] A. Hasegawa, F. Tappert, Appl. Phys. Lett. **45**, 142 (1973).
[13] L. F. Mollenauer, R. H. Stolen, J. P. Gordon, Phys. Rev. Lett. **45**, 1095 (1980).
[14] V. E. Zakharov, A. B. Shabat, Sov. Phys. JETP **34**, 62 (1972).
[15] M.D. Levenson, R.M. Shelby, J. Mod. Optics **34**, 775 (1987).

Nonlinear and Quantum Dynamics of an Optical Cavity Containing Cold Atoms

E. Giacobino, J.M. Courty, and A. Lambrecht

Labortoire Kastler Brossel, Université Pierre et Marie Curie, 4 place Jussieu, F-75252 Paris Cedex 05, France

This paper presents the nonlinear dynamics and the quantum properties of laser cooled and trapped cesium atoms placed inside an optical cavity and interacting with a probe light beam.

1. Introduction

Soon after the first implementations of magneto-optic traps[1], the strong nonlinear properties of laser cooled atoms were recognized. It was shown that a probe beam going through the cloud of cold atoms could experience a strong gain due to Raman transitions involving the trapping beams[2]. When the atoms are placed in a resonant optical cavity without a probe beam, laser action corresponding to that gain feature was demonstrated [3].

On the other hand, when cold atoms interact with a probe laser beam inside an optical cavity, bistability is very easily observed at very low light powers (as low as 5μW)[4]. This high nonlinear dispersion comes from the fact that, all atoms being virtually motionless, the Doppler width is much smaller than the natural linewidth and the probe frequency can be set close to atomic resonance.

Such a nonlinear behaviour indicates that the system is capable to strongly modify the quantum fluctuations of the probe beam, and should lead to squeezing.

To understand the nonlinear dynamics as well as the quantum properties of a collection of cold atoms in a cavity in more details, we suppressed the trapping beams during the measurements. In such conditions, new features such as instabilities have been found. We describe the experiment in section 2, in section 3, we give a model explaining the dynamical behaviour of the system and in section 3, we indicate the perspectives for quantum noise reduction.

2. Bistability and instabilities with cold atoms

In the experiments described in the following, we prepare a cloud of cold cesium atoms in a cell using the background pressure to fill the trap. The set-up includes three orthogonal trapping beams generated by a Ti:Sapphire laser and an inhomogeneous magnetic field. The laser is detuned by a few linewidths on the low frequency side of the $6S_{1/2}$ F=4 to $6P_{3/2}$ F=5 transition. We obtain a cloud of

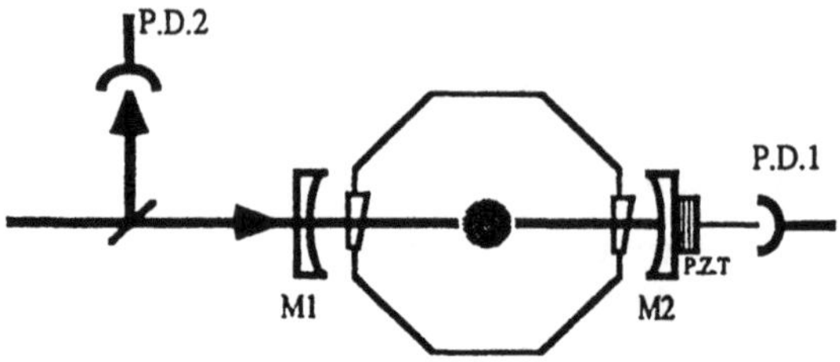

Fig.1 Experimental set-up with the cell containing the cold atom cloud; M1, M2, mirrors, PD1, PD2, photodiodes.

cesium atoms the typical temperature of which is of the order of 1 mK, which gives a Doppler width much smaller than the natural width. The number of atoms contained in the volume of the cavity fundamental mode is of the order of 10^8.

Because the cell has optical quality antireflecting windows, we can build a good finesse optical cavity around the atomic cloud (Fig.1). It is a 50 cm-long linear cavity, with a waist of about 400 μm. The losses due to the two windows are of the order of 1%. Mirrors with different transmission coefficients have been used. The bistability parameter can be as high as 300. The cavity is in the symmetry plane of the trap, making a 45° angle with the two trapping beams that propagate in this plane.

After the trap is loaded, we turn off the trapping laser beams in order to get unperturbed atoms. We have about 10 ms to perform the measurements before the atoms have escaped out of the interaction region. To look for a bistable behaviour of the optical cavity containing cold atoms, we send a *circularly polarized* probe beam into the cavity. It can detuned by 0 to 150 MHz on either side of the $6S_{1/2}$ F=4 to $6P_{3/2}$ F=5 atomic transition.

We scan the cavity length for a fixed value of the input intensity, as shown in fig.2. The recording shows the characteristic hysteresis cycle due to the bistable behaviour, where the output power switches abruptly between low to high values when the length is scanned.

In a rather broad range of experimental parameters, we also observed instabilities that exhibit unusal features. They are present for detunings ranging from -10 Γ to 10 Γ, where Γ is the linewidth of the excited state, and for input powers ranging from 50 to 300 μW. Depending on the detuning at which the cavity is set, the frequency of these instabilities varies from 100 kHz to a few MHz.

Let us point out that similar oscillations had been observed previously when bistability was investigated in the presence of trapping beams but only at high intensity or small atomic detuning. Here, on the contrary, they are very easy to observe.

Figure 3 shows a set of recordings of the output power of the cavity when the length is scanned for different values of the input power. It is clear here that oscillations appear in at some low power and that diasppear at high input power, where only bistability is observed. It is not possible to keep the optical cavity length

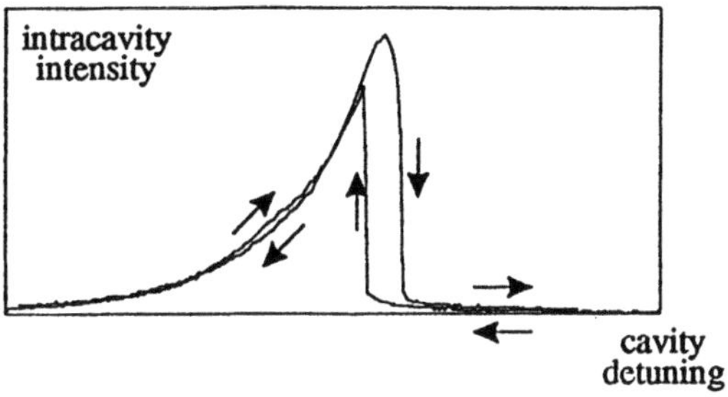

Fig.2 Recording showing the bistable switching from low to high transmission when the cavity length is scanned across Fabry-Perot resonance

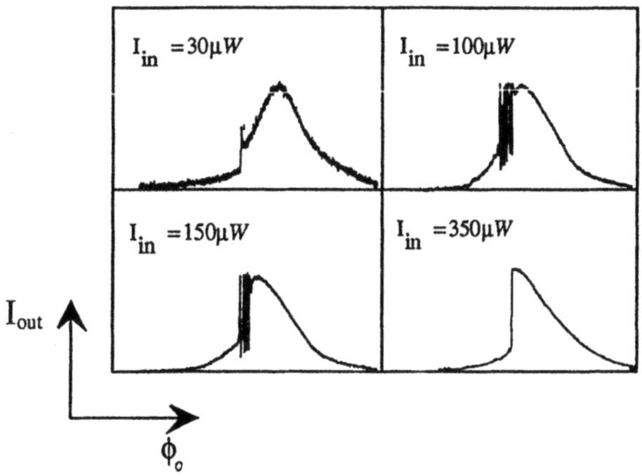

Fig.3 Recording of the output power of the cavity containing cold atoms when the cavity length is scanned, for 4 different values of the input power I_{in}.

fixed to observe the behaviour of the output power because the atoms slowly escape from the original cloud, thus changing the index of refraction. However, in such conditions, the effective cavity scan is very slow and the oscillations are observed on long time durations. Enlarging these oscillations show that they look like limit cycles.

3. Model for instabilities

Owing to the well defined properties of the cold atom medium, we have been able to develop a simple model to understand the origin of the observed oscillations. Roughly, they result from the competition between a fast nonlinear process, the saturation of the optical transition and the much slower optical pumping. With circularly polarized light, optical pumping tends to empty out all the magnetic sublevels of low magnetic number to accumulate all the atoms in the sublevels with highest magnetic number.

The system can be modelled with only 4 variables fulfilling 3 equations. First, we introduce the ground state orientation, p, which increases with the intracavity intensity I and decays at some rate γ due to magnetic precession in residual fields and to radiative transitions to other hyperfine sublevels:

$$dp / dt = - \gamma p + \beta I (1 - p) \ . \tag{1}$$

The round trip phase shift in the cavity Φ is the sum of three contributions, the linear phase shift Φ_0, the Kerr-like phase shift due to the non-linearity of the atoms χI and the phase shift αp coming from the change in the populations of the ground state sublevels due to optical pumping:

$$\Phi = \Phi_0 + \alpha p + \chi I \ . \tag{2}$$

The Kerr-like non-linearity has been assumed to have an instantaneous response.

The third equation describes the evolution of the field E inside the cavity (with $|E|^2$=I). The change of the field on a round trip of duration τ is due to the driving field E_{in} entering through the coupling mirror of transmission t, to the mirror decay coefficient $\varkappa$ (with $\varkappa = t^2/2$) and to the round trip phase shift Φ:

$$\tau \, dE / dt = t \, E_{in} + (\varkappa + i\Phi) \, E \ . \tag{3}$$

The various rates involved in these equations are of very different orders of magnitude. In our experimental situation, the relaxation γ is much slower than the optical pumping rate βI, which is itself much slower than the cavity evolution time $\tau\varkappa$. The atomic non-linearity, which evolves with a time constant of the order of the atomic detuning is even larger and the response has been considerd instantaneous.

These equations have been used to numerically calculate the motion of the system. In some range of initial conditions, oscillations and limit cycles are found in the intracavity intensity (and in the output intensity). To compare these results with the experimental data, we have calculated the output intensity when Φ_0 (i.e. the cavity length is scanned slowly) for several values of the input intensity.

The result is shown in fig.4. It can be seen that the curves reproduce the experimental recordings in a satisfactory way. In particular, they show oscillations setting in for intermediate powers. When the input power is strong enough, the oscillations disappear, due to the fact that the optical pumping is fast enough to bring all the atoms in the sublevels of high magnetic number before the oscillation can start.

This behaviour can be understood by considering that the optical pumping changes the optical length of the cavity. It can then scan the cavity length back and for in the vicinity of the bistable regime, causing periodic abrupt changes in the cavity transmission. A similar behaviour has been observed with thermal effects[5].

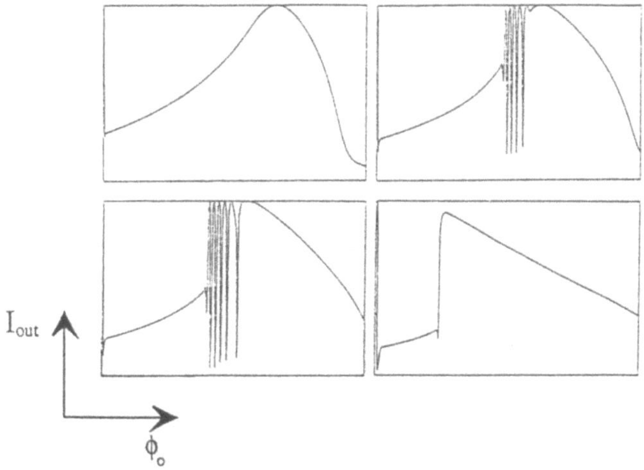

Fig.4 Calculated output power of the cavity containing cold atoms when the cavity length is scanned for 4 values of the input power corresponding to those of fig.3

A more complete treatment of the non-linearity and of the optical pumping involving saturation effects has been done. This gived only minor changes in the results, the general behaviour remaining the same.

Let us note that for high powers, the probe laser tends to push the atoms out of the beam. This phenomenon could give rise to a bistable behaviour due to the change of the effective linear index of refraction of the medium with the number of atoms in the interaction zone. We have shown that such mechanical effects were negligible under our experimental conditions.

4. Quantum noise measurements

Nearly perfect squeezing is expected from a cavity containing a pure Kerr medium[6-7-8]. When motionless two-level atoms are used, the quantum noise reduction is predicted to persist in spite of the added fluorescence noise. Specific theories have to be used to take these effects into account. The usual method consists in writing a Fokker-Planck equation for the coupled atomic and field variables and deriving an equivalent set of Langevin equations[9-10-11-12]. An alternative method, which relies on the derivation of Langevin equations for the field only, has been introduced recently[13-14]. In this treatment, the atomic fluctuations are incorporated in a linearized input output theory similar to the semiclassical one[8], which applies to the parametric media only.

Squeezed vacuum was first observed in a four-wave mixing experiment using an atomic beam as a nonlinear medium[15]. Later, squeezing in a beam of finite

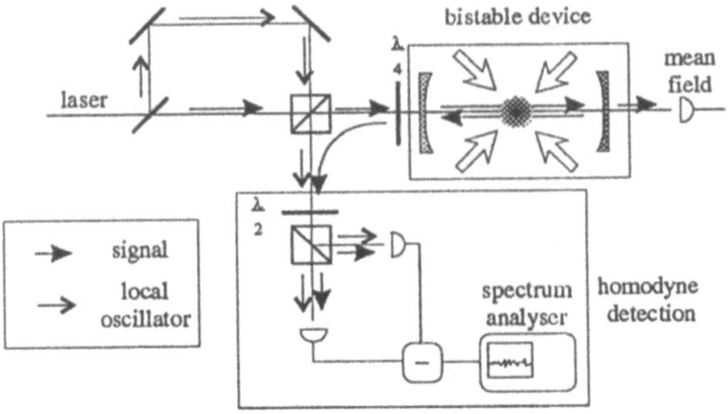

Fig. 5. Experimental set-up designed to study non-linearities and squeezing with cold atoms. The bistability experiments have been done using the photodiode placed behind the end mirror of the cavity to measure the mean signal field. The homodyne detection shown here is used for the study of quantum noise.

mean intensity (bright squeezing) was demonstrated at the output of a bistable cavity traversed by an atomic beam[16-17].

Cold atoms seem to be good candidates to achieve quantum noise reduction, since the motion is much better controlled than in an atomic beam.

To experimentally study the quantum fluctuations at the output of the cavity containing the cold atoms, we use a linear cavity having only one coupling mirror. The coupling mirror has a high transmission, so as to be in the "bad cavity" case, where the decay time of the cavity is larger than that of the atoms, and for which the theoretically predicted squeezing is the best. The other mirror is a highly reflecting one. (fig. 5.)
The field reflected from the cavity is separated by an optical circulator. The polarizing beamsplitter is also used to mix a local oscillator beam with the probe field going out of the cavity. The signal is detected with a balanced heterodyne detector and sent to a spectrum analyser.

The fluctuations of the field are detected at a fixed analysis frequency, while the phase of the local oscillator is rapidly varied with apiezoelectric transducer and the length of the optical cavity is slowly scanned through the resonance. When the cavity is close to resonance, the presence of the nonlinear medium manifests itself by a phase dependent noise as can be seen in fig.6. Excess noise due to atomic fluorescence is observed. Investigation of the parameter space to look for squeezing is under progress.

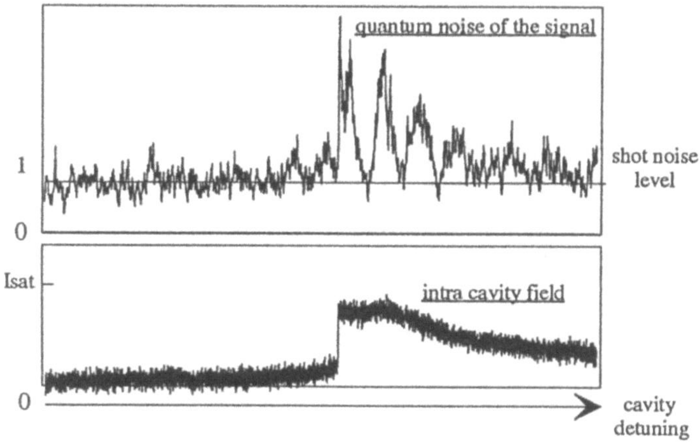

Fig.6 Recording of the noise in the output field homodyned with the local oscillator as a function of the cavity lenght for a fixed noise freqency of 5 MHz. The lower trace gives the mean field as recorded behind the highly reflecting mirror.

5. Conclusion

The recent development of the magneto-optic trap, which enables to get clouds of motionless atoms with a density comparable to that of the atomic beams is of great interest for nonlinear optics and for squeezing. In such traps, one can get large non-linearities by setting the driving field close to resonance without having much absorption. The observation of dispersive bistability in such media makes it possible to do a detailed evaluation of the properties of the cold trapped atoms as a device to manipulate the quantum noise.

Acknowledgements :This work has been supported in part by the C.E.C. contract ESPRIT BRA 6934 and the CNRS ULTIMATECH programme.

References

1 E. Raab, M. Prentiss, A. Cable, S. Chu, D. Pritchard, Phys. Rev. Lett. **59**, 2631 (1987); C. Monroe, W. Swann, H. Robinson, C. Wieman, Phys. Rev. Lett. **65**, 1571, (1990)

2 D. Grison, B. Lounis, C. Salomon, J.Y. Courtois, G. Grynberg, Europhys. Lett. **15**, 149 (1991); J. Tabosa, G. Chen, Z. Hu, R. Lee, H.J. Kimble, Phys. Rev. Lett. **66**, 3254 (1991)

3 L. Hilico, C. Fabre, E. Giacobino, Europhys. Lett. **18**, 685 (1992)

4 E. Giacobino, J.M. Courty, C. Fabre, L. Hilico, A. Lambrecht, in "Laser Spectroscopy XI", eds L. Bloomfield, T. Gallagher, D. Larson, AIP Press (1994); L. Hilico, Thesis, Université Paris VI, 1992

5 S. L. McCall, Appl. Phys. Lett. **32**, 284 (1978)

6 L. Lugiato, G. Strini, Optics Commun. **41**, 67 and 374 (1982).

7 M.J. Collett, D.F. Walls, Phys. Rev. **A32**, 2887 (1985).

8 S. Reynaud,C. Fabre, E. Giacobino, A. Heidmann Phys. Rev. **A40**, 1440 (1989).

9 H.J. Carmichael, Phys. Rev. **A33**, 3262 (1986)

10 L.A. Orozco, M.G. Raizen, Min Xiao, R.J. Brecha, H.J. Kimble, J. Opt. Soc. Am. **B 4**, 1490 (1987)

11 F. Castelli, L.A. Lugiato, M. Vadacchino, Nuovo Cimento B **10**, 183 (1988)

12 M.D. Reid, Phys. Rev. **A37**, 4792 (1988)

13 J.M. Courty, P. Grangier, L. Hilico, S. Reynaud, Opt. Commun. **83**, 251 (1991)

14 L. Hilico, C. Fabre, S. Reynaud; E. Giacobino, Phys. Rev. **A46**, 4397 (1992)

15 R.E. Slusher, L. Hollberg, B. Yurke, J. Mertz, J.F. Valley, Phys. Rev. Letters **55**, 2409 (1985).

16 M.G. Raizen, L.A. Orozco, Min Xiao, T.L. Boyd, H.J. Kimble, Phys. Rev. Lett. **59**, 198 (1987)

17 D.M. Hope, H.A. Bachor, P.J. Manson, D.E. McClelland, P.T.H. Fisk, Phys. Rev. **A46**, R1181, (1992)

Quantum Optics with Mirrors

J.G. Rarity[1], D.M. Atkin[1], P.R. Tapster[1], P.C.M. Owens[1], T. Herzog[2], H. Weinfurter[2], and A. Zeilinger[2]

[1]DRA Malvern, St. Andrews Rd., Malvern, Worcs. WR14 3PS, UK
[2]Institut für Experimentalphysik, Universität Innsbruck, Technikerstr. 25, A-6020 Innsbruck, Austria

Abstract. The close presence of a mirror can profoundly modify the spontaneous emission process. Here I describe some preliminary experiments with electrically pumped micro-cavities where we have shown enhancement and suppression of spontaneous emission into the cavity modes. We have also demonstrated modification of the spontaneous emission of photon pairs in parametric down-conversion using suitably placed mirrors. In contrast with the single photon case the mirrors can be placed at large distances from the non-linear crystal source due to the energy conservation inherent in the down-conversion of narrow band pump beam photons.

1 Introduction

Modification of the spontaneous emission of an atom by its surrounding environment was first suggested some years ago [1] but it is only recently that the phenomenon has been investigated in various elegant experiments involving atoms close to mirrors[2, 3]. In a simple description, light propagating from an atom placed close to a mirror can reach a detector either directly or via reflection in the mirror. When the two paths cannot be distinguished, constructive and destructive interference effects occur between reflected and direct spontaneous emission. This interference effect decays away with increasing atom-mirror distance due to the limiting bandwidth (hence coherence length) associated with the emission process or predetection filter. A similar effect occurs when an atom is located in a short cavity. Emission into the cavity mode can be strongly enhanced when the atom is placed at an anti-node and strongly suppressed at a node. Strong enhancement of spontaneous emission is only seen when the emission peak is within the cavity resonance. For broad band emission we see a narrowing of the emission and enhanced spectral density at the cavity resonance. Below (section 2) we present preliminary results of electro-luminescence in $In_{0.2}Ga_{0.8}As$ quantum well resonant cavity light emitting diodes demonstrating these simple effects.

Another spontaneous emission process occurs in parametric downconversion in the weak pump limit. In this case there are three modes involved, the pump, signal and idler modes and there is a finite probability of spontaneous conversion of a pump photon to a signal-idler photon pair satisfying energy and momentum conservation rules. We can enhance and suppress this spontaneous emission process by suitably arranging three mirrors such that there are two indistinguishable routes for the creation of the downconverted photon pairs[5] (section 3).

Single atom experiments can be described by semi-classical models where the light can be thought of as a classical field interacting with a quantum mechanical oscillator, the atomic transition. In contrast, suppression of two-photon spontaneous emission is a truly non-classical interference effect because the correlation of detection events in pairs allows us to gate detection of one-photon states conditionally on detection of partner photons while still observing the effect.

2 One-photon spontaneous emission

We have designed and built resonant cavity light emitting diodes based on two $In_{0.2}Ga_{0.8}As$ quantum wells placed at various positions within a one wavelength ($1\mu m$) long GaAs cavity. A schematic diagram of the three device structures we have grown is shown in figure 1. They are grown by molecular beam epitaxy (MBE) on an n^+ GaAs substrate. Below the depleted GaAs cavity region is a $6\frac{1}{2}$ element quarter wave stack made from alternating layers of n-type GaAs and AlAs which is grown on an n-type GaAs buffer layer on the GaAs substrate. The top contact of the device is made using a thin layer of AlAs and a stabilising layer of GaAs both of which are heavily p-type doped (about $10^{19}cm^{-3}$). This layer is topped by a 120nm thick layer of silver which acts both as a reflector and an electrical connection. The thickness of the p-type layers is chosen to compensate the phase change at the silver mirror calculated using the known values of silver and GaAs refractive indices.

One can use the method of images to calculate the spontaneous emission modification in such a structure but it is more simple to use vacuum field modes[4] as described below. Given a plane wave incident on a layered dielectric medium we can work out the fields within the media by matching the boundary conditions at every dielectric interface. Spontaneous emission can be modelled by emission stimulated by vacuum fluctuations. Associating the standing wave amplitude with the amplitude of vacuum fluctuations within the sample we can predict the variation of spontaneous emission. Emission into many directions is simply mapped by superposing and integrating many plane wave propagation directions with suitable normalisation[4]. Such a map of spontaneous emission enhancement factor is overlayed on our device design in figure 1.

Integration is over an external cone angle of 1/8 radians to represent the finite collection angle of our light measurement system but corresponds to a single wavelength ($1\mu m$). Strong enhancements and suppressions are clearly seen within the cavity region where the expected positions of our quantum wells are marked.

We can also vary the wavelength and predict the emission spectra at fixed positions within the cavity. Predicted spectra for our devices are shown in

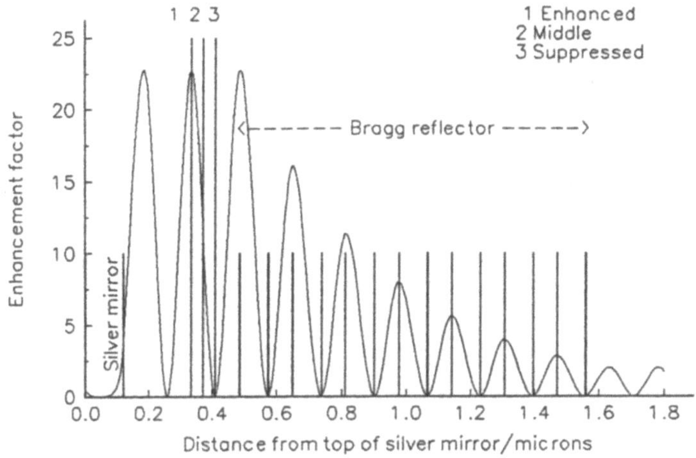

Figure 1: Device design showing silver mirror, one wavelength cavity and $6\frac{1}{2}$ period AlAs/GaAs quarter wave stack. The device emits through the GaAs substrate. Three wafers were grown with quantum wells in the positions (1,2,3) shown. Also shown is the enhancement factor (solid curve) for spontaneous emission into an external cone angle of 1/8 radians plotted throughout the device

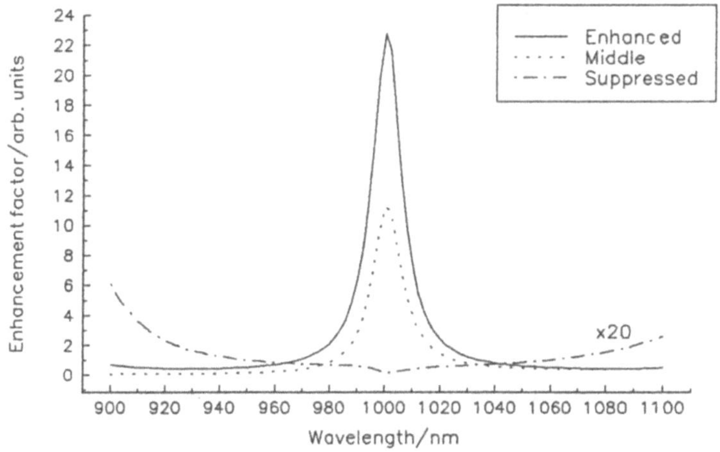

Figure 2: Predicted spontaneous emission enhancement spectra for the three quantum well positions shown in figure 1. Note that the suppressed spectrum is magnified by a factor of 20

figure 2. With the emitter at the antinode, at line centre we see an enhancement factor of 22 above emission intensity outside the cavity. For the node we see suppression by a similar factor although there is some probability of emission away from line centre. When the emitter is midway between node and antinode

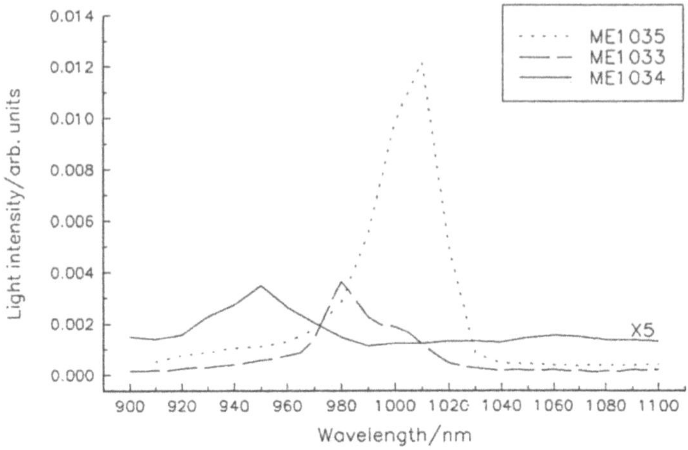

Figure 3: Measured emission spectra from the three wafers with cavity resonance matching the emission peak. Quantum well positions within the cavity were: ME1035 at the antinode (position 1), ME1033 between node and antinode (position 2) and ME1034 at the node (position 3). Note that the suppressed spectrum is multiplied by 5 thus increasing the background light level.

the enhancement is roughly halved. The spectral widths reflect the width of the cavity resonance and we have assumed a flat broadband emission spectrum. This is not exactly the situation for InGaAs as we see from the experimental measurements.

Light output from the three devices varied considerably showing a factor of >100 difference in output between the enhanced and supressed samples. Typical light intensity spectra measured from selected devices on the three wafers are shown in figure 3. The devices were driven by a current of 60mA. Cavity thickness changes by 5% across the wafer and these devices were selected near to the edge where the cavity resonance (as estimated from transmission measurements) matches the peak output from the $In_{0.2}Ga_{0.8}As$ quantum wells. We see a good qualitative agreement between the measured spectra and those predicted in figure 2. Experimental spectral widths appear a little broader than those predicted possibly due to cavity losses and non-ideal Bragg reflectors.

3 Inhibition of two-photon spontaneous emission

Consider the process of parametric down-conversion as shown in figure 4. A UV pump beam is reflected back on itself to pass through a non-linear crystal two times. This provides two possibilities for spontaneous emission of a pair of

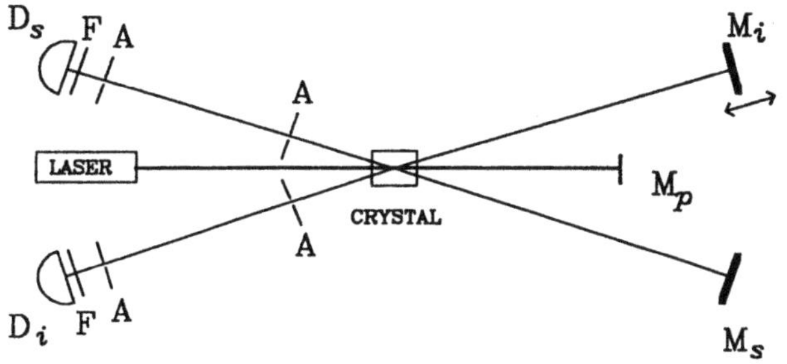

Figure 4: Schematic of the two-photon spontaneous emission experiment. Two possible ways of creating a photon pair in a non-linear crystal arise because the pump laser beam is retroreflected at mirror M_p. When these two possibilities are made to overlap with a further pair of mirrors (M_s, M_i) we see interference between them. The diaphragms (A) serve to define single spatial modes and filters (F) select out energy matched pairs. Pairs are detected in single photon counting detectors (D_s, D_i

red photons into modes propogating in opposite directions. We then reflect one set of signal-idler modes back through the crystal such that they overlap with the other signal-idler modes and detect pairs in photon counting detectors D_s, D_i. Viewed from the detectors there are two indistinguishable routes for any event to occur which leads to an interference effect. In the weak pump limit the wavefunction can be written

$$|\Psi\rangle = \alpha \left(e^{i\phi_p} + e^{i(\phi_s+\phi_i)}\right) |1\rangle_s |1\rangle_i \ , \tag{1}$$

where $|\rangle_s|\rangle_i$ represent the signal and idler modes, ϕ_s and ϕ_i are the phases accumulated by the first signal and idler, respectively, to their mirror and back, ϕ_p is the phase accumulated by the reflected pump and $\alpha \ll 1$ is the probability amplitude for creation of a signal-idler pair for each passage of the pump beam. Equation 1 makes the remarkable prediction that the signal, idler and (as a result) coincidence count rates all vary as

$$I_c \propto I_i = I_s = 2I_0 \left\{1 + \cos(\phi_i + \phi_s - \phi_p)\right\} \ , \tag{2}$$

where $I_0 \propto |\alpha|^2$ is the rate of photon emission into either mode without mirrors present.

Intensities I_i and I_s are varying in an identical way. Thus varying, say the position of the idler mirror, varies both intensities with period given by the idler wavelength. Clearly this can be interpreted as enhancement and suppression of two-photon spontaneous emission.

A more complete theoretical approach has to take into account the finite coherence lengths of the pump beam (meters) and of the down-converted radiation ($\approx 260\mu m$). Energy conservation confines the sum of signal and idler photon energies to that of the pump light. When considered in unison signal-idler photon pairs have a coherence length equal to that of the pump beam thus the signal and idler mirrors can be placed a large distance from the crystal without destroying the effect. We must only *equalise* the crystal-mirror path lengths of signal and idler beams to within their $260\mu m$ coherence lengths. Similarly the crystal-mirror distance of the pump, and signal/idler mirrors should not differ by more than the pump coherence length. Recall in contrast that in the experiments with atomic sources it was necessary to place the radiating sample close to the mirrors (within the coherence length of the spontaneously emitted light) in order to see a modification of the spontaneous emission. Here it is only the *relative* position of signal and idler mirror which has to lie within the coherence length of the spontaneously emitted down-converted light.

In our experiment[5] the source of the two photons is a non-linear crystal of $LiIO_3$ optically pumped by the 351.1 nm line of a single-mode argon-ion laser. Apertures are placed to satisfy the phase matching conditions for signal and idler photons at wavelengths 632.8 nm and 788.7 nm respectively. This down-conversion light is further limited to a bandwidth of 5 nm by employing interference filters. The two beams are detected using Silicon avalanche photodiodes operating in the Geiger mode [6].

Singles count rates and the coincidence count rate are recorded on a personal computer which is also used to control the positions of the mirrors. The two signal (idler) modes are made indistinguishable by placing two small diaphragms separated by 90cm in both the signal and idler beam between the crystal and detectors, which are about 1.2 m from the crystal. These apertures are closed to their smallest diameter ($\approx$ 0.8mm) to roughly define a single spatial mode.

By careful adjustment of the aperture positions the mode overlap can be improved to give high visibility interference in the singles count rates. Figure 5 shows the signal count rate, idler count rate and coincidence count rate measured as a function of the idler mirror position. In agreement with equation 2 the fringe period is given by half the wavelength of the beam whose mirror is scanned. The slight quantitative discrepancy is due to drift of the mirror mounts. All three measured curves vary in the same way both in period and phase which again confirms the prediction of equation 2. Furthermore, the fringe visibilities can be quite high in the experiment. The maximum visibilities we observed were 51.7% for the idler beam and 86% for the coincidences. A lower visibility in the signal beam arose because of poorer mode selectivity. When the signal mirror is removed there is no variation of the idler intensity with any mirror position clearly indicating that we are observing a two-photon effect.

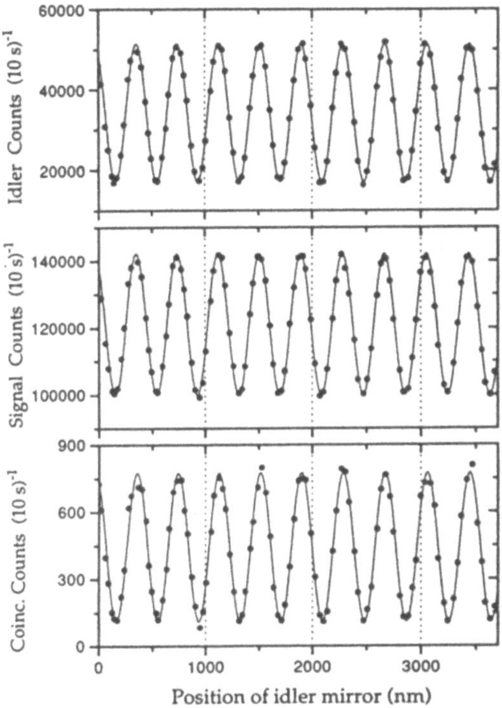

Figure 5: Simultaneous measurement of signal count rate, idler count rate and coincidence count rate as a function of the displacement of the idler mirror. The phase agreement of these oscillations shows clearly that the boundary condition controlling the emission of both photons in the entangled state can be controlled by changing the position of just one of the mirrors.

4 Conclusions

In conclusion we have reported enhancement and inhibition of spontaneous emission of light in resonant microcavity LED's. The devices show narrower emission spectra than conventional InGaAs LED's and enhanced spectral density. Such devices may have advantages over conventional sources in fibre optic communications[7]. Again we emphasise that the effects demonstrated are semi-classical where the field can in principle be described classically in these devices. When higher finesse cavities are used other interesting cavity quantum electrodynamics effects[8] occur and eventually we expect fully quantum mechanical descriptions will be necessary[9].

We have for the first time observed the suppression and enhancement of the spontaneous creation of an entangled photon pair. The effect may be fully understood as interference between two possible ways for emission to occur rather than interference of the photons themselves. The fact that the mirrors can be placed a large distance from the crystal allows the conception of some interesting 'Welcher Weg' experiments. The existence of a measurable field between the crystal and mirrors when the emission is completely suppressed is can now be probed along with inevitable propagation delays associated blocking of the crystal mirror path.

J.G.R. wishes to acknowledge the hospitality and support of the Institut für Experimentalphysik, Innsbruck where part of this work was carried out. D.M.Atkin's present address is Optical Research Centre, Southampton University, Southampton SO9 5NH, UK

References

[1] E. M. Purcell, Phys. Rev. **69**, 681 (1946).

[2] K. H. Drexhage, Progress in Optics **12**, 163 (1974).

[3] D. G. Deppe, J. C. Campbell, R. Kuchibhotla, T. J. Rogers, and B. G. Streetman, Electron. lett. **26**, 1666 (1990)

[4] H Khosravi and R Loudon, Proc. Roy. Soc. Lon. A 436, 373 (1992).

[5] T Herzog, J G Rarity, H Weinfurter and A Zeilinger Phys. Rev. Lett. **72**, 629 (1994).

[6] R G W Brown, J G Rarity and K D Ridley, Applied Optics **25**, 4122 (1986).

[7] N E J Hunt, E F Schubert, R A Logan and G J Zydzik, Appl Phys Letts 61, 2287 (1992).

[8] C Weisbuch, N Nishioka, A Ishikawa and Y Arakawa, Phys. Rev. Lett. **69**, 3314 (1992).

[9] G Rempe, R J Thompson, R J Brecha, W D Lee and H J Kimble, Phys. Rev. Lett. **67**, 1727 (1991).

Decoherence: Is Dissipation All It Takes?

H.J. Carmichael

Department of Physics, Chemical Physics Institute
and Institute of Theoretical Science, University of Oregon,
Eugene, OR 97403-1274, USA

Abstract. I discuss the relationship between the decoherence of a quantum ensemble seen at the level of the density operator and the complementary evolutions revealed by quantum trajectory theory for individual members of the ensemble.

The idea of decoherence has become quite popular as a possible way to explain how classical dynamical behavior emerges from the formalism of quantum mechanics [1]. The central result is that for a system interacting with an environment, certain types of superposition states decay very rapidly into mixtures. The decay occurs at a rate governed by the separation of the superposition state components in phase space, a rate very much faster than the rate of loss of energy from the system. The standard example in quantum optics is a lossy electromagnetic cavity mode prepared in a superposition of coherent states. Let the amplitudes of the coherent states be $+A$ and $-A$; the components of the superposition state then describe an electromagnetic field that oscillates with phase 0 or π. Further, let $A^2 = 10^7$ photons and let the photon decay rate be $2\kappa = 10^7\,\mathrm{s}^{-1}$. The energy then leaks out of the cavity in a time on the order of $(2\kappa)^{-1} = 10^{-7}\,\mathrm{s}$; however, solving the master equation for the density operator of the damped field mode, we find that the superposition of coherent states is convert into a mixture after a time on the order of $(2\kappa A^2)^{-1} = 10^{-14}\,\mathrm{s}$ [2,3].

The result of decoherence is quite clear, but the interpretation of this result must be approached with considerable caution. It is tempting to conclude that the environmental interaction "contaminates" the system in some objective sense (in the way in which thermal noise can scramble a phase for example), so that after the $10^{-14}\,\mathrm{s}$ the diagonal density operator may be transcribed into a probability distribution describing a classical ensemble of cavities, each supporting an electromagnetic field with *either* the phase 0 *or* the phase π. In this interpretation the environment objectively, and locally (through an interaction at the cavity output mirror), destroys the quantum coherence and realizes a definite phase for each member of the ensemble—in each individual run of an experiment. The fallacy of an interpretation along these lines comes with the leap from a density operator, which provides the statistical information required to calculate ensemble averages, to statements, or implied statements, about realistic "happenings" occurring for individual members of the ensemble.

Springer Proceedings in Physics, Vol. 77
Editors: D.F. Walls · J.D. Harvey
Quantum Optics VI

In making this leap, a dynamics belonging to individual members of an ensemble is visualized without ever introducing a well-defined formalism that permits the construction of the ensemble evolution out of an ensemble *of* evolutions. "Evolutions" in the latter sense are not defined in the standard formulation of quantum mechanics. The stated interpretation imports the notion, nonetheless, (by implication at least) in a manner that is at best ad hoc.

It is possible to illustrate how such ill-defined constructions can be misleading by analyzing the decay of a superposition of coherent states using quantum trajectory theory. Quantum trajectory theory is a stochastic formulation of quantum mechanics which describes a complete experimental scenario, including the realization of stochastic measurement records (time series of real numbers) [4]. The theory defines an ensemble of individual evolutions for an open (observable) quantum system, where the system is described at all times by a pure quantum state satisfying a stochastic dynamic. The environmental interaction is accounted for by the parallel, and coupled, evolution of the measurement record. The record is provided by ideal detectors (real or imagined) which convert every quantum lost from the system into classically accessible information. The quantum state in this formalism is a *conditioned* state; it describes the state of the open system at time t, conditioned on the history of the accompanying measurement record prior to t. In this theory the environment is monitored, and from this fact it is meaningful to speak of "happenings." The "happenings" in the records are by definition observable "realities," and these "realities" are accompanied by inferred "happenings" in the conditioned quantum state of the open system.

The stochastic part of quantum trajectory theory follows from the attitude that "happenings" can only "happen" if we are prepared to realize the probabilistic predictions of quantum theory by sampling the probability distributions the theory gives to us to generate stochastic measurement records. (Remembering of course that the outcomes of the sampling must feed back into the dynamics through some sort of quantum state collapse. The "collapse" employed by quantum trajectory theory is always applied at the one-quantum level. It is the elementary collapse associated with realizing a single photoelectron through the photoelectric effect [4].) The stochastic character is not, however, the focus in this paper. The principal point here is that the inferred "happenings" only become well-defined after we specify *how* the environment is to be monitored. They are not defined solely by the local interaction at the cavity output mirror which is responsible for coupling energy into the environment. More precisely, if we follow individual quantum trajectories by monitoring the electromagnetic field leaking out of the cavity using homodyne detection, the conditioned quantum state of the field inside the cavity behaves in completely different ways depending on the choice of phase for the homodyne detector. The setting of this phase is an adjustment to a local oscillator, and may be made, in principal, at a location quite remote from the leaky cavity. The point is that the behavior of the decaying superposition of coherent states has *complementary appearances* which lead to observable differences in measurement records. The ambiguity that comes with the complementarity is not automatically resolved in an objective, mechanistic way through the mere presence of an interaction with an environment and the resultant decoherence. It remains until we define a complete physical scenario which permits observation (reading of the environment) through some recording apparatus (be it real or imagined). Given such definition, the ambiguity is then resolved into a choice of one appearance over others, assuming we are prepared to introduce something into the theory to enable the recording apparatus to record—that is, assuming we are prepared to include a sampling of probability distributions as part of the dynamics.

The stochastic quantum trajectory equations for this problem can be solved analytically. The details are reported elsewhere [5]. The principal results are the following: The local oscillator phase may be given a setting $\theta = 0$ for which individual trajectories for the (unnormalized) conditioned quantum state of the intracavity electromagnetic field take the form

$$|\bar{\psi}_c(t)\rangle = e^{+\phi(t)}|+Ae^{-\kappa t}\rangle + e^{-\phi(t)}|-Ae^{-\kappa t}\rangle, \tag{1}$$

where $\phi(t)$ is a real stochastic variable which diffuses in the time-dependent potential

$$V[\phi, \lambda(t)] \propto -\ln[\cosh(2\phi)e^{2[A^2-\lambda(t)]} + 1], \tag{2}$$

with

$$\lambda(t) = A^2(1-e^{-2\kappa t}). \tag{3}$$

The measurement record that conditions the state is the stochastic charge $Q(t)$ accumulated at the balanced detector registering the homodyne current; $Q(t)$ is proportional to $\phi(t)$; therefore the statistics of the measurement records follow those of $\phi(t)$.

The local oscillator phase may alternatively be set to $\theta = \pi/2$. Then the conditioned quantum states take the form

$$|\bar{\psi}_c(t)\rangle = e^{+i\phi(t)}|+Ae^{-\kappa t}\rangle + e^{-i\phi(t)}|-Ae^{-\kappa t}\rangle\,, \tag{4}$$

where the time-dependent potential is

$$V[\phi, \lambda(t)] \propto -\ln[1+\cos(2\phi)e^{-2[A^2-\lambda(t)]}]. \tag{5}$$

Both of these trajectory evolutions are consistent with the solution to the master equation for the damped cavity mode; both give pure state ensembles which satisfy

$$\rho(t) = \overline{|\psi_c(t)\rangle\langle\psi_c(t)|}\,, \tag{6}$$

where $\rho(t)$ is the solution to the master equation, $|\psi_c(t)\rangle$ is the normalized conditioned quantum state, and the overbar denotes the ensemble average. Decoherence occurs therefore, in both cases, for the ensemble as a whole. But the conclusion that the initial superposition of probability amplitudes is replace, for individual members of the ensemble, by the either-or choice of classical probability does not automatically follow. This naive conclusion is correct when the local oscillator phase is $\theta = 0$ [Eqs. (1) and (2)]. It is not correct when $\theta = \pi/2$ [Eqs. (4) and (5)].

When $\theta = 0$ the potential defined in Eq. (2) rapidly splits the initial probability distribution $\Phi(\phi) = \delta(\phi)$ into two peaks. Individual realizations of $\phi(t)$ then choose, during the decoherence time, to be either positive (and quite large) or negative, in which case one piece of the superposition state in Eq. (1) dominates over the other. The time-scale for this to happen is determined by the time, on average, for one quantum to escape the leaky cavity, and from the quantum trajectory viewpoint the decoherence is seen to follow from the phase shift that accompanies the elementary collapse associated with this photon escape [4].

This behavior is contrasted dramatically by the individual trajectory evolution for a local oscillator phase $\theta = \pi/2$. In this case the potential defined by Eq. (5) is initially flat, and it remains that way for a time on the order of 10^8 decoherence times (for the chosen A^2 and 2κ). After this time, the cosine term in Eq. (5) becomes important. Then the potential develops deep corrugations which confine the realizations of $\phi(t)$, and hence the measurement record $Q(t)$, to a periodic distribution. Thus, the asymptotic distribution $Q(\infty)$ records the phase-space interference fringe associated with the initial superposition of coherent states [5]. Throughout this evolution the quantum state along each individual trajectory is always a superposition state [Eq. (4)], but a superposition state with a diffusing relative phase; it is this diffusion of the relative phase that ensures that the ensemble as a whole decoheres while individual trajectories remain as superpositions.

The initially surprising feature about this behavior is that the quantum interference fringe does not even begin to be written into the record until some 10^8 decoherence times have elapsed. The reason for this is that the information necessary to identify the initial state of the cavity field as a superposition state is contained in the entire state. Every last photon is needed to recognize the superposition. This can be appreciated by noting that the two superpositions, $|+A\rangle + |-A\rangle$ and $|+A\rangle - |-A\rangle$, differ in the property that the first contains only even numbers of photons, and the second, only odd numbers of photons. To distinguish these states when $A^2 = 10^7$ a discrimination between 10^7 photons and $10^7 - 1$ photons must be made. The 10^8 decoherence times translates, therefore, into the time for which the record making must continue in order that, on average, the very last photon has leaked from the cavity and been recorded.

Of course, this last observation also explains why actual experiments to observe quantum interference fringes are so exquisitely sensitive to the efficiency of the recording process when large numbers of quanta are involved. Certainly, for practical efficiencies the complementary behaviors presented here are not equally observable (the second is even difficult to observe for $A^2 \sim 2$). This fact may resolve the ambiguous nature of the quantum dynamics *for all prac-*

tical purposes. Such a resolution is not, however, the objective, mechanistic process that is all too easily deduced from the notion of decoherence. It relies on nonmechanistic issues related to the gathering of information (the making of records). Decoherence does not provide an escape from the need to define the way of observing (in principal if not in practice) in a manner that goes deeper than the choice of an environmental interaction before an unambiguous meaning can be given to a quantum mechanical evolution.

Acknowledgements

Work supported by the National Science Foundation under Grant No. 9214501.

References

[1] W. H. Zurek, *Physics Today*, October 1991, pp. 36-44; Letters to the Editor, *Physics Today*, April 1993, pp. 13-15, 81-90.
[2] A. O. Caldeira and A. J. Legget, *Phys. Rev.* A **31**, 1059 (1985).
[3] D. F. Walls and M. G. Milburn, *Phys. Rev.* A **31**, 2103 (1985).
[4] H. J. Carmichael, *An Open Systems Approach to Quantum Optics*, Lecture Notes in Physics, New Series m: Monographs, Vol. m18 (Springer-Verlag, Berlin, 1993).
[5] H. J. Carmichael, P. Kochan, and L. Tian, "Coherent States and Open Quantum Systems: A Comment on the Stern-Gerlach Experiment and Schrödinger's Cat," in the proceedings of the International Symposium on Coherent States: Past, Present, and Future, Oak Ridge, Tennessee, June 14-17, 1993, edited by J. R. Klauder (World Scientific, Singapore, in press).

Part IV

New Theoretical Techniques

Quantum Optical Correlations: A Green's Function Approach

C.J. Mertens and T.A.B. Kennedy

School of Physics, Georgia Institute of Technology, Atlanta, GA 30332, USA

Abstract. We outline a Green function method for the calculation of quantum optical correlations, paying particular attention to nonlinear effects beyond the domain of linearized noise theory.

1. Introduction

In the last decade a great deal of progress has been made in the understanding of quantum statistical properties of the electromagnetic field. At optical frequencies, photodetection can be used to measure a variety of correlation functions, and thereby obtain detailed information about quantum dissipative dynamics. As experiments in the optical regime begin to probe the quantum limit, where dissipation is reduced to a level comparable or less than the coherent coupling [1], electromagnetic field states far from the familiar coherent, squeezed and thermal states are envisaged, and traditional classical notions about phase transitions in lasers and parametric oscillators breakdown.

To describe dynamical phenomena in this regime is a difficult task, by any approach, as linearized noise perturbation theory is insufficient. In the present work we summarize a nonequilibrium Green function approach, which attacks the problem of calculating correlation functions head on. While this method is not new [2]-[3], it is unfamiliar in the context of quantum optics. The Green function approach provides a compact way of deriving and organizing the hierarchy of equations for correlation functions which typically arise. Moreover, it presents a systematic way to truncate the hierarchy for self-consistent perturbation theory [4]. The latter turns out to be a very efficient means of practical calculation, although we will not discuss this aspect further here. Perhaps most importantly, this rather direct approach retains the flavor of the elementary dynamical processes in its structure. These can be visualized with Feynman diagrams, a particularly appealing feature from a physical standpoint.

Here we describe the Green function approach in the context of the optical parametric oscillator (OPO) [5]. In the threshold region, linearized correlation functions diverge, a signature that the system senses its own size: specifically the threshold photon number. More generally, if this number is "small" for example, in an ultra-low threshold OPO, or "thresholdless" laser one expects nonperturbative dynamical behavior. In section 2 we consider Markovian dis-

Springer Proceedings in Physics, Vol. 77 Quantum Optics VI

Editors: D.F. Walls · J.D. Harvey

sipation, and extend the theory to include nonlinear parametric interactions in section 3. Truncation of the hierarchy of correlation functions is discussed in section 4. Closed form Dyson equations and their physical interpretation is treated in section 5.

2. Cavity dissipation, input-output and Green functions

Here we summarise how Markovian dissipation may be incorporated into the Green function formalism. Consider the quantum Langevin equation for an empty-cavity mode

$$\frac{d}{dt}a^\dagger(t_\zeta) = -a^\dagger(t_\zeta) - \sqrt{2}\,a^\dagger_{in}(t_\zeta) + J^{\zeta *}(t), \tag{1}$$

with input noise $[a_{in}(t), a^\dagger_{in}(t)] = \delta(t - t')$, and an external source $J^{\zeta *}(t)$, $\zeta = \pm 1$. Time is scaled to the cavity lifetime, and is considered to lie on the upper or lower lip of the Keldysh contour, Fig.1 [6]. For example, a time ordered correlation function is evaluated on the upper lip $\zeta = 1$, and anti-time ordered functions on the lower lip $\zeta = -1$. Different orderings are prescribed by the Keldysh path ordering operator T_c. The source is introduced as a formal device to enable all correlation functions to be generated by functional differentiation of (1), and is set to zero at the end. This leads to the Green function equations [7]

$$\begin{aligned}\frac{d}{dt}D^{\zeta\zeta'}(t,t') &= -D^{\zeta\zeta'}(t,t') + i\zeta'\,\delta_{\zeta\zeta'}\,\delta(t-t') \\ &\quad + \sqrt{2}i\left\langle T_c\left(a^\dagger_{in}(t_\zeta), a(t'_{\zeta'})\right)\right\rangle, \end{aligned}\tag{2}$$

$$\frac{d}{dt}\mathcal{D}^{\zeta\zeta'}(t,t') = -\mathcal{D}^{\zeta\zeta'}(t,t') - \sqrt{2}\left\langle T_c\left(a_{in}(t_\zeta), a(t'_{\zeta'})\right)\right\rangle, \tag{3}$$

where the *normal* propagator $iD^{\zeta\zeta'}(t,t') = \left\langle T_c\left(a^\dagger(t_\zeta), a(t'_{\zeta'})\right)\right\rangle$, the *anomalous* propagator $\mathcal{D}^{\zeta\zeta'}(t,t') = \left\langle T_c\left(a(t_\zeta), a(t'_{\zeta'})\right)\right\rangle$, and $< a, b >=< ab > - < a >< b >$. For squeezed quantum white noise inputs [8]

$$< a^\dagger_{in}(t)a_{in}(t') >\equiv N\,\delta(t-t'), \quad < a_{in}(t)a^\dagger_{in}(t') >\equiv (N+1)\,\delta(t-t'), \tag{4}$$

$$< a_{in}(t)a_{in}(t') >\equiv M\,\delta(t-t'), \quad < a^\dagger_{in}(t)a^\dagger_{in}(t') >\equiv M^*\,\delta(t-t'), \tag{5}$$

Figure 1: The Keldysh contour, $\zeta = 1, -1$ on the upper and lower lip, respectively.

where M determines the degree of squeezing, the input-internal correlations in equations (2) and (3) may be computed using the method of [8]. The result is a closed set of equations for the normal and anomalous propagators, the Dyson equations. In Fourier space $(f(\omega) = \int d\tau e^{i\omega\tau} f(t, t+\tau))$,

$$\left[(i\omega + 1)\underline{\hat{1}} + \underline{\Lambda}\right] \underline{D}(\omega) = i\underline{\sigma}_z - (\underline{\Upsilon})^* \underline{\mathcal{D}}(\omega), \tag{6}$$

$$\left[(i\omega + 1)\underline{\hat{1}} + \underline{\Xi}\right] \underline{\mathcal{D}}(\omega) = \underline{\Upsilon}\, \underline{D}(\omega), \tag{7}$$

where the matrices are defined

$$\underline{D}(\omega) = \begin{pmatrix} D^{++}(\omega) & D^{+-}(\omega) \\ D^{-+}(\omega) & D^{--}(\omega) \end{pmatrix} \tag{8}$$

and similarly for $\underline{\mathcal{D}}$; ω is the frequency offset from resonance in units of the cavity bandwidth, $\sigma_z = diag(1, -1)$, and the damping matrices are given by

$$\underline{\Lambda} = 2\begin{pmatrix} -(N+1) & (N+1) \\ -N & N \end{pmatrix}, \quad \underline{\Xi} = 2\begin{pmatrix} N & -N \\ (N+1) & -(N+1) \end{pmatrix},$$
$$\underline{\Upsilon} = 2Mi\begin{pmatrix} 1 & -1 \\ 1 & -1 \end{pmatrix}. \tag{9}$$

The solutions of (6) and (7) are readily obtained. The Green functions for squeezed white noise are given by

$$D^{++}(\omega) = \frac{\omega(\omega^2+1) - i[(\omega^2+1)(2N+1) - 4|M|^2]}{(\omega^2+1)^2}, \tag{10}$$

$$D^{+-}(\omega) = \frac{-2i[(\omega^2+1)(N+1) - 2|M|^2]}{(\omega^2+1)^2}, \tag{11}$$

$$D^{-+}(\omega) = \frac{-2i[(\omega^2+1)N - 2|M|^2]}{(\omega^2+1)^2}, \tag{12}$$

$$D^{--}(\omega) = \frac{-\omega(\omega^2+1) - i[(\omega^2+1)(2N+1) - 4|M|^2]}{(\omega^2+1)^2}, \tag{13}$$

and

$$\mathcal{D}^{\alpha\beta}(\omega) = \frac{2M}{\omega^2+1} \tag{14}$$

for all α and β. The squeezed input drives anomalous correlations which are otherwise zero.

3. Nonlinear optical parametric oscillator theory

In the OPO a pump photon (mode 2) splits into two subharmonic photons (mode 1) intracavity. The pump is driven by a laser and non-thermal equilibrium is established. Following the approach of [9], we define the following vectors for subharmonic (mode 1) and pump (mode 2):

$$\underline{\alpha} = \begin{pmatrix} \alpha_1 \\ \alpha_2 \end{pmatrix} = \begin{pmatrix} a_1 \\ a_1^\dagger \end{pmatrix}, \quad \underline{\beta} = \begin{pmatrix} \beta_1 \\ \beta_2 \end{pmatrix} = \begin{pmatrix} a_2 \\ a_2^\dagger \end{pmatrix}. \tag{15}$$

The commutation relations are given by $[\alpha_i, \alpha_j] = [\beta_i, \beta_j] = \tau_{ij}$, where $\underline{\tau} = i\underline{\sigma}_y$. Below threshold, the coherent interactions are described by the Hamiltonian, [7].

$$H = i\hbar \frac{p}{2}\left(\alpha_2^2 - \alpha_1^2\right) + i\hbar \frac{1}{2\sqrt{n_{th}}}\left(\alpha_2^2\beta_1 - \alpha_1^2\beta_2\right). \tag{16}$$

The pump amplitude p is normalized so that $p = 1$ at threshold, and n_{th}, the threshold pump photon number, is the system-size parameter in the sense that $n_{th} \to \infty$ reduces equation (16) to the standard linearized theory [7].

We write down quantum Langevin equations for the modes in the presence of external sources $\underline{j}^\zeta$ and $\underline{k}^\zeta$ which perturb the system according to

$$H_J^\zeta = i\hbar\left(j_1^\zeta\,\alpha_1 + j_2^\zeta\,\alpha_2\right) + i\hbar\left(k_1^\zeta\,\beta_1 + k_2^\zeta\,\beta_2\right), \tag{17}$$

and generate equations of motion for the Green functions by functional differentiation: $G_{il}^{\zeta\zeta'}(t,t') = \delta < \alpha_i^\zeta(t) > /\delta j_l^{\zeta'}(t')$, and $H_{il}^{\zeta\zeta'}(t,t') = \delta < \beta_i^\zeta(t) > /\delta k_l^{\zeta'}(t')$ for subharmonic G and pump H, respectively. The correlation function hierarchy that results is given by

$$\begin{aligned}
\left[\frac{d}{dt} + 1\right] G_{ij}^{\zeta\zeta'}(t,t') &= -M_{(ij)}^{\zeta\zeta_1}\, G_{ij}^{\zeta_1\zeta'}(t,t') + \tau_{ij}\,\delta_{\zeta\zeta'}\,\delta(t-t') \\
&\quad + \left(p\,(\sigma_x)_{ik} + \frac{<\beta_{il}^{(\zeta)}(t)>}{\sqrt{n_{th}}}\,(\sigma_x)_{lk}\right) G_{kj}^{\zeta\zeta'}(t,t') \\
&\quad + \frac{\zeta}{\sqrt{n_{th}}}\,\frac{\delta}{\delta\, K_{il}^{(\zeta)}(t)}\,(\sigma_x)_{lk}\, G_{kj}^{\zeta\zeta'}(t,t'), && (18) \\
\left[\frac{d}{dt} + 1\right] H_{ij}^{\zeta\zeta'}(t,t') &= -\gamma\, M_{(ij)}^{\zeta\zeta_1}\, H_{ij}^{\zeta_1\zeta'}(t,t') + \tau_{ij}\,\delta_{\zeta\zeta'}\,\delta(t-t') \\
&\quad - \frac{\zeta}{2\sqrt{n_{th}}}\,\frac{\delta}{\delta\, k_j^{(\zeta')}(t')}\, G_{ii}^{\zeta\zeta}(t,t), && (19)
\end{aligned}$$

where γ is the pump to subharmonic decay rate, time is scaled to the subharmonic lifetime, and the damping matrices $\underline{M}_{1j} = \underline{\Xi}$ and $\underline{M}_{2j} = \underline{\Lambda}$, (we set $N = M = 0$) (9). We have taken the limit $j(t) \to 0$, and used $< \underline{\alpha} >= 0$ in steady state. The Green functions decompose by mode according to

$$
\begin{aligned}
\underline{G}_{11}(t,t') &= \underline{\mathcal{D}}_1(t,t')\underline{\sigma}_z\,, \\
\underline{G}_{21}(t,t') &= i\underline{D}_1(t,t')\underline{\sigma}_z\,, \\
\underline{G}_{12}(t,t') &= i\begin{pmatrix} D_1^{++}(t',t) & D_1^{-+}(t',t) \\ D_1^{+-}(t',t) & D_1^{--}(t',t) \end{pmatrix}\underline{\sigma}_z\,, \\
\underline{G}_{22}(t,t') &= \begin{pmatrix} \mathcal{D}_1^{--}(t,t') & \mathcal{D}_1^{-+}(t,t') \\ \mathcal{D}_1^{+-}(t,t') & \mathcal{D}_1^{++}(t,t') \end{pmatrix}^* \underline{\sigma}_z\,,
\end{aligned} \tag{20}
$$

and similarly for mode two. In addition, we have defined 2x2 matrices

$$
\left(\frac{\delta}{\delta\,\underline{K}(t)}\right)^{(\zeta)}_{ij} = \frac{\delta}{\delta\,k_i^{\zeta}(t)}\delta_{ij}, \quad \langle\underline{\beta}(t)\rangle^{(\zeta)}_{ij} = \left\langle\beta_i^{\zeta}(t)\right\rangle\delta_{ij}. \tag{21}
$$

4. The self-energy and truncation of the hierarchy

Define the inverse subharmonic Green function G^{-1} such that [3]

$$
\int_{-\infty}^{+\infty} dt_2\, \left(G^{-1}\right)^{\zeta\zeta_2}_{ik}(t,t_2) G^{\zeta_2\zeta'}_{kj}(t_2,t') = \delta_{ij}\delta_{\zeta\zeta'}\delta(t-t') \tag{22}
$$

and

$$
\left(G^{-1}\right)^{\zeta\zeta'}_{ij}(t,t') \equiv \left(\hat{G}^{-1}\right)^{\zeta\zeta'}_{ij}(t,t') - \Sigma^{\zeta\zeta'}_{ij}(t,t'), \tag{23}
$$

where $\hat{G}^{-1}$ is the inverse empty-cavity Green function, and Σ is the self-energy. From (18) and (19) the self-energy is given by

$$
\begin{aligned}
\Sigma^{\zeta\zeta_2}_{ik}(t,t_2) &= \tau_{il}^{-1}\left[p(\sigma_x)_{lk} + \frac{1}{\sqrt{n_{th}}}\langle\underline{\beta}(t)\rangle^{(\zeta)}_{lm}(\sigma_x)_{mk}\right]\delta_{\zeta\zeta_2}\delta(t-t_2) \\
&\quad + \frac{\zeta}{\sqrt{n_{th}}}\tau_{il}^{-1}(\sigma_x)_{mn}\int_{-\infty}^{+\infty} dt_1\, G^{\zeta\zeta_1}_{np}(t,t_1)\frac{\delta\,\Sigma^{\zeta_1\zeta_2}_{pk}(t_1,t_2)}{\delta\,K^{(\zeta)}_{lm}(t)}.
\end{aligned} \tag{24}
$$

This enables self-consistent approximations to the self-energy to be generated iteratively [4]. For example, $n_{th} \to \infty$ yields the simple self-energy of linearized theory: the lowest truncation. Finite system-size effects are included by incorporating terms in n_{th}.

A good way to see this is to introduce the vertex function Γ by means of the definition below:

$$
\begin{aligned}
\frac{\delta\,\Sigma^{\zeta_1\zeta_2}_{pk}(t_1,t_2)}{\delta\,K^{(\zeta)}_{lm}(t)} &= \int_{-\infty}^{\infty} dt_3\left[\frac{\delta\,\Sigma^{\zeta_1\zeta_2}_{pk}(t_1,t_2)}{\delta < \beta_q^{\zeta_3}(t_3) >}\,\frac{\delta < \beta_q^{\zeta_3}(t_3) >}{\delta\,k_l^{(\zeta)}(t)}\right]\delta_{lm} \\
&\equiv \int_{-\infty}^{+\infty} dt_3\, \Gamma^{\zeta_1\zeta_2;\zeta_3}_{pk;q}(t_1,t_2;t_3)\, H^{\zeta_3\zeta}_{qm}(t_3,t).
\end{aligned} \tag{25}
$$

Figure 2: The nonlinear part of the exact self-energy is illustrated by the loop, and the vertex function by the heavy dot. In the bare vertex approximation only the lowest order vertex contribution is retained as shown.

The lowest order approximation to Σ, proportional to $< \beta >$, yields the bare vertex

$$\Gamma^{\zeta_1\zeta_2;\zeta_3}_{pk;q}(t_1,t_2;t_3) = \frac{1}{\sqrt{n_{th}}}\tau^{-1}_{pq}(\sigma_x)_{qk}\delta_{\zeta_1\zeta_2}\delta_{\zeta_1\zeta_3}\delta(t_1-t_3)\delta(t_1-t_2), \tag{26}$$

upon functional differentiation.

The bare vertex approximation, (25), truncates the self-energy expansion at order $1/n_{th}$. This is illustrated in Fig. 2. Inserted into the Dyson equations this yields self-consistent solutions to all orders by the summation of classes of terms at each order. This approximation is expected to be good for $n_{th} \gg 1$ but can be systematically improved by the above procedure, vertex renormalisation.

5. Dyson equations

That the elementary dynamical processes are reflected in the structure of the Dyson equations can be seen from Fig. 3. This illustrates that the propagation of subharmonic photons involves three elementary processes: free propagation in the empty cavity, resonant parametric scattering from either the pump mean field, or quantum fluctuations of the pump. We must account for large numbers of the latter processes in the nonperturbative regime. The loop diagram represents polarization of subharmonic photons in which both subharmonic and pump play a dynamical role. This is the origin of the nonlinearity, and the system-size effects, since the polarization diagrams scale with n_{th}. The closed Dyson equations sum classes of these elementary scattering processes to all orders. Using the decomposition (20) and the bare vertex approximation we find the Dyson equations

$$\begin{aligned}
\underline{D}_1(\omega) &= \hat{\underline{D}}_1(\omega) + \hat{\underline{D}}_1(\omega)\,(-\underline{\sigma}_1(\omega))\,\underline{\mathcal{D}}_1(\omega) + \hat{\underline{D}}_1(\omega)\underline{\Pi}_1(\omega)\underline{D}_1(\omega)\ , & (27)\\
\underline{\mathcal{D}}_1(\omega) &= (\hat{\underline{D}}_1(-\omega))^T(-\underline{\sigma}_1'(\omega))\underline{D}_1(\omega) + (\hat{\underline{D}}_1(-\omega))^T(\underline{\Pi}_1(-\omega))^T\underline{\mathcal{D}}_1(\omega) & (28)\\
\underline{D}_2(\omega) &= \hat{\underline{D}}_2(\omega) + \hat{\underline{D}}_2(\omega)\underline{P}_2'(\omega)\underline{\mathcal{D}}_2(\omega) + \hat{\underline{D}}_2(\omega)\underline{\Pi}_2(\omega)\underline{D}_2(\omega)\ , & (29)\\
\underline{\mathcal{D}}_2(\omega) &= (\hat{\underline{D}}_2(-\omega))^T(\underline{\Pi}_2(-\omega))^T\underline{\mathcal{D}}_2(\omega) - (\hat{\underline{D}}_2(-\omega))^T\underline{P}_2(\omega)\underline{D}_2(\omega)\ , & (30)
\end{aligned}$$

where the empty-cavity Green functions are denoted by $\hat{\underline{D}}_k(k=1,2)$. These are

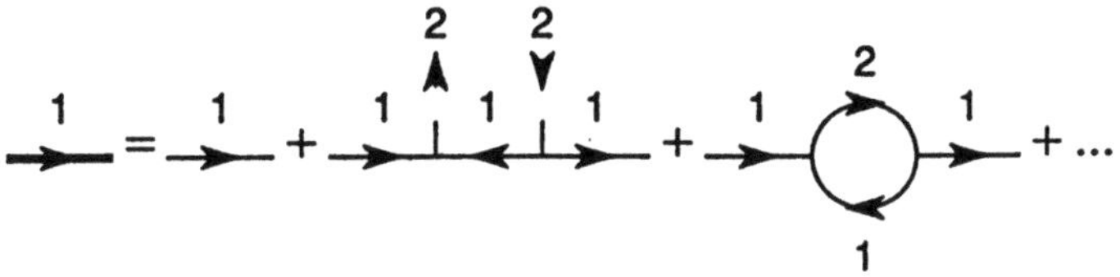

Figure 3: Elementary parametric scattering processes which perturb the subharmonic evolution, see text for a description.

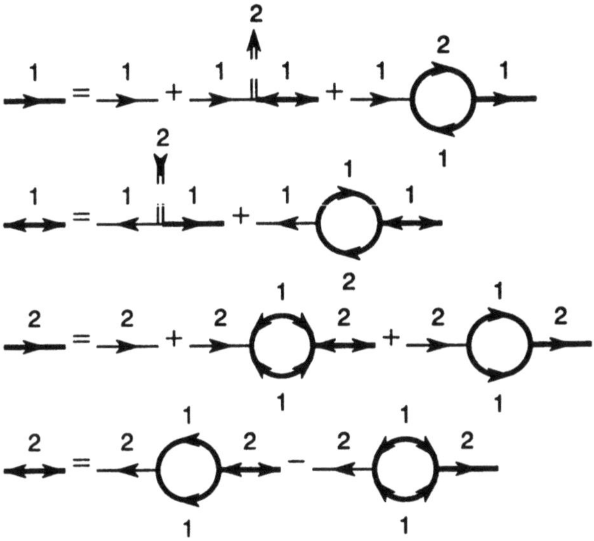

Figure 4: Graphical representation of the Dyson equations (27-30). The meaning of the parts can be read off directly from the equations.

illustrated in Fig. 4. The subharmonic and pump *normal polarization* functions are given by

$$-i\Pi_1^{\alpha\beta}(\omega) = \frac{\alpha\beta}{2\pi n_{th}} \int_{-\infty}^{+\infty} d\omega'\, D_2^{\alpha\beta}(\omega') D_1^{\beta\alpha}(\omega' - \omega)\,, \tag{31}$$

$$-\,i\Pi_2^{\alpha\beta}(\omega) = \frac{\alpha\beta}{4\pi n_{th}} \int_{-\infty}^{+\infty} d\omega'\, D_1^{\alpha\beta}(\omega') D_1^{\alpha\beta}(\omega - \omega'). \tag{32}$$

The subharmonic and pump *anomalous polarization* functions are given by

$$P_1^{\alpha\beta}(\omega) = \frac{\alpha\beta}{2\pi n_{th}} \int_{-\infty}^{+\infty} d\omega'\, \mathcal{D}_1^{\alpha\beta}(\omega') \left(\mathcal{D}_2^{\alpha\beta}(\omega - \omega')\right)^*, \alpha \neq \beta\,, \tag{33}$$

$$P_1^{\alpha\alpha}(\omega) = \frac{\alpha\alpha}{2\pi n_{th}} \int_{-\infty}^{+\infty} d\omega'\, \mathcal{D}_1^{\alpha\alpha}(\omega') \left(\mathcal{D}_2^{-\alpha-\alpha}(\omega - \omega')\right)^*, \tag{34}$$

$$P_2^{\alpha\beta}(\omega) = \frac{\alpha\beta}{4\pi n_{th}} \int_{-\infty}^{+\infty} d\omega' \mathcal{D}_1^{\beta\alpha}(-\omega')\mathcal{D}_1^{\alpha\beta}(\omega-\omega') \ , \tag{35}$$

and

$$\underline{P}_1'(\omega) = \begin{pmatrix} P_1^{--}(\omega) & P_1^{+-}(\omega) \\ P_1^{-+}(\omega) & P_1^{++}(\omega) \end{pmatrix}^* \ , \tag{36}$$

and similarily for $\underline{P}_2'(\omega)$. The σ functions play the role of the pump amplitude renormalized by the anomalous propagator and polarization. This may be compared to condensate renormalization in a partially condensed quantum gas [9].

$$\underline{\sigma}_1(\omega) = \bar{p}^* \underline{\sigma}_z - \underline{P}_1(\omega) \ , \tag{37}$$

$$\underline{\sigma}_1'(\omega) = \bar{p}\underline{\sigma}_z + \underline{P}_1'(\omega) \ , \tag{38}$$

$$\bar{p} = p + \frac{\langle a_2 \rangle}{\sqrt{n_{th}}} = \left(p - \frac{1}{2\gamma\sqrt{n_{th}}} \int_{-\infty}^{+\infty} \frac{d\omega}{2\pi} \mathcal{D}_1^{\alpha\beta}(\omega)\right) \tag{39}$$

for any α and β.

The convolution structure of the photon polarization integrals enables fast Fourier transform methods to be used in an iterative solution of the Dyson equations. We have found this to be an efficient, practical procedure with good convergence properties.

6. Conclusion

We have given an analysis of the Green function approach to quantum optical systems with Markovian damping. The method's advantages are that it enables derivation and self-consistent truncation of correlation function hierarchies, it retains the dynamical structure of the elementary processes under consideration, and provides a practical numerical method for calculation beyond the limits of linearized quantum noise theory.

Acknowledgement

We thank our colleague S. Swain for his collaboration, and NATO for travel support.

References

[1] R.J. Thompson, G. Rempe, and H.J. Kimble Phys. Rev. Lett. **67**, 1132 (1992).

[2] E.M. Lifshitz and L.P. Pitaevskii, *Physical Kinetics*, Course of Theoretical Physics Vol. 10 (Pergamon, Oxford, 1981).

[3] V. Korenman, Ann. Phys. (N.Y.) **39**, 72 (1966).

[4] L.P. Kadanoff and G. Baym, *Quantum Statistical Mechanics*, (Benjamin, California, 1962).

[5] H.R. Zaidi, Can. J. Phys. **66**, 164 (1988); **66**, 854 (1988).

[6] L.V. Keldysh, Zh. Eksp. Teor. Fiz. **47**, 1515 (1964), [Sov. Phys. JETP **20**, 1018 (1965)].

[7] C.J. Mertens, T.A.B. Kennedy, and S. Swain, Phys. Rev. A **48**, 2374(1993); Phys. Rev. Lett. **71**, 2014(1993).

[8] C.W. Gardiner and M.J. Collett, Phys. Rev. A **31**, 3761(1985).

[9] C. De Dominicis and P.C. Martin, J. Math. Phys. **5**, 14 (1964).

Wigner Distributions for Quantum States with Minimum Phase Uncertainty

B. Daeubler, H. Risken, and L. Schoendorff

Abteilung für Theoretische Physik, Universität Ulm, D-89069 Ulm, Germany

We investigate the Wigner distributions for those states which lead to a minimum phase uncertainty for a fixed mean intensity. In contrast to the work of Bandilla, Paul, and Ritze (BPR), who use approximately a variance, we employ the Süssmann measure for minimizing the phase, leading to a phase distribution which is about twice as peaked than the BPR states. Our Wigner–functions show negative values for an appreciable mean intensity. These negative values occur when the curvature of the surface of constant energy in the classical phase space gets positive.

1. Introduction

Various investigations have been devoted to the problem of defining a phase operator for a single-mode optical field. Five years ago Pegg and Barnett [1] have given a nice prescription for a hermitian phase operator, which has the required properties of a phase operator, see for instance [2]. In the limit of a very large number of states their description leads to the continuous phase distribution [see Eq. (6.12) of [3]

$$P(\phi) = \frac{1}{2\pi} \sum_{n,m=0}^{\infty} \langle m|\rho|n\rangle \exp[i(n-m)\phi] \, , \tag{1}$$

where $\langle m|\rho|n\rangle$ are the matrixelements of the density operator in the Fock states. The phase distribution is normalized on the interval $(-\pi, \pi)$. For a pure state $\rho = |\psi\rangle\langle\psi|$, (1) reduces to the phase distribution (4,6) of Ref. [4] or (3.4,3.5) of Ref. [5].

For minimal uncertainty in the phase the distribution should be made as peaked as possible. Without any restrictions the phase distribution can be made as narrow as possible. Therefore we use as additional constraints a fixed mean intensity. The question now arises what is a ideal measure for the phase uncertainty $\Delta\phi$. Bandilla, Paul and Ritze (BPR) have used $(\Delta\phi)^2_{\mathrm{BPR}} = 1 - |\langle\cos\phi\rangle|^2$ [6], which reduces to the

Springer Proceedings in Physics, Vol. 77
Editors: D.F. Walls · J.D. Harvey
Quantum Optics VI

variance, $(\Delta\phi)^2_{\rm BPR} = (\Delta\phi)^2$ for peaked functions, if the 4'th, 6'th, etc. moments are much smaller than the second one. The effect of different measures of phase uncertainties is discussed in [7].

Here we use a measure of the uncertainty $(\Delta\phi)_S$ introduced by Süssmann, which for the normalized distribution (1) is defined by [8],[9]

$$(\Delta\phi)_S = \left[\int_{-\pi}^{\pi} P^2(\phi)\, d\phi \right]^{-1} . \tag{2}$$

This measure is small for high narrow peaks. It does not care very much about the wings of the distribution. On the other hand a small variance is only possible if the wings are very small. For a rectangular distribution of width a and height $1/a$ the Süssmann measure is given by a whereas the square root of the variance is given by $a/\sqrt{12}$. For a Gaussian function with variance σ^2 the Süssmann measure is $2\sqrt{\pi}\,\sigma$. Thus for these two functions, which fall off very rapidly, the Süssmann measure is about three times the value of the square root of the variance. For the Lorentz distribution $\{\pi a[1 + (x/a)^2]\}^{-1}$ (the integration runs from $-\infty$ to $+\infty$) the Süssmann measure is given by $2\pi a$, i.e., by two times the inverse of the peak height. Because the wings of the Lorentz distribution do not fall off very rapidly its variance is infinitely large. As we will see the phase distribution of the Süssmann measure is about twice as peaked and the halfwidth is about half that of the BPR states.

The Wigner-function is calculated here for the first time. It shows some negative values for large intensities. By comparing it to the distribution in coordinate space $\psi(x)$ it can be shown that the potential $V(x)$ of this coordinate space distribution shows some anomalies in its curvature.

The present paper is organized as follows: In Sect. 2 we derive the minimal states for the Süssmann measure and give a simple explicit expression for the phase distribution. In Sect. 3 we discuss the Wigner-function for these states. Finally, in Sect. 4, we summarize our results.

2. Derivation of the Minimal Uncertainty States

It seems reasonable to believe that minimal uncertainty states only exist for pure states. Without loss of generality we further assume that the maximum in the phase distribution is at $\phi = 0$ and that these states are symmetric with respect to the $\phi = 0$ line. Then the coefficients c_n of the pure state in the Fock representation

$$|\psi\rangle = \sum_{n=0}^{\infty} c_n |n\rangle \tag{3}$$

are real. For this special case (1) simplifies to

$$P(\phi) = \frac{1}{2\pi} \left[a_0 + 2 \sum_{\nu=1}^{\infty} a_\nu \cos(\nu\phi) \right] , \tag{4}$$

with real coefficients a_ν, where the a_ν takes the form

$$a_\nu = \sum_{n=0}^{\infty} c_n c_{n+\nu} \ . \tag{5}$$

Inserting (4) into the Süssmann measure (2) we obtain

$$(\Delta\phi)_S = 2\pi / H \ , \tag{6}$$

where H is given by

$$H = a_0^2 + 2 \sum_{\nu=1}^{\infty} a_\nu^2 \ . \tag{7}$$

Obviously H must be maximized under the constraints of normalization and mean intensity I, i.e.,

$$1 = \sum_{n=0}^{\infty} c_n^2 \ , \quad I = \sum_{n=0}^{\infty} n \ c_n^2 \ . \tag{8}$$

Using Lagrange parameters α and β and inserting (5) one thus has to find the free maximum of

$$\begin{aligned} F(c_0, c_1, c_2, \ldots, \alpha, \beta) \ &= \ \left[\sum_{n=0}^{\infty} c_n^2 \right]^2 + 2 \sum_{\nu=1}^{\infty} \left[\sum_{n=0}^{\infty} c_n c_{n+\nu} \right]^2 \\ &+ \ \frac{\alpha}{2} \left[I - \sum_{n=0}^{\infty} n \, c_n^2 \right] + \frac{\beta}{2} \left[1 - \sum_{n=0}^{\infty} c_n^2 \right] \ . \end{aligned} \tag{9}$$

To find the extrema of $F(c_0, ..., \alpha, \beta)$ one has to solve the infinite set of equations

$$\partial F / \partial c_k = 0 \quad \text{for} \quad k = 0, 1, 2, \ldots \ . \tag{10}$$

It is easy to check that the coefficients

$$c_n = \delta \gamma^n \quad \text{with} \quad \delta = 1/\sqrt{1+I} \quad \text{and} \quad \gamma = \sqrt{I} \ / \sqrt{1+I} \tag{11}$$

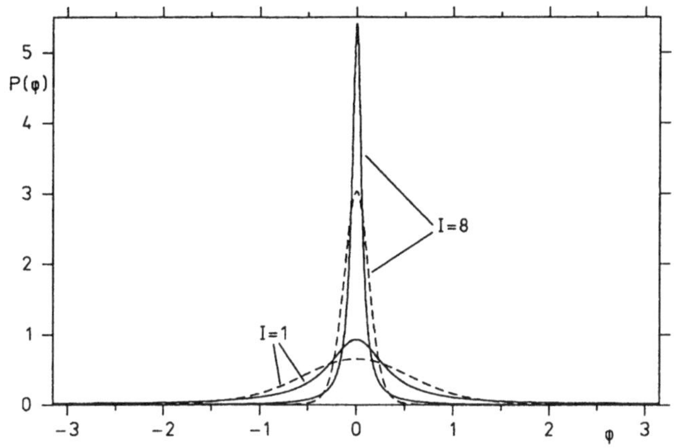

Figure 1: A comparison of our phase distribution (14) (full lines) with the BPR minimal phase distribution (dashed lines) for $I = 1$ and $I = 8$.

solve the infinite set of equations (9) with the Lagrange parameters given by

$$\alpha = 4\ , \quad \beta = 4 + 4I\ . \tag{12}$$

Furthermore the coefficients (11) also satisfy the constraints (8). Thus (3) with (11) are the proper minimal states of the phase uncertainty for the Süssmann measure [8]. The states (3,11) have been termed "coherent phase states" by Shapiro et al. [10]. Sometimes they are also called "thermal phase states" because the photon number measurement $p_n = c_n^2$ on these states leads exactly to the Bose–Einstein distribution for thermal light.

Using the states (3,11) it is now an easy matter to obtain the phase distribution $P(\phi)$. First, the coefficients (5) take the form

$$a_\nu = \gamma^\nu = (\sqrt{I}\ /\sqrt{1+I}\,)^\nu \tag{13}$$

as may be easily seen. Inserting these coefficients in (4) we can sum up the infinite series exactly leading to

$$P(\phi) = \frac{1}{2\pi}\ \frac{1}{1 + 2I - 2\sqrt{I}\ \sqrt{1+I}\ \cos\phi}\ . \tag{14}$$

In Fig. 1 we have shown some values of the phase distribution. In addition we have also added those of the BPR states. As can be seen clearly both distributions only agree for small I, otherwise they differ. The maximum of our distribution is about twice as large as the one of BPR. Furthermore the halfwidth of our distribution is only 1/2 of the one for the BPR states. Thus by looking at the plots in particular for large I our distributions are more peaked than those of BPR.

3. Wigner-function

The Wigner-function can best be obtained from the $\psi(x)$ function. Using the eigenfunction

$$\psi_n(x) = \frac{1}{\sqrt{2^n n! \sqrt{\pi}}} H_n(x) e^{-x^2/2} \tag{15}$$

we obtain

$$\psi(x) = \delta \sum_{n=0}^{\infty} \gamma^n \psi_n(x) . \tag{16}$$

The Wigner-function can be evaluated by the expression

$$W(\alpha) = \frac{1}{\sqrt{2\pi}} \int_{-\infty}^{\infty} \psi^* \left(\sqrt{2}\,\alpha_r + \frac{\xi_r}{\sqrt{2}}\right) \psi \left(\sqrt{2}\,\alpha_r - \frac{\xi_r}{\sqrt{2}}\right) e^{i\alpha_i 2\xi_r} d2\xi_r . \tag{17}$$

For large I the γ value is nearly 1, which means that lots of functions have to be taken into account. Because of the bad convergence properties of the sum for $I = 64$, for instance, 1500 functions have to be taken into account with an accuracy of at least 35 decimals. In Fig. 2 shows the Wigner-function for $I = 2$, $I = 4$, $I = 16$, and $I = 64$. It is worth mentioning that the Q-function has zeros at certain points, i.e. near $\gamma\alpha \approx 1.7 \pm i3.2$, $2.9 \pm i4.1$, at least for our numerical precision. Thus the Wigner function has always negative values for $I \neq 0$ but pronounced negative values show up for $I \approx 4$ and for larger values.

Like the Q-function, which had to be maximal for $\sqrt{I/2}$ see [9] for large I, the maximum value for the Wigner-function occurs near but nevertheless larger than this value $\sqrt{I/2}$.

In Fig. 3 we have plotted the functions $\psi(x)$ together with their potentials

$$V(x) = \frac{1}{\psi(x)} \frac{d^2\psi(x)}{dx^2} . \tag{18}$$

The dominant behavior of $V(x)$ for large $|x|$ can approximated as

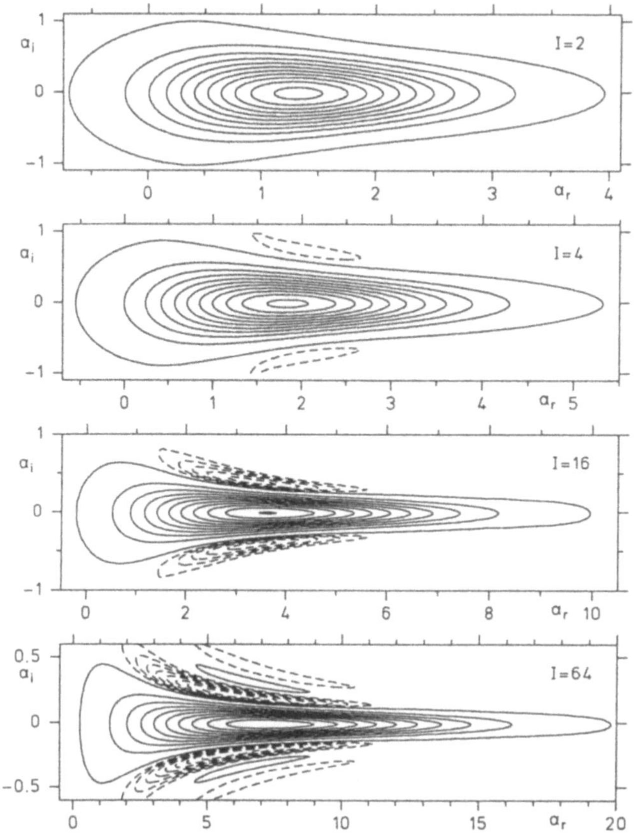

Figure 2: Contour lines of the Wigner–function (17) for $I = 2$, $I = 4$, $I = 16$, and $I = 64$. The dashed lines are negative values.

$$V(x) \approx \begin{cases} \frac{1}{2}\left(\frac{1-\gamma^2}{1+\gamma^2}\right)^2 x^2 , & x \gg 1 \\ \frac{1}{2}x^2 , & x \ll -1 . \end{cases} \tag{19}$$

As follows from a work by Berry [11] there is a deep relation between the Wigner function and classical trajectories in the phase space. It can be shown, that as one moves off the surface of constant energy, resp. the trajectory corresponding to our phase state, the behavior of the Wigner function can be approximated by an Airy function. Thus the Wigner function behaves oscillatory in one direction and exponential in the other. In Fig. 4 we have plotted the contour lines of constant energy

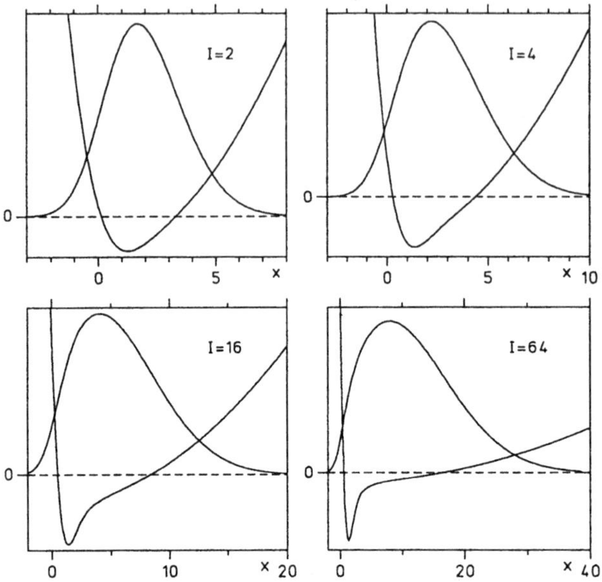

Figure 3: Wave–functions (16) and potentials (18) for $I = 2$, $I = 4$, $I = 16$, and $I = 64$.

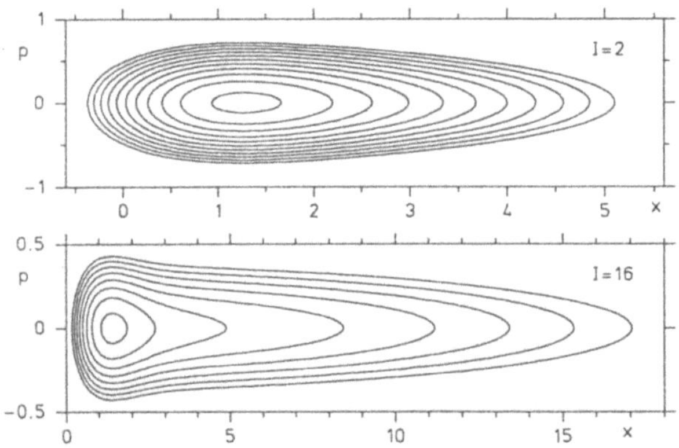

Figure 4: Energie surfaces $p^2/2 + V(x) = \text{const.}$ in the classical phase space for $I = 2$, and $I = 16$.

in the classical phase space. As can be seen the trajectories for large intensities have a clear tendency to change the sign of their curvature in the region where the Wigner function oscillates in the outward direction [11]. This is due to the change of the curvature of the potential for small x.

4. Summary

We have found an analytical solutions of the form $|\psi\rangle \propto \sum \gamma^n |n\rangle$ by maximizing the phase uncertainty of the Süssmann measure for a given norm and a given mean intensity. These states are the thermal phase states already discussed by Shapiro et al. [10]. Though they lead to the same photon probability distribution as the thermal states, nevertheless they are pure states with a sharp phase distribution, in contrast to the ordinary mixed thermal states with a flat phase distribution. Furthermore we have shown that the phase distribution of the thermal phase states is narrower and more peaked than the distribution following from the phase uncertainty states of Bandilla et al. [6]. One of the main aims was to discuss the Wigner–function. It is shown that the Wigner-function for large mean intensities has regions with pronounced oscillatory behavior. This effect is related to the detailed behavior of the potential, i.e. higher derivatives for small x resp. the change of the sign of curvature of the classical trajectories in phase space.

Acknowledgments

We would like to thank Arno Bandilla, Iwo Bialynicki–Birula, Matthias Freyberger, and Wolfgang Schleich for stimulating discussions.

References

1. D. T. Pegg and S. M. Barnett: Europhys. Lett. **6**, 483 (1988).

2. P. Carruthers and M. M. Nieto: Rev. Mod. Phys. **40**, 411 (1968).

3. J. Eiselt and H. Risken: Phys. Rev. A **43**, 346 (1991).

4. W. Schleich, R. J. Horowicz, and S. Varro: Phys. Rev. A **40**, 7405 (1989).

5. W. Schleich, R. J. Horowicz, and S. Varro: in: *Quantum Optics V*, edited by J. D. Harvey and D. F. Walls, Springer, Berlin, 1989, page 133.

6. A. Bandilla, H. Paul, and H. H. Ritze: Quantum Opt. **3**, 267 (1991).

7. I. Bialynicki–Birula, M. Freyberger, and W. Schleich: Physica Scripta **T48**, 113, (1993).

8. G. Süssmann: In: Sudarshan Festschrift, Conference Proceedings, Austin, Texas, Sept. 1991, to be published by World Scientific Singapore.

9. B. Daeubler, Ch. Miller, H. Risken, and L. Schoendorff: Physica Scripta **T48**, 119, (1993).

10. J. H. Shapiro, S. R. Shepard, and N. C. Wong: in *Coherence and Quantum Optics VI*, edited by J. H. Eberly, L. Mandel, and E. Wolf, Plenum Press, New York, 1990, page 1077.

11. M.V. Berry: Proc. R. Soc. Lond. A **423**, 219 (1989).

Spatial and Temporal Optical Field Reconstruction Using Phase-Space Tomography

M.G. Raymer, M. Beck, and D. McAlister

Department of Physics and Chemical Physics Institute,
University of Oregon, Eugene, OR 97403, USA

Abstract. Phase-space tomography has recently been developed for the experimental determination of the amplitude and phase structure of a multimode optical field in either the space or time domain. The methods can be applied to the precise characterization of optical fields, in the cases of fully or partially coherent sources.

1. Introduction

While it is relatively easy to measure the mean intensity distribution of an optical field, often there is hard-to-get information contained in the phase structure of the field. For example, a great deal of effort has gone into trying to fully characterize light from ultrashort-pulse lasers. Another problem of general interest is "phase retrieval," that is, the determination of the transverse phase structure of an optical field.

This paper reviews new concepts for the characterization of light fields, including spatial and temporal phase structure. The methods can be applied to multimode optical fields, in the cases of fully or partially coherent sources. The methods are based on "phase-space tomography," using the Wigner distribution function as a primary analysis tool. The phase structure is obtained purely from intensity measurements, using no reference fields. An advantage of the tomographic method is that the data analysis requires no deconvolution, and is noniterative, in contrast to established phase-retrieval techniques.[1] Furthermore, it is linear in the signal field, and so can be applied to weak fields.

Both "classical" and "quantum" fields can be treated in this manner. The one-or-two-mode quantum case has been reviewed previously, and so will be treated only briefly here.[2,3] In the classical case the fields may be multimode (transverse or temporal), and the two-point (or two-time), complex-field correlation function is obtained. This case will be discussed in detail.

2. Wigner Representation of Wave Fields

The Wigner function provides a way of simultaneously representing information about two mutually incompatible variables. In quantum physics the incompatibility arises through noncommutation of conjugate variables. In classical physics it arises through Fourier-transform uncertainty relations.

Define a wave field $\psi(\chi)$, where the amplitude ψ may be classical electric-field strength (or, where appropriate, quantum mechanical wave function). The (dimensionless) independent variable χ can be transverse position (in the classical, spatial case), or frequency (in the classical, time-frequency case), or quadrature amplitude (in the quantum, single-mode case).[4] In the transverse spatial case we consider only quasi-monochromatic fields. In the time-frequency case we consider only quasi-plane-wave fields. In the case of partial coherence, where an ensemble of similarly prepared systems is under consideration, a useful characterization is provided in the classical case by the two-point correlation function, also called the mutual intensity in the spatial case, which is equal to

$$\Gamma(\chi,\chi')=\langle\psi(\chi)\psi^*(\chi')\rangle \ , \tag{1}$$

where the brackets indicate an average over the set of realizations of the functions $\psi(\chi)$. In the case of a field obeying gaussian statistics this function provides its complete statistical characterization.[5,6]
The wave field may also be represented in the Fourier space in terms of the (dimensionless) conjugate variable η, by

$$\tilde{\psi}(\eta)=(2\pi)^{-1/2}\int_{-\infty}^{\infty}\psi(\chi)\exp(i\eta\chi)d\chi \quad . \tag{2}$$

A more general representation can be obtained by considering the χ,η variables to make up a quasi-phase space, and defining a distribution function on this space. We might want this distribution to correspond to an intuitive weighting function, or joint distribution, for simultaneous presence of the magnitudes of χ and η. Unfortunately for this naive viewpoint, the quasi-distribution can sometimes take on negative values, due to the intrinsic incompatibility of the two variables. Only after convolving the quasi-distribution with an "instrument resolution function" does the resulting distribution have positive values guaranteed.[7,8]

The Wigner quasi-distribution is defined as [9,10]

$$W(\chi,\eta)=\frac{1}{\pi}\int_{-\infty}^{\infty}\langle\psi(\chi+\chi')\psi^*(\chi-\chi')\rangle\exp(-i2\chi'\eta)d\chi' \quad . \tag{3}$$

It has a one-to-one relation to the mutual intensity function (density matrix). This distribution has the nice property that its standard deviations of χ and η obey the inequality $\Delta\chi\Delta\eta\geq 1/2$, for any possible wave field (state). So in the χ,η plane the "stuff" represented by the Wigner function, "W-stuff, is more or less like an incompressible liquid."[11] It cannot be squeezed into arbitrary shape. This property embodies the uncertainty principle, both in classical [12] and quantum [9] cases.

Even though $W(\chi,\eta)$ can be negative, and so is not a real joint probability, its integrated (marginal) distributions are always positive, and are simply related to the wave functions:

$$P_\chi(\chi)=\int_{-\infty}^{\infty}W(\chi,\eta)d\eta =\langle|\psi(\chi)|^2\rangle\geq 0 \quad , \tag{4a}$$

$$P_\eta(\eta)=\int_{-\infty}^{\infty}W(\chi,\eta)d\chi =\langle|\tilde{\psi}(\eta)|^2\rangle \geq 0 \quad . \tag{4b}$$

The marginal distributions for single variables are directly measurable distributions. In special cases, when the wave fields are nonstochastic (pure states in QM), the knowledge of the marginals is sufficient to determine uniquely (using an iterative numerical algorithm [13,1]) the wave field.

In the case of stochastic fields, a more powerful approach is needed, and this is provided by Wigner-space rotations and "fractional Fourier transforms."

3. Wigner-Space Rotations and Fractional-Order Fourier Transforms [14]

If we define a new pair of "rotated" independent variables according to $\chi_\theta = \chi\cos\theta + \eta\sin\theta$, $\eta_\theta = -\chi\sin\theta + \eta\cos\theta$, then the Wigner function can be expressed in these variables, and a set of marginal distributions can be found for these new variables:

$$P_\theta(\chi_\theta) = \int_{-\infty}^{\infty} W(\chi_\theta\cos\theta - \eta_\theta\sin\theta, \chi_\theta\sin\theta + \eta_\theta\cos\theta)\, d\eta_\theta \quad . \tag{5}$$

The basis of the method is to find ways to measure a sufficiently large set of these distributions to allow the Wigner function to be determined by inverting Eq.(5).

This goal is facilitated by finding a new transformed wave field $\psi_\theta(\chi_\theta)$ which is related to the Wigner function by

$$\begin{aligned} &W(\chi_\theta\cos\theta - \eta_\theta\sin\theta, \chi_\theta\sin\theta + \eta_\theta\cos\theta) = \\ &\quad \frac{1}{\pi}\int_{-\infty}^{\infty} \langle\psi_\theta(\chi_\theta + \chi_\theta')\psi_\theta^*(\chi_\theta - \chi_\theta')\rangle \exp(-i2\chi_\theta'\eta_\theta)d\chi_\theta' . \end{aligned} \tag{6}$$

It is straightforward to verify that this is achieved by the integral

$$\psi_\theta(\chi_\theta) = A_\theta \int_{-\infty}^{\infty} \exp\left[-i\left(\chi'\chi_\theta\csc\theta - \tfrac{1}{2}(\chi'^2 + \chi_\theta^2)\cot\theta\right)\right]\psi(\chi')d\chi' \quad , \tag{7}$$

where $A_\theta = (2\pi i\,|\sin\theta\,|)^{-1/2}$. It is clear that for $\theta = \pi/2$ this is proportional to the Fourier transform of $\psi(\chi)$. Also, for $\theta = 0$ it is equal to $\psi(\chi)$. For θ between 0 and $\pi/2$, Eq.(7) in some sense interpolates between $\psi(\chi)$ and its Fourier transform. For this reason it has been referred to as the "fractional-order Fourier transform" (FOFT).[15,16] This operation corresponds to rotation in the Wigner phase space.[16] (In the case that $\psi(\chi)$ is a QM wave function, Eq.(7) corresponds to a change of basis, or representation, from the χ variable to the χ_θ variable.[17]

The generalized marginal distributions can be expressed more simply as

$$P_\theta(\chi_\theta) = \langle|\,\psi_\theta(\chi_\theta)\,|^2\rangle. \tag{8}$$

This gives a clue about how to measure the marginal distributions -- find a physical mechanism that effects the FOFT for arbitrary "rotation" angle. We will return to this in Sec. 5.

4. Phase-Space Tomography on the Wigner Function

It was first pointed out by Bertrand and Bertrand that, because the marginal distributions in Eq.(8) amount to a set of projection integrals of the Wigner function along parallel lines for a range of angles, it should be possible to use the well-known inverse-Radon transform to reconstruct $W(\chi,\eta)$ from knowledge of a sufficiently large set of the $P_\theta(\chi_\theta)$ functions.[18] This is illustrated in Fig.1. The inverse Radon transform is a linear, integral transform, and is well studied in medical imaging.[19] The filtered-back projection algorithm reconstructs uniquely the Wigner function to within a filtering bandwidth imposed by the necessarily finite number of marginal distributions measured.

Vogel and Risken first proposed a concrete method for measuring a set of marginal distributions, in the case of quadrature amplitudes for a single-mode field.[20] This proposal, which used balanced optical homodyne detection, was realized experimentally in Refs.[21,22] . A method for measuring two-mode quantum fields has also been proposed.[2] The present discussion reviews recent proposals for measuring marginal distributions in the cases of (classical) multimode fields, in either time or spatial domains.

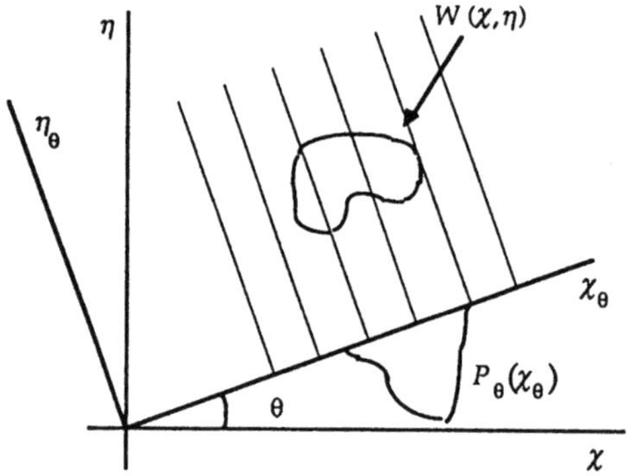

Fig.1 Projection integrals of Wigner function (shown as contour) gives the observed distributions.

5. Spatial Field Reconstruction using Phase-Space Tomography [23]

Notice the remarkable fact that the FOFT in Eq.(7) is identical in form to the well-known Fresnel integral that describes spatial diffraction (in one transverse dimension) through an optical system with quadratic-phase elements. If we identify $\chi = x/x_0$, where x is transverse position and x_0 is a characteristic scaling length, then the transform variable η is identified as $\eta = k_x x_0$, where k_x is the transverse propagation constant (momentum). The total propagation constant of the quasi-monochromatic wave is k. Refer to Fig.2, which shows the wave front to be reconstructed, $E(x,0)$, located in the 0 plane. After propagating through a lens with focal length f in the plane at d, the field in the plane at D is [5]

$$E(x,D) = C(x)\int_{-\infty}^{\infty} \exp\left[ik\left(\frac{xx'}{L} - \frac{x^2 + x'^2}{2R}\right)\right]E(x',0)dx' \quad , \tag{9}$$

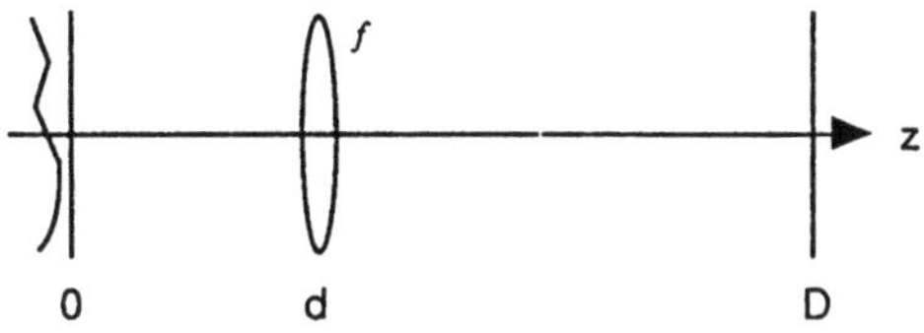

Fig.2 Set up for diffraction measurements for phase-space tomography.

where the curvature radius is given by $R = R_0 + d$, where

$$\frac{1}{R_0} = \frac{1}{D-d} - \frac{1}{f} \quad , \tag{10}$$

and the effective propagation length is given by $L = (D-d)(1 + d/R_0)$. The prefactor is a constant times a phase function, $C(x) = C_0 \exp[ih(x)]$, and is not important when measuring only normalized intensity distributions.

The connection between Fresnel diffraction and the FOFT is rather intuitive, since it is well known that the Fresnel integral interpolates between the original function at the input plane and its Fourier transform in the focal plane of the lens ($D = d + f$). Let us now see how this interpolation corresponds to rotation in the Wigner space, thereby providing a new geometrical view of diffraction. [16]

Compare Eqs.(7) and (9), with x' replaced by $\chi' x_0$. Define a new variable by $\tilde{\chi}_\theta = -\chi_\theta (L \csc\theta / k x_0^2)$, which is a scaled version of χ_θ. Then equating coefficients of χ' shows that we should identify $\cot\theta = -k x_0^2 / R$,as determining the phase-space rotation angle θ. This angle is controlled by varying R, which depends on the position and focal length of the lens, and on the position of the observation plane. With this scaling, then, Eq.(9) is identical to Eq.(7) to within an unimportant x-dependent phase factor. Measuring the intensity $\langle | E(x,D) |^2 \rangle$ for a number of different values of R (equivalently θ) thus determines $P_\theta(\chi_\theta)$, according to Eq.(8). This, in turn, is equal to the projection integral through the Wigner function, as in Eq.(5). Using the inverse Radon transform, then, one can reconstruct this Wigner function. Equation (3) can then be inverted using a Fourier transform to yield the complex correlation function $\langle \psi(\chi) \psi^*(\chi') \rangle$.

In the case that the field is spatially coherent, the mutual intensity factorizes, and the complex wave field can be recovered (up to an unimportant, constant phase). It is interesting that this method uses only refractive optics (lenses), in contrast to older methods which use interference with a coherent reference wave. Other noninterferometric methods have recently been developed for coherent fields.[24] Such methods have not been developed for the case of partial coherence.

The spatial field reconstruction just discussed can be generalized to allow the reconstruction of quasi-monochromatic fields with 2-D transverse structure, by using "4-D Phase-Space Tomography."[23] The spatial Wigner function $W(\vec{r}, \vec{k})$ has arguments which are 2-D transverse position $\vec{r} = (x_1, x_2)$ and 2-D transverse spatial frequency $\vec{k}$ in a plane normal to the propagation direction.[10] It is uniquely related to the two-point spatial correlation function, $\Gamma(\vec{r}, \vec{r}') = \langle E(\vec{r}) E^*(\vec{r}') \rangle$, by a generalization of Eq.(3).

Figure 2 is generalized to have a pair of cylindrical lenses, with independently variable focal lengths and positions, and oriented with orthogonal axes. The intensity

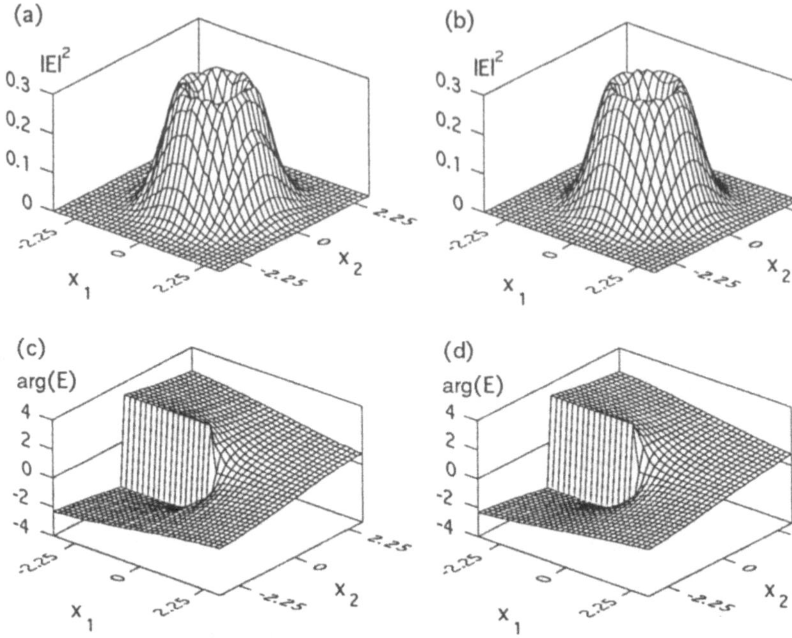

Fig.3 Plots of a vortex wave field, $E(x_1, x_2) \propto (x_1 + i x_2)\exp[-(x_1^2 + x_2^2)/1.3]$. (a) and (b) Amplitude squared; (c) and (d) Phase. (b) and (d) Initial analytic form. (a) and (c) Reconstructed using phase-space tomography. Note 2π phase jump at branch cut. The amplitude goes to zero at the origin, and so is a good test case for the method.

distribution at a position beyond the lenses is recorded for many different combinations of lens parameters. Each observed distribution corresponds to a 4-D projection of the Wigner function, generalizing the 2-D projection shown in Fig.1. The full set of distributions can then be numerically inverted using a series of inverse Radon transforms. The procedure corresponds to doing computed tomography in the 4-D phase space defined by two transverse coordinates and their conjugate spatial frequencies (or propagation constants). The transformation on the field as it propagates from the source plane to the detection plane is given by a 2-D Fresnel integral, generalizing Eq.(9).

As an example, we have numerically simulated the reconstruction of a coherent wave field with a known analytic form, which is plotted in Fig.3, which shows both the amplitude and phase structure. From this function we analytically determined the 961 measured intensity distributions that would be recorded by the camera for a well defined set of lens parameters. We treated these distributions numerically as if they were experimentally measured data files, and reconstructed the 4-D Wigner function using the filtered back-projection algorithm for the inverse Radon transform. From the reconstructed Wigner function we obtained the correlation function $\Gamma(x_1, x_2, 0, x_1', x_2', 0)$, from which we obtained the reconstructed wave field, shown.

6. FOFTs and Chronocyclic Phase-Space Tomography for Measuring Temporal Structure of Optical Pulses [25]

If the field of interest is broad band temporally, but diffraction-limited spatially, then the two-time correlation function $\langle E(t)E^*(t')\rangle$ carries useful information about its temporal structure. This function can be measured by phase-space tomography if we find a way to implement the FOFT "rotation" given by Eq.(10). If in Eq.(1) we identify $\chi = \omega/\omega_0$, where ω is frequency and ω_0 is a characteristic scaling frequency, then the transform variable η in Eq.(2) is identified as scaled time, $\eta = t\omega_0$. Then the FOFT in Eq.(10) is identical in form to the well-known convolution integral that describes pulse propagation through an optical system with time- or frequency-quadratic-phase elements,[26]

Consider an ensemble of short optical pulses passing through a grating-pair based dispersive delay line with dispersion ϕ_ω'', and then through a temporal phase modulator. If each pulse passes through the phase modulator during a maximum of the sinusoidal modulation, it sees approximately a quadratically varying phase shift with rate Φ_t''. Then the output pulse from this system is

$$\tilde{E}_{out}(\omega) = (i/2\pi\Phi_t'')^{1/2}\exp(-i\omega^2/2\Phi_t'')\times \int_{-\infty}^{\infty}\exp\left\{-i[-\omega\omega'/\Phi_t''+\tfrac{1}{2}(\phi_\omega''+1/\Phi_t'')\omega'^2]\right\}\tilde{E}(\omega')\,d\omega'. \tag{11}$$

This transform is identical in form to Eq.(7) for the FOFT, to within an unimportant phase function. This shows how to achieve the FOFT in the time domain, and also provides a different way to visualize temporal pulse propagation as a rotation in the Wigner phase space. For a particular choice of DDL dispersion and temporal phase chirp, a measurement of the spectrum $\langle|\tilde{E}_{out}(\omega)|^2\rangle$ of the pulses, after proper scaling of the frequency variable, yields the average time-dependent intensity of the input pulse, $\langle|\tilde{E}_{out}(\omega)|^2\rangle$.[27,28] In the present language, this works because the quadratic-phase system performs the FOFT in Eq.(7), rotating the Wigner-space variables by an angle $\theta = \pi/2$, thereby converting time into frequency.

In the proposed technique, one makes measurements of the spectra not just at one setting of the dispersion and phase modulation, but at many different combinations of settings. By performing computer-assisted tomography on the set of measured spectra, the time-frequency domain ("chronocyclic"[12]) Wigner function $W(\omega,t)$ of the pulse can be reconstructed, using the arguments discussed above. Details are given in Beck, et al.[25] In the case of a reproducible pulse, the Wigner function can be used to determine the amplitude and phase of the pulse, or the complex field $E(t)$; otherwise two-time correlation function $\langle E(t)E^*(t')\rangle$ is determined.

The main advantage of the technique is that the data inversion does not require deconvolution or iterative algorithms, which are subject to error or ambiguity when the signal-to-noise ratio is not high enough. The temporal imaging system which will be used in these measurements also has the advantage that the detected field is a linear functional of the input field, and thus is not subject to the same restrictions on input pulse energy and wavelength as schemes that use nonlinear devices.

In Fig. 4 we show a numerical simulation of the reconstruction of a (nonstochastic) double-peaked pulse with a quadratic phase structure.[25] It can be seen that the technique works quite well for reconstructing the amplitude and phase of this complicated pulse.

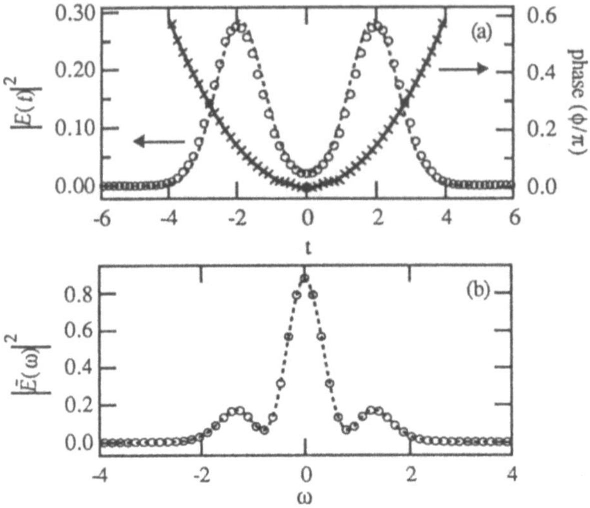

Fig. 4. (a) The pulse intensity $|E(t)|^2$ is plotted vs. time t for the input pulse (dashed line) and the reconstructed pulse (circles). Also plotted is the phase of the input pulse (solid line) and of the reconstructed pulse (X's). (b) The pulse-energy spectrum $|\tilde{E}(\omega)|^2$ is plotted vs. frequency ω for the original pulse (dashed line) and the reconstructed pulse (circles).

Acknowledgments. We thank J. Cooper, A. Faridani, and D. Smithey for helpful contributions to this work. Supported by the National Science Foundation.

References

1. For review of phase retrieval see *Inverse Source Problems in Optics*, H. P. Baltes, ed., (Springer, Heidelberg, 1978); and J. R. Fienup, Appl. Opt. **21**, 2758 (1982).
2. "Measurement of the Wigner function in quantum optics," M. G. Raymer, D. T. Smithey, M. Beck, M. Anderson, and D. McAlister, Proceedings of 3rd International Wigner Symposium (Oxford, Sep. 1993, to be published 1994).
3. "Quantum states and number-phase uncertainty relations measured by optical homodyne tomography," M. G. Raymer, D. T. Smithey, M. Beck, and J. Cooper, to appear in Acta Physica Polonica (1994).
4. χ can also be transverse position in the case of a quantum mechanical, masssive particle. See refs. 2 and 23.
5. J. W. Goodman, *Statistical Optics* (Wiley, New York, 1985).
6. In the QM case, the analogous quantity is the density matrix, also given by Eq.(1). It provides a complete description for any one-particle state (ignoring spin).
7. K.-H. Brenner and K. Wodkiewicz, Opt. Commun. **43**, 103 (1982); Y. Lai and H. A. Haus, Quantum Opt. **1**, 99 (1989);

8. J. H. Eberly and K. Wodkiewicz, J. Opt. Soc. Am. **67**, 1252 (1977).
9. M. Hillery, R. F. O'Connell, M. O. Scully and E. P. Wigner, Phys. Rep. 106, 121 (1984), for the QM case.
10. M. J. Bastiaans, J. Opt. Soc. Am. A, **3**, 1227 (1986), for the classical case.
11. Adolf W. Lohmann (private communication, 1994).
12. J. Paye, IEEE J. Quant. Electron. **QE-28**, 2262 (1992).
13. R. W. Gerchberg and W. O. Saxon, Optik **35**, 237 (1972).
14. The results in the present section were developed in different contexts and published independently in refs.16, 18, and in 2 and 25.
15. V. Namias, J. Inst. Math Appl. **25**, 241 (1980); H. M. Ozaktas and D. Mendlovic, Opt. Commun. **101**, 163 (1993).
16. A. W. Lohmann, J. Opt. Soc. Am. A, **10**, 2181 (1993).
17. B. Yurke, W. Schleich and D. F. Walls, Phys. Rev. A **42**, 1703 (1990).
18. J. Bertrand and P. Bertrand, Found. Phys. **17**, 397 (1987); P. Bertrand and J. Bertrand, Rech. Aérosp. **1985-5**, 277 (1985).
19. Herman, G. T., *Image reconstruction from projections : the fundamentals of computerized tomography* (Academic Press, New York, 1980).
20. K. Vogel and H. Risken, Phys. Rev. A **40**, 2847 (1989).
21. D. T. Smithey, M. Beck, M. G. Raymer, and A. Faridani, Phys. Rev. Lett. **70**, 1244 (1993).
22. D. T. Smithey, M. Beck, J. Cooper, M. G. Raymer, and A. Faridani, Physica Scripta, **T48**, 35 (1993).
23. "Complex wave-field reconstruction using phase-space tomography," M. G. Raymer, M. Beck, and D. McAlister, Phys. Rev. Lett., **72**, 1137, (1994).
24. N. Streibl, Opt. Commun. **49**, 6 (1984); K. Ichikawa, A. W. Lohmann, and M. Takeda, Appl. Opt. **27**, 3433 (1988).
25. M. Beck, M. G. Raymer, I. A. Walmsley and V. Wong, Opt. Lett. **18**, 2041 (1993).
26. E. B. Treacy, IEEE J. Quant. Electron. **QE-5**, 454 (1969).
27. B. H. Kolner and M. Nazarathy, Opt. Lett. **14**, 630, (1989).
28. M. T. Kauffman, A. A. Godil, B. A. Auld, W. C. Banyai and D. M. Bloom, Electron. Lett. **29**, 268 (1993); M. T. Kauffman, W. C. Banyai and D. M. Bloom, CLEO '93 (May, 1993, Baltimore, MD), paper CPD33 .

Driving Atoms with Light of Arbitrary Statistics

C.W. Gardiner[1] *and A. Parkins*[2]

[1]Physics Department, University of Waikato, Hamilton, New Zealand
[2]Fakultät für Physik, Universität Konstanz, D-7834 Konstanz, Germany

1. The Problem

Since it is now possible to produce reliable sources of squeezed, antibunched and sub-Poissonian light, it is relevant to consider what effects these might have when impinged on quantum systems. Two questions immediately arise:

- Given a source of non-classical light, how does one work out the effect of coupling it to a quantum system?
- In a theoretical context, how does one specify a non-classical light beam? With thermal or squeezed light beams this is normally done by assuming that they are Gaussian, which is theoretically true, though in principle practical squeezed beams may deviate from being exactly Gaussian. For Gaussian beams the means and second order correlation functions specify all higher correlation functions, so that one need only specify these to specify the beam exactly. There is no such simple method for non-Gaussian beams, such as antibunched beams.

2. The Solution

The first problem has been solved independently by three authors. Kolobov and Sokolov [1] gave the appropiate formula in 1987, though not in this context. Carmichael [2] and Gardiner [3] gave independent derivations last year. All derivations are quite different, though I think my own is simplest.

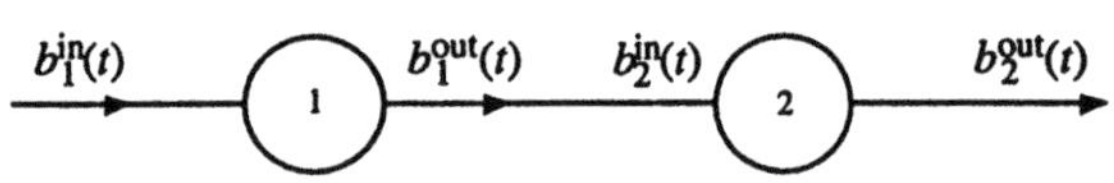

FIG. 1. Schematic diagram indicating how system 2 is driven by the output from system 1.

The basic idea is very simple. We consider a quantum system which can be decomposed into two subsystems, as in Fig.1. Using the notation of [4, 5], a driving field $b_{in}(1, t)$ drives the first system, and gives rise to an output $b_{out}(1, t)$ which, after a propagation delay τ, becomes the input field $b_{in}(2, t)$ to the second system: let a_1, a_2 be operators

Springer Proceedings in Physics, Vol. 77 Quantum Optics VI
Editors: D.F. Walls · J.D. Harvey

for the two systems, and let $H_{\rm sys}$ be the sum of their otherwise independent system Hamiltonians. The quantum Langevin equations for these are

$$\begin{aligned}\dot{a}_1 &= -\frac{i}{\hbar}[a_1, H_{\rm sys}] \\ &-[a_1, c_1^\dagger]\left\{\frac{\gamma_1}{2}c_1 + \sqrt{\gamma_1}b_{\rm in}(1,t)\right\} + \left\{\frac{\gamma_1}{2}c_1^\dagger + \sqrt{\gamma_1}b_{\rm in}^\dagger(1,t)\right\}[a_1, c_1], \end{aligned}\tag{1}$$

$$\begin{aligned}\dot{a}_2 &= -\frac{i}{\hbar}[a_2, H_{\rm sys}] \\ &-[a_2, c_2^\dagger]\left\{\frac{\gamma_2}{2}c_2 + \sqrt{\gamma_2}b_{\rm in}(2,t)\right\} + \left\{\frac{\gamma_2}{2}c_2^\dagger + \sqrt{\gamma_2}b_{\rm in}^\dagger(2,t)\right\}[a_2, c_2]. \end{aligned}\tag{2}$$

The output field from the first system is given by

$$b_{\rm out}(1,t) = b_{\rm in}(1,t) + \sqrt{\gamma_1}c_1(t). \tag{3}$$

If it takes a time τ for light to travel from system 1 to system 2, then we can get the effect of feeding the output of system 1 into the input of system 2 by writing

$$b_{\rm in}(2,t) = b_{\rm out}(1,t-\tau) = b_{\rm in}(1,t-\tau) + \sqrt{\gamma_1}c_1(t-\tau). \tag{4}$$

If now a is an operator from either of the systems, the resulting two quantum Langevin equations can be written as one equation for a (using the abbreviated notation $b_{\rm in}(1,t) \to b_{\rm in}(t)$):

$$\begin{aligned}\dot{a} &= -\frac{i}{\hbar}[a, H_{\rm sys}] \\ &-[a, c_1^\dagger]\left\{\frac{\gamma_1}{2}c_1 + \sqrt{\gamma_1}b_{\rm in}(t)\right\} + \left\{\frac{\gamma_1}{2}c_1^\dagger + \sqrt{\gamma_1}b_{\rm in}^\dagger(t)\right\}[a, c_1] \\ &-[a, c_2^\dagger]\left\{\frac{\gamma_2}{2}c_2 + \sqrt{\gamma_1\gamma_2}c_1(t-\tau) + \sqrt{\gamma_2}b_{\rm in}(t-\tau)\right\} \\ &+\left\{\frac{\gamma_2}{2}c_2^\dagger + \sqrt{\gamma_1\gamma_2}c_1^\dagger(t-\tau) + \sqrt{\gamma_2}b_{\rm in}^\dagger(t-\tau)\right\}[a, c_2]. \end{aligned}\tag{5}$$

In developing these equations it has been required that the output from the first atom can be connected to the input of the second atom without there being a corresponding connection from the second atom back into the first. Experimentally, this is routinely achieved by appropriate isolation techniques, and it is possible [3] to write such a one-way coupling in terms of a Hamiltonian.

The major technical difficulty is the fact that operators at the two times t and $t-\tau$ both turn up. Because we are considering only the case in which the driving is one way, it is clear that τ may be chosen arbitrarily—its only effect is to shift the origin of the time axis for the second atom. In [6] it is shown how we can in fact simply set $\tau \to 0$, with the understanding that the origin of time is simply delayed by τ in the second atom.

From this it follows the a master equation can be derived for the density operator $\rho(t)$ by usual technique [4,5] of setting $\langle da(t)\rho\rangle \equiv \langle da\rho(t)\rangle$, and this master equation takes the form

$$\frac{d\rho}{dt} = \frac{i}{\hbar}[\rho, H_{\rm sys}] + \frac{\gamma_1}{2}\left\{2c_1\rho c_1^\dagger - \rho c_1^\dagger c_1 - c_1^\dagger c_1\rho\right\}$$

$$
\begin{aligned}
&+\frac{\gamma_2}{2}\left\{2c_2\rho c_2^\dagger - \rho c_2^\dagger c_2 - c_2^\dagger c_2\rho\right\} \\
&-\sqrt{\gamma_1\gamma_2}\left\{[c_2^\dagger, c_1\rho] + [\rho c_1^\dagger, c_2]\right\} \\
&+\frac{\bar{N}}{2}\left[[\sqrt{\gamma_1}c_1 + \sqrt{\gamma_2}c_2, \rho], \sqrt{\gamma_1}c_1^\dagger + \sqrt{\gamma_2}c_2^\dagger\right] \\
&+\frac{\bar{N}}{2}\left[[\sqrt{\gamma_1}c_1^\dagger + \sqrt{\gamma_2}c_2^\dagger, \rho], \sqrt{\gamma_1}c_1 + \sqrt{\gamma_2}c_2\right] \\
&-\left[\mathcal{E}_{\text{in}}(t)(\sqrt{\gamma_1}c_1^\dagger + \sqrt{\gamma_2}c_2^\dagger) - \mathcal{E}_{\text{in}}^*(t)(\sqrt{\gamma_1}c_1 + \sqrt{\gamma_2}c_2), \rho\right] .
\end{aligned} \tag{6}
$$

3. Imperfect coupling

The previous derivation assumes perfect coupling of the first output into the second input. Using the input-output formulation it is straightforward to generalise to include multiple inputs and outputs, coupling of only part of the output into the next input and losses in transmission between the two atoms. The second two are not the same. For details see [6].

4. Applications

4.1 Linear dependence of a two photon transition rate using squeezed light

Squeezed light, as produced by parametric oscillators, is characterized by correlated pairs of photons. Therefore, it is quite natural to expect that the behaviour of two photon transitions in atoms should be influenced in a significant way by squeezed light. That this is indeed the case has been pointed out in studies of the effect of squeezed light on the two photon transition rate in a three-level "ladder" system [7].

With coherent or thermal light as the driving field, two photon excitation to the upper excited state occurs primarily, in the weak field limit, via a two step process, with each step proportional to the driving field intensity. Hence, the two photon transition rate exhibits a quadratic dependence on driving field intensity.

If, however, the driving field consists of correlated pairs of photons (coincident upon the atom within a certain correlation time), then two photon excitation can occur essentially as a single step process proportional to the incident intensity. This leads to a linear dependence of the two photon transition rate on intensity [7].

To highlight this fundamental difference between conventional and squeezed light excitation, we set up a model describing the following situation, which is also depicted in Fig.2. A three-level "ladder" atom is driven by the output from a *nondegenerate* parametric oscillator (this could be either frequency or polarization nondegenerate: in the case that it is frequency nondegenerate the two cavity modes are assumed to be well separated in frequency).

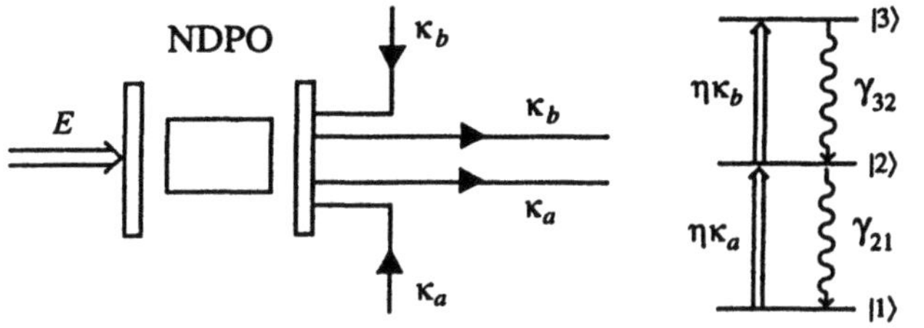

FIG. 2. Schematic diagram of the coupling between the output modes of a non-degenerate parametric oscillator (NDPO) and a three-level atom.

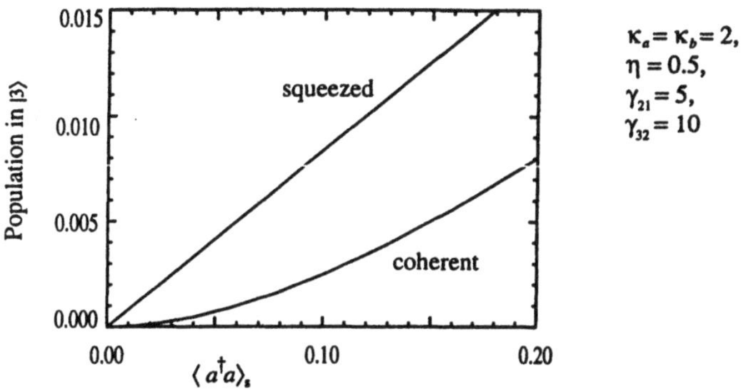

FIG. 3. Population of the upper atomic state $|3\rangle$ as a function of the cavity mode photon number $\langle a^\dagger a\rangle_s$, which is a direct measure of the incident intensity. Parameters for this figure are $\kappa_a = \kappa_b = 2$, $\eta = 0.5$, $\gamma_{21} = 5$, $\gamma_{32} = 10$. The case of excitation with correlated squeezed light is contrasted with the case of independent coherent excitation of the two transitions.

With an appropriate unitary transformation to a rotating frame, we can write the master equation describing this configuration as

$$\begin{aligned}\frac{\partial \rho}{\partial t} &= \frac{1}{2}\left[E a^\dagger b^\dagger - E^* ab, \rho\right] \\ &+\frac{1}{2}\kappa_a\left(2a\rho a^\dagger - a^\dagger a\rho - \rho a^\dagger a\right) \\ &+\frac{1}{2}\kappa_b\left(2b\rho b^\dagger - b^\dagger b\rho - \rho b^\dagger b\right) \\ &-\sqrt{\eta\kappa_a\gamma_{21}}\left\{[\sigma_{12}^+, a\rho] + \text{h.c.}\right\} - \sqrt{\eta\kappa_b\gamma_{32}}\left\{[\sigma_{23}^+, b\rho] + \text{h.c.}\right\} \\ &+\frac{1}{2}\gamma_{21}\left(2\sigma_{12}^-\rho\sigma_{12}^+ - \sigma_{12}^+\sigma_{12}^-\rho - \rho\sigma_{12}^+\sigma_{12}^-\right) \\ &+\frac{1}{2}\gamma_{32}\left(2\sigma_{23}^-\rho\sigma_{23}^+ - \sigma_{23}^+\sigma_{23}^-\rho - \rho\sigma_{23}^+\sigma_{23}^-\right). \qquad (7)\end{aligned}$$

The two cavity modes are damped at the rate κ_a and κ_b, and the atomic transitions $|2\rangle \to |1\rangle$ and $|3\rangle \to |2\rangle$ have linewidths γ_{21} and γ_{32} respectively. The fraction of input to the atom contributed by the parametric oscillator output is given by the parameter η ($\eta \leq 1$).

In Fig.3 we plot the population of state $|3\rangle$ as a function of $\langle a^\dagger a\rangle_s = \langle b^\dagger b\rangle_s$ (which is directly proportional to the intensity). The linear dependence exhibited for squeezed light excitation is contrasted with the quadratic dependence obtained when the two cavity modes are uncoupled and driven independently by weak coherent fields, in such a way as to give $\langle a^\dagger a\rangle_s = \langle b^\dagger b\rangle_s$.

4.2 Two level atom driven by antibunched light from another atom

The problem of an atom driven by antibunched light is one for which solutions were unknown until the coupled systems approach was applied to it [3]. As in that work, we begin by taking our source of antibunched light to be a single two-level atom. Given that our source of antibunched light is a two-level atom, a pumping mechanism must be provided to excite this atom and induce fluorescence from which the source field is derived. The two distinct possibilities that we shall consider are

- Coherent pumping provided by incident light from a laser.
- Incoherent pumping provided by finite bandwidth thermal light.

Coherent excitation of the source atom: The configuration that we shall consider is shown schematically in Fig.4 where we illustrate the various input, output, and coupling channels.

For the case of coherent excitation of the source atom the master equation describing this situation takes the form

$$\begin{aligned}\frac{\partial\rho}{\partial t} &= \frac{1}{2}\gamma_1\left(2\sigma_1^-\rho\sigma_1^+ - \sigma_1^+\sigma_1^-\rho - \rho\sigma_1^+\sigma_1^-\right) - \sqrt{\epsilon_1\gamma_1}[E\sigma_1^+ - E^*\sigma_1^-,\rho] \\ &\quad -\sqrt{(1-\epsilon_1)(1-\epsilon_2)\gamma_1\gamma_2}\left([\sigma_2^+,\sigma_1^-\rho] + [\rho\sigma_1^+,\sigma_2^-]\right) \\ &\quad +\frac{1}{2}\gamma_2\left(2\sigma_2^-\rho\sigma_2^+ - \sigma_2^+\sigma_2^-\rho - \rho\sigma_2^+\sigma_2^-\right). \end{aligned} \tag{8}$$

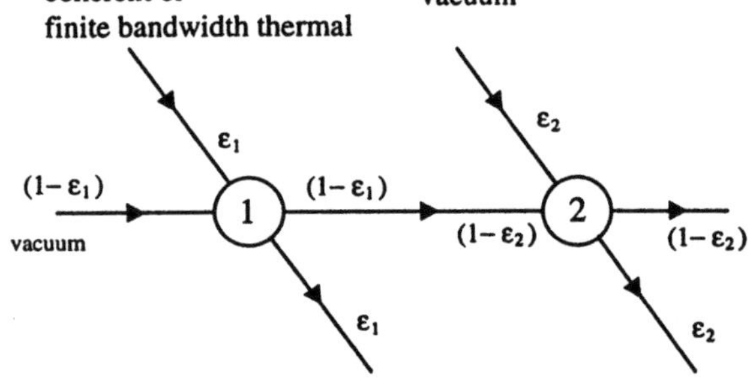

FIG. 4. Schematic diagram of the coupling between a coherently-driven atom (1) and a second atom (2). The various input, output, and coupling channels are discussed in the text.

The (classical) coherent electric field E is incident upon atom 1 through the channel labelled ϵ_1. This coherent field is not incident upon atom 2. Antibunched light emitted from atom 1 into channel $(1-\epsilon_1)$ provides the light source for atom 2 with which it is coupled through channel $(1-\epsilon_2)$. The (vacuum input) channel ϵ_2 provides an observation channel for the light emitted from atom 2.

To solve the master equation we use direct numerical integration on a computer. Application of the quantum regression theorem yields all of the correlation functions and spectra of interest. More details of our computational technique are given in [6].

We compute the photon counting correlation functions

$$\langle : I_i(t)I_j(0) : \rangle_s \doteq \beta_i\beta_j \mathrm{Tr}\left\{\sigma_i^+\sigma_i^- V(t)\left\{\sigma_j^-\rho_s\sigma_j^+\right\}\right\}, \tag{9}$$

where $i,j=1,2$ (first or second atom), $V(t)$ is the (two-sided) evolution operator, such that $V(t)\rho(t')=\rho(t+t')$, and ρ_s is the stationary solution of the master equation. These functions can be interpreted as the conditional probability of counting a photon from atom i a time t after counting a photon from atom j. The most significant qualitative features of the intensity correlation functions are similar for coherent and incoherent excitation of the source atom, and are illustrated in Fig.5 for both cases, where we plot the normalized intensity correlation functions,

$$g_{ij}^{(2)}(t)=\frac{\langle : I_i(t)I_j(0) : \rangle_s}{\langle I_i\rangle_s\langle I_j\rangle_s}, \tag{10}$$

for two different kinds of driving. These plots present a typical set of results.

Incoherent excitation of the source atom:

We would like to test to what extent the antibunched nature of the light emitted from the first atom characterizes the nature of the photon counting from the second atom. To do this we shall show that we can produce a light beam from the first atom which has the same intensity and $g^{(2)}(t)$ as that emitted when the light incident is coherent, but is otherwise quite different.

Let us consider excitation of atom 1 by a thermal light field with finite bandwidth. The corresponding master equation is given by

$$\begin{aligned}\frac{\partial\rho}{\partial t} &= \kappa(\bar{N}+1)\left(2a\rho a^\dagger - a^\dagger a\rho - \rho a^\dagger a\right) + \kappa\bar{N}\left(2a^\dagger\rho a - aa^\dagger\rho - \rho aa^\dagger\right)\\ &\quad -\sqrt{2\kappa\eta_1\gamma_1}\left([\sigma_1^+, a\rho] + [\rho a^\dagger, \sigma_1^-]\right)\\ &\quad +\frac{1}{2}\gamma_1\left(2\sigma_1^-\rho\sigma_1^+ - \sigma_1^+\sigma_1^-\rho - \rho\sigma_1^+\sigma_1^-\right)\\ &\quad -\sqrt{\eta_2\gamma_1\gamma_2}\left([\sigma_2^+, \sigma_1^-\rho] + [\rho\sigma_1^+, \sigma_2^-]\right)\\ &\quad +\frac{1}{2}\gamma_2\left(2\sigma_2^-\rho\sigma_2^+ - \sigma_2^+\sigma_2^-\rho - \rho\sigma_2^+\sigma_2^-\right).\end{aligned} \tag{11}$$

The finite bandwidth thermal field is provided by the output from a cavity driven (through the input channel denoted by ϵ_κ) by a broadband, or "white noise", thermal field. The cavity output through channel $(1-\epsilon_\kappa)$ is coupled to atom 1 through channel

ϵ_1. Again, this driving field is not incident upon atom 2. The coupling parameters are given by $\eta_1 = (1 - \epsilon_\kappa)\epsilon_1$ and $\eta_2 = (1 - \epsilon_1)(1 - \epsilon_2)$.

Comparison with coherent excitation: different kinds of antibunched light: We want to compare the cases of purely coherent (Eq.(8)) and purely incoherent (Eq.(11)) thermal excitation of atom 1 when the intensity of the light driving atom 2,

$$\eta_2\gamma_1\langle\sigma_1^+\sigma_1^-\rangle_s \,, \tag{12}$$

($\eta_2 = (1 - \epsilon_1)(1 - \epsilon_2)$) and the intensity correlation function,

$$g_{11}^{(2)}(t) = \frac{\langle\sigma_1^+\sigma_1^+(t)\sigma_1^-(t)\sigma_1^-\rangle_s}{\langle\sigma_1^+\sigma_1^-\rangle_s^2} \,, \tag{13}$$

are as close as possible the same for both situations. The two different driving fields for atom 2 can then be regarded as having identical antibunching properties.

Such a comparison is presented in Fig.5 where we plot the functions $g_{ij}^{(2)}(t)$ for the two cases. By a suitable choice of parameters, we are able to closely match $g_{11}^{(2)}(t)$ and $\eta_2\gamma_1\langle\sigma_1^+\sigma_1^-\rangle_s$ for the two cases (such that they would quite certainly be indistinguishable in an experiment).

The curves describing $g_{ij}^{(2)}(t)$ for $(ij) = (12), (21), (22)$ are clearly different for the two cases, as are the respective intensities of light emitted from atom 2—values of the excited state population for atom 2 are $\langle\sigma_2^+\sigma_2^-\rangle_s = 0.0651$ for thermal driving and 0.0929 for coherent driving.

These results demonstrate that the photon counting properties of the light emitted by an atom driven with antibunched light are not determined by the photon counting properties of the incident light—i.e., those properties specified by $g_{11}^{(2)}(t)$ and $\eta_2\gamma_1\langle\sigma_1^+\sigma_1^-\rangle_s$.

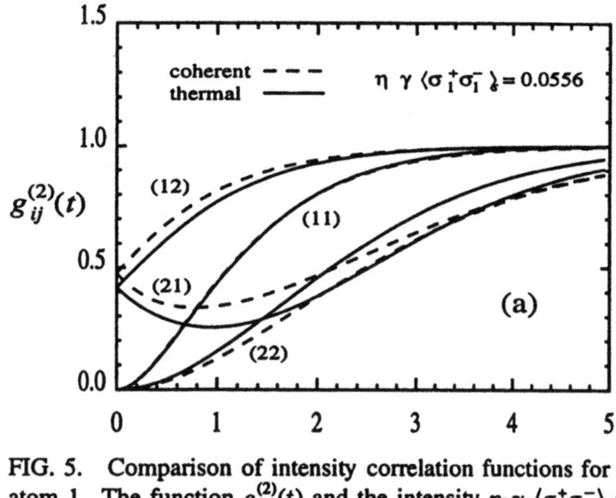

FIG. 5. Comparison of intensity correlation functions for coherent and thermal excitation of atom 1. The function $g_{11}^{(2)}(t)$ and the intensity $\eta_2\gamma_1\langle\sigma_1^+\sigma_1^-\rangle$ ($\eta_2 = (1 - \epsilon_1)(1 - \epsilon_2)$) have been matched for the two cases. Parameters for the two figures, referring to the master equations (11) and (8), are: Thermal: $\kappa = 1.15$, $\bar{N} = 0.207$, $\eta_1 = 0.25$, $\gamma_1 = 0.8$, $\eta_2 = 0.3125$, $\gamma_2 = 1$; Coherent: $\gamma_1 = 2$, $\sqrt{\epsilon_1\gamma_1}E = 0.25$, $\epsilon_2 = 0.5$, $\gamma_2 = 1$.

Indeed, the spectral properties of the antibunched driving field are quite different for the two cases. Properties displayed in such spectra must evidently be taken into consideration when characterizing an antibunched light field with respect to its influence on the purely photon counting properties of light emitted by the atoms upon which it is incident.

5. Summary

The method introduced here is sufficient in principle to treat any kind of non-classical light acting on any quantum system. However, to do so, we must have a full description of the system being used to produce the non-classical light, so that it can be modelled accurately, and inserted into the formalism.

The conclusion on characterisation of light field is quite simple—a light field can be completely characterised by the system used to produce it. This is a far more practical characterisation than a requirement that all correlation functions be known. While this is correct, there is as yet no way of either knowing which correlation functions are more important than others or of using such knowledge in any computatiional algorithm.

The method is clearly more adapted to systems for which the resultant density matrix (truncated if necessary) is not of too large a dimension. However, larger problems could be tackled by wavefunction stochastic differential equation methods.

Acknowledgements: I wish to thank Peter Zoller and Howard Carmichael for stimulating and fruitful discussions, and Andrea Eschmann for help in calculations.

References

[1] M.I. Kolobov and I.V. Sokolov, Opt. Spektrosk. **62**, 112 (1987).
[2] H.J. Carmichael, Phys. Rev. Lett. **70**, 2273 (1993).
[3] C.W. Gardiner, Phys. Rev. Lett. **70**, 2269 (1993).
[4] C.W. Gardiner and M.J. Collett, Phys. Rev. A **31**, 3761 (1985).
[5] C.W. Gardiner, *Quantum Noise* (Springer, Heidelberg, 1991).
[6] C.W. Gardiner, A.S. Parkins, Phys. Rev. A, (1994)
[7] J. Gea-Banacloche, Phys. Rev. Lett. **62**, 1603 (1989); J. Javanainen and P.L. Gould, Phys. Rev. A **41**, 5088 (1990); Z. Ficek and P.D. Drummond, Phys. Rev. A **43**, 6247 (1991); **43**, 6258 (1991).
[8] B.R. Mollow, Phys. Rev. **175**, 1555 (1968).

Quantum Trajectories and Stochastic Schrödinger Equation for Nonlinear Optical Processes

P. Goetsch and R. Graham

Fachbereich Physik, Universität GH Essen, D-45117 Essen, Germany

Abstract. The stochastic Schrödinger equation in a linear and equivalent nonlinear forms arises naturally by the back action of a broad-band continuous-in-time measurement of a quantum observable. It represents the classical stochastic outcomes of single measurements as stochastic quantum expectation values performed with a stochastic wave-function. Applications of this description to some nonlinear optical processes such as subharmonic amplification, nonlinear absorption and second-harmonic generation are discussed.

1 Introduction

Measurements on quantum systems performed continuously in time can be modelled by a system-environment interaction. Recently the list of methods available for the treatment of such problems has received a new entry – the stochastic Schrödinger equation, or, equivalently, the stochastic master equation. Originally introduced in a phenomenological way (see e.g. [1-5]) it received meanwhile a microscopic foundation in models of quantum measurements ([7-9]). Here we wish to discuss the foundations of the method for Markoffian quantum processes, its close links to the input-output formalism [10], its application to some exactly soluble cases (harmonic oscillator, subharmonic amplification) and present numerical results for some nonlinear optical processes (nonlinear absorber, second-harmonic generation) [11].

We are indebted to Martin Schlautmann for useful remarks concerning section 2 [12]. After completing this manuscript we were informed about unpublished work by M. Collet also linking the input-output formalism and the stochastic Schrödinger equation [13].

2 The measurement process and the stochastic Schrödinger equation

Let us consider the origin of the stochastic Schrödinger equation in the quantum theory of continous-in-time measurements (see e.g. [6,7], [9]). For concreteness we shall assume that a system variable X with the canonical conjugate P is measured by the emission of a quantized meter field, say an electromagnetic field, into the vacuum. The input-output formalism [10] is perfectly adapted to this situation: The interaction of the system with the input field $\Xi(t)$ generates an output field

Springer Proceedings in Physics, Vol. 77 Quantum Optics VI
Editors: D.F. Walls · J.D. Harvey

$\Theta(t)$ depending on the observable X via a coupling constant

$$d\Theta(t) = 2\sqrt{\gamma}X(t)dt + d\Xi(t). \tag{2.1}$$

In sections 2 and 3 we use upper case notation to indicate operators and lower case for the corresponding eigenvalues. Eq. (2.1) holds in this non-retarded form for broad band measurements with a time-resolution much better than the characteristic time-scales of the system, and we shall restrict ourselves to this Markoffian case here. (For a non-Markoffian generalization of the input-output formalism see [14]). Then the incoming vacuum fluctuations are described by a free field Heisenberg operator $\Xi(t)$ representing a quantized Wiener process with uncorrelated and causally independent future-directed increments $d\Xi(s)$, $0 \leq s < t$, satisfying

$$\begin{aligned} &[d\Xi(s),\, d\Xi(s')] = 0\,, \qquad d\Xi(s)d\Xi(s) = ds\,, \\ &[d\Xi(t),\;\; Y(s)] = 0\,, \qquad \text{for } t \geq s \end{aligned} \tag{2.2}$$

for any system variable Y. As a consequence of (2.2) there is a representation in which all $d\Xi(s)$ are diagonal and represented by a classical Wiener process $d\xi(s)$, $\xi(t) = \int_0^t d\xi(s)$. In terms of negative and positive frequency parts of creation and annihilation operators of the white noise [15]

$$\begin{aligned} &[dB(t), dB(t')] = 0 = [dB^\dagger(t), dB^\dagger(t')]\,, \\ &[dB(t), dB^\dagger(t')] = 0 \quad \text{for} \quad t \neq t', \\ &dB(t)dB^\dagger(t) = dt, \quad dB^\dagger(t)dB(t) = 0\,, \end{aligned} \tag{2.3}$$

the input meter field is represented as

$$d\Xi(t) = dt \int \frac{d\omega}{2\pi} \left(B(\omega)e^{-i\omega t} + B^\dagger(\omega)e^{i\omega t}\right) = dB(t) + dB^\dagger(t). \tag{2.4}$$

The output field $\Theta(t)$ in the Heisenberg picture is generated from the input field by the unitary evolution operator $U_{\text{int}}(t)$ of the total system in the interaction representation $\Theta(t) = U^\dagger_{\text{int}}(t)\Xi(t)U_{\text{int}}(t)$. In a quantum optical context the variables $X(t)$, $P(t)$ will be given by $X(t) = \frac{1}{2}(A(t) + A^\dagger(t))$, where the mutually adjoint operators $A(t)$, $A^\dagger(t)$ describe the quantized slowly varying complex amplitude of some oscillation at large frequency ω_0. The frequency ω_0 is split-off in the interaction picture and thus specifies a new natural origin of the frequencies ω in (2.4). The frequencies ω can then not only be positive or negative, but can even be extended, without harm, to the range $-\infty < \omega < \infty$. The white-noise and Markoff assumptions therefore actually refer to the flatness of the frequency spectrum in the neighborhood of ω_0. The interaction Hamiltonian H_{int} of the measured system in the rotating wave and Markoff approximations is of the form

$$H_{\text{int}}(t)dt = i\sqrt{\gamma}\left(AdB^\dagger(t) - dB(t)A^\dagger\right). \tag{2.5}$$

Hence $U_{\text{int}} = [\exp(-i\int_0^t dsH_{\text{int}}(s))]_+$ where $[\ldots]_+$ denotes time-ordering. Eq. (2.1) for the output process is then a direct consequence of the definition of $\Theta(t)$. The output process has properties corresponding to (2.2) namely

$$[d\Theta(s), d\Theta(s')] = 0\,; \quad [d\Theta(s), X(t)] = 0 \quad \text{for}\, t \geq s. \tag{2.6}$$

The first of these equations follows from the interaction Hamiltonian with the assumed white-noise properties [7], the second is a general causality relation which even holds for non-Markoffian processes [14]. We note that eqs. (2.1), (2.2) and (2.3) imply that also $[d\Xi(s), d\Theta(s')] = 0$ for $s \geq s'$. Therefore, in the representation where $d\Xi(s)$ is diagonal all the $d\Theta(s')$ with $s' \leq s$ can also be diagonalized and treated as a classical stochastic process $d\theta(s')$. We can go even further and consider normalized time-evolutions of the total system in which all eigenvalues $d\theta(s)$ for $0 \leq s < t$ have been measured and can be considered as given and fixed. Taking the expectation value of eq. (2.1) in such states we obtain

$$d\theta(t) = 2\sqrt{\gamma}\langle X(t)\rangle_\theta dt + d\xi(t) \tag{2.7}$$

for the increment of the output process conditioned on all of its earlier increments $d\Theta(s)$ for $0 \leq s < t$. Note that, at least within the limited class of models considered here, this is the fulfillment of an old dream — the direct description of the random results of individual quantum measurements by a stochastic process.

It remains to derive the stochastic Schrödinger equation. The generator of the total time-evolution is defined as $Ldt = dU_{\rm int} \cdot U_{\rm int}^\dagger$ where $dU_{\rm int}(t)$ is calculated from $U_{\rm int}(t)$ by using the Ito calculus for the stochastic differential $dB(t)$ and its adjoint. The result is

$$L(t)dt = -\frac{\gamma}{2}A^\dagger A dt + \sqrt{\gamma}(AdB^\dagger(t) - dB(t)A^\dagger). \tag{2.8}$$

Initially the total wave function $|\psi(0)\rangle \otimes |\{0\}\rangle$ is in the vacuum state of the field and we may represent the total wave-function at time t as the time-ordered product over discretized time

$$\left[\prod_{t_i=0}^{t} (1 + L(t_i)dt_i)\right]_+ |\psi(0)\rangle \otimes |\{0\}\rangle. \tag{2.9}$$

Then we may use the fact $[dB(t), dB^\dagger(t')] = 0$ for $t \neq t'$, implied by eq. (2.3), to bring all $dB(t_i)$ to the right where they are annihilated when hitting $|\{0\}\rangle$. By the same token we may in (2.9) replace $dB^\dagger(t_i)$ by $dB^\dagger(t_i) + dB(t_i) = d\Xi(t_i)$, which may be further replaced by $d\xi(t_i)$ in the ξ-representation. This leaves us with a Schrödinger equation for the total wave-function of the form

$$\left(d + \frac{\gamma}{2}A^\dagger A dt - \sqrt{\gamma}A \cdot d\xi(t)\right) |\psi_\xi(t)\rangle \otimes \varphi_0(\{\xi\}) = 0 \tag{2.10}$$

where $\varphi_0(\{\xi\})$ is the field vacuum in the ξ-representation. This field vacuum may now be factored out (i.e. we do not have to take a trace) and we obtain the linear stochastic Schrödinger equation in the sense of Ito for $|\psi_\xi\rangle$ alone [7]

$$d|\psi_\xi\rangle = -\frac{\gamma}{2}A^\dagger A|\psi_\xi\rangle dt + \sqrt{\gamma}A|\psi_\xi(t)\rangle \cdot d\xi(t). \tag{2.11}$$

We note in passing that there is also a complex noise version of (2.11). It arises e.g. in heterodyne measurements [9] where the observable $X(t)$ has the form $X(t) = \frac{1}{2}(A^\dagger e^{i\Omega t} + Ae^{-i\Omega t})$ with some detuning Ω small compared to ω_0 but large compared to all other relevant inverse time scales. Then A in (2.11) is replaced by $Ae^{-i\Omega t}$. Averaging over many cycles of Ω the complex noise $e^{-i\Omega t}d\xi(t) = d\xi_1(t) + id\xi_2(t)$ has the properties $(d\xi_1)^2 = (d\xi_2)^2 = \frac{1}{2}dt$, $d\xi_1 d\xi_2 = 0$. For simplicity we continue to discuss the case of real noise only.

By construction expectation values of system variables $Y(t)$ with the total wave-function in the interaction picture are obtained from $|\psi_\xi\rangle$ via

$$\langle Y(t)\rangle = \int d\mu_t^W[\xi]\langle\psi_\xi(t)|Y|\psi_\xi(t)\rangle\,, \tag{2.12}$$

where $d\mu_t^W[\xi] = |\varphi_0(\{\xi\})|^2 D\xi$ is the classical Wiener measure of $\xi(s)$ in the interval $[0,t]$. Taking $Y = 1$ the norm $\langle\psi_\xi(t)|\psi_\xi(t)\rangle$ is recognized as a density of $\xi(s)$, $0 \leq s < t$, which is normalized with respect to the Wiener measure. Thus the expression

$$d\mu_t[\xi'] = d\mu_t^W[\xi']\langle\psi_{\xi'}(t)|\psi_{\xi'}(t)\rangle \tag{2.13}$$

is also a normalized probability measure. The increment $d\xi'$ of the process with this measure is distributed [5] according to

$$p(d\xi') = p^W(d\xi')\left(1 + \frac{d\langle\psi_{\xi'}|\psi_{\xi'}\rangle}{\langle\psi_{\xi'}|\psi_{\xi'}\rangle}\right), \tag{2.14}$$

where $p^W(d\xi')$ is the Gaussian distribution of the Wiener increment. With

$$\frac{d\langle\psi_{\xi'}|\psi_{\xi'}\rangle}{\langle\psi_{\xi'}|\psi_{\xi'}\rangle} = \sqrt{\gamma}\frac{\langle\psi_{\xi'}|A + A^\dagger|\psi_{\xi'}\rangle}{\langle\psi_{\xi'}|\psi_{\xi'}\rangle}d\xi' \tag{2.15}$$

following from (2.11) we see that $p(d\xi')$ is a shifted Gaussian, or, equivalently, that $d\xi'$ may be expressed as

$$d\xi'(t) = d\xi(t) + 2\sqrt{\gamma}\frac{\langle\psi_{\xi'}(t)|X|\psi_{\xi'}(t)\rangle}{\langle\psi_{\xi'}(t)|\psi_{\xi'}(t)\rangle}dt\,, \tag{2.16}$$

where $d\xi(t)$ is the Wiener increment. A comparison with eq. (2.7) shows that $d\xi'(t)$ coincides with the increment $d\theta(t)$ of the output process, whose measure is therefore given by (2.13).

The normalized stochastic wave-function

$$|\phi_\xi(t)\rangle = |\psi_\xi(t)\rangle/\langle\psi_\xi(t)|\psi_\xi(t)\rangle^{1/2} \tag{2.17}$$

is distributed according to the measure (2.13), as can be seen by rewriting (2.12) in terms of $|\phi_\xi(t)\rangle$; i.e. the ξ in $|\phi_\xi\rangle$ is distributed according to this measure, i.e. it is in fact the output field and better renamed as θ. Reexpressing this θ in terms of the Wiener process of the input field ξ one finally obtains the *nonlinear* stochastic Schrödinger equation for the non-normalized $|\tilde{\psi}_\xi\rangle \equiv |\psi_\theta\rangle$

$$d|\tilde{\psi}_\xi\rangle = \left(-\frac{\gamma}{2}A^\dagger A dt + \sqrt{\gamma}A(d\xi + 2\sqrt{\gamma}\langle X\rangle_\xi dt)\right)|\tilde{\psi}_\xi\rangle \tag{2.18}$$

and an equivalent version for the normalized $|\tilde{\phi}_\xi\rangle \equiv |\phi_\theta\rangle = |\tilde{\psi}_\xi\rangle/\langle\tilde{\psi}_\xi|\tilde{\psi}_\xi\rangle^{1/2}$. Expectation values are then given by

$$\langle Y\rangle = \int d\mu_t^W[\xi]\langle\tilde{\phi}_\xi|Y|\tilde{\phi}_\xi\rangle = \int d\mu_t[\theta]\langle Y\rangle_\theta\,, \tag{2.19}$$

with the notation $\langle Y \rangle_\theta = \langle \phi_\theta | Y | \phi_\theta \rangle$, which has already been used in eq. (2.7). We note that according to (2.17) $\langle Y \rangle_\theta$ can be calculated from the linear form (2.11) simply renaming ξ as θ. However, the measure $d\mu_t[\theta]$ of the output process and of the measured value $\langle X \rangle_\theta$ is given by (2.13) and more complicated than the Wiener measure. Therefore, the form (2.11) is most convenient for analytical solutions, while eq. (2.18) is best suited for numerical simulations.

3 Exactly solvable simple examples

The linear form (2.11) of the stochastic Schrödinger equation can be solved analytically in simple cases. E.g. if A, $A^\dagger$ describes a harmonic oscillator with a coherent state as initial state $|\psi_\xi(0)\rangle = f(\alpha)|\alpha\rangle$, $A|\alpha\rangle = \alpha|\alpha\rangle$, the solution of (2.11) is easily obtained as

$$|\psi_\xi(t;\alpha)\rangle = \tag{3.1}$$
$$f(\alpha) \exp\left[\frac{\alpha^2 + |\alpha|^2}{2}(e^{-\gamma t} - 1) + \sqrt{\gamma}\alpha \int_0^t e^{-\frac{\gamma}{2}s} d\xi(s)\right] |\alpha e^{-(\gamma/2)t}\rangle.$$

As any initial state can be expanded in coherent states, with $f(\alpha) = \frac{1}{\pi}\langle \alpha | \psi_\xi(0)\rangle$, and eq. (2.11) is linear, the general solution of (2.11) in the present case is simply

$$|\psi_\xi(t)\rangle = \int d^2\alpha |\psi_\xi(t;\alpha)\rangle. \tag{3.2}$$

We note that by normalizing $|\psi_\xi\rangle$ according to (2.17), and replacing in (3.2) $d\xi(s)$ by $d\xi'(s) = 2\sqrt{\gamma}\langle X \rangle_{\xi'} dt + d\xi$, we can even construct a 'general' solution of the nonlinear equation (2.18); but it is highly implicit as it still contains $\langle X(s) \rangle_{\xi'}$, which must be calculated from (3.2) with (3.2), replacing ξ by ξ'. In general this will give rise to a nonlinear integral equation for $\langle X(s) \rangle_{\xi'}$. Only in the special case of a single coherent state will this solution be simple, since then $\langle X(s) \rangle_{\xi'} = \frac{1}{2}(\alpha + \alpha^*)e^{-\frac{\gamma}{2}s}$. We note however, that even in the general case the solution of the linear stochastic Schrödinger equation is sufficient to obtain the complete measure of the output process and the measured values of X conditioned on the output process. In the simplest case of a coherent state we see that the measurement by radiation into the vacuum does not perturb the measured amplitude $\langle X(s) \rangle_{\xi'}$ in a random way, after all, and, as is well known, the output field is simply the superposition of the incoming vacuum fluctuations and a coherently radiated exponentially damped field.

In order to consider yet another example of eq. (2.11) we turn to subharmonic generation where the right hand side of eq. (2.11) must include the additional term

$$-\frac{\kappa}{2}(A^2 e^{-2i\varphi} - (A^\dagger)^2 e^{2i\varphi})|\psi_\xi(t)\rangle dt \tag{3.3}$$

with a coupling constant κ. The arbitrary phase φ is introduced to allow for an arbitrary phase-shift between the pump and the measured quadrature X. Let the initial state be the vacuum, $|\psi_\xi(0)\rangle = |0\rangle$. The ansatz

$$|\psi_\xi(t)\rangle = \exp\left(g(t) + \alpha(t)A^\dagger + \frac{1}{2}\beta(t)(A^\dagger)^2\right)|0\rangle \tag{3.4}$$

solves eq. (2.11), provided three stochastic differential equations for g, α, β are satisfied with initial conditions $\alpha(0) = \beta(0) = g(0) = 0$. Their integration yields

$$\begin{aligned}
\beta(t) &= \frac{e^{2i\varphi}}{\kappa + \gamma e^{2i\varphi}}\left(-\frac{\gamma}{2} + \sqrt{C}\tanh\sqrt{C}(t-t_0)\right), \\
\alpha(t) &= \sqrt{\gamma}\int_0^t e^{-\frac{\gamma}{2}(t-\tau)-(\gamma+\kappa e^{-2i\varphi})\int_\tau^t \beta(s)ds}\cdot\beta(\tau)d\xi(\tau), \\
g(t) &= -\frac{1}{2}\int_0^t d\tau[(\kappa e^{-2i\varphi}+\gamma)\alpha^2(\tau)+\beta(\tau)] + \sqrt{\gamma}\int_0^t \alpha(\tau)d\xi(\tau),
\end{aligned} \tag{3.5}$$

with $C = \gamma^2/4 + \kappa^2 + \kappa\gamma e^{2i\varphi}$. The coefficient $\beta(t)$ describes squeezing. It contains the constant of integration t_0, which is fixed by $\beta(0) = 0$. The squeezing coefficient $\beta(t)$ is affected by the measurement through γ, but unlike the amplitude coefficient $\alpha(t)$ it is not affected by the measurement *noise*. It is easy to calculate the stochastic prediction for the measured value $\langle X\rangle_\xi$ using the state (3.4). We obtain

$$\langle X(t)\rangle_\xi = \frac{\mathrm{Re}\,(\alpha + \beta\alpha^*)}{1 - |\beta|^2} \tag{3.6}$$

and the predicted output process follows from eq. (3.6).

4 The nonlinear absorber

This model is well-known and has been investigated recently by Gardiner [15] and by Schack and Schenzle [16] using the positive P representation and by Gisin and Percival [4] using a nonlinear stochastic Schrödinger equation of the form

$$\begin{aligned}
|d\psi\rangle &= \epsilon(a^\dagger - a)|\psi\rangle dt + \frac{\gamma}{2}\left[2\langle(a^\dagger)^2\rangle_\psi a^2 - (a^\dagger)^2 a^2\right. \\
&\quad \left.-\langle(a^\dagger)^2\rangle_\psi\langle a^2\rangle_\psi\right]|\psi\rangle dt + \sqrt{\frac{\gamma}{2}}\left(a^2 - \langle a^2\rangle_\psi\right)|\psi\rangle d\xi,
\end{aligned} \tag{4.1}$$

with $d\xi d\xi = 0 = d\xi^* d\xi^*$, $d\xi^* d\xi = 2dt$. Here a, $a^\dagger$ are Boson operators and in the notation $|\psi\rangle$ we now supress the dependence on ξ. This form, which we shall also adopt in the following, in principle corresponds to (2.18) but assumes a nonlinear 2-photon heterodyne measurement, where a local oscillator near frequency 2ω beats against the 2-photon amplitude a^2, $(a^\dagger)^2$. We solved this equation numerically in an N-dimensional Fock state basis [11]. The time is scaled with γ and we choose the dimensionless parameters $\epsilon/\gamma = 0.05$, which permits us to set the dimension of the Hilbert space to the modest value three. The solution evolving from the vacuum at $t = 0$ is shown in Fig. 1. One can clearly see a systematic oscillation with a period $T \simeq 30$, with stochastic fluctuations superimposed which depend on the realization of the Wiener process. In the same figure we also show an ensemble average over 1000 stochastic realizations. Our results agree with those of Gisin and Percival [4]. They also agree with an approximate analytical

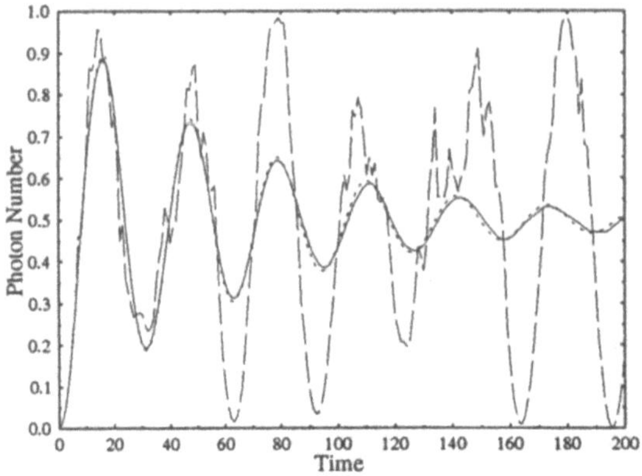

Figure 1: Nonlinear absorber: photon number $\langle\psi(t)|a^\dagger a|\psi(t)\rangle$ for a single quantum trajectory (dashed line) and mean photon number (solid line: for an ensemble of 1000 quantum trajectories; dotted line: from the approximate solution of the master equation (adiabatic elimination)). Parameter: $\epsilon/\gamma = 0.05$.

solution based on the fact that $\epsilon/\gamma \ll 1$; this means that the damping of the oscillator is much larger than the external pumping and permits us to adiabatically eliminate the second excited level of the mode. The analytical solution gives a period of $\approx \pi/2\epsilon$; it is shown as a dotted line in Fig. 1. The result disagrees with a numerical result obtained by using the positive P function [16] and reveals the defects of that method.

5 Second-harmonic generation

A study of second-harmonic generation is of some interest because the classical dynamics of the amplitudes (see e.g. Savage and Walls [17], Dörfle and Schenzle [18]) contains instabilities leading to period doubling and chaos. This system was studied in the semiclassical limit in [19], using the Wigner function, and in [16], [18-20] using the positive P representation.

The Hamiltonian is given by

$$H = \sum_{i=1}^{2} \hbar\omega_i a_i^\dagger a_i + i\hbar\left(\frac{\chi}{2}(a_1^\dagger)^2 a_2 + a_1^\dagger F e^{-i\omega_p t} - \text{h. c.}\right) \tag{5.1}$$

with the second order nonlinear susceptibility χ, and in the following we shall consider the case of nonvanishing detunings $\delta_1 = \omega_1 - \omega_p$, $\delta_2 = \omega_2 - 2\omega_p$. We have solved the stochastic nonlinear Schrödinger equation corresponding to (2.18) and a heterodyne measurement of a quadrature component of the fundamental and the second harmonic mode [11]. The wave-function is expanded in the Fock state basis of the two modes. The semiclassical limit is approached for $\chi/\gamma \ll 1$, increasing the value of the external force F appropriately. Unfortunately the

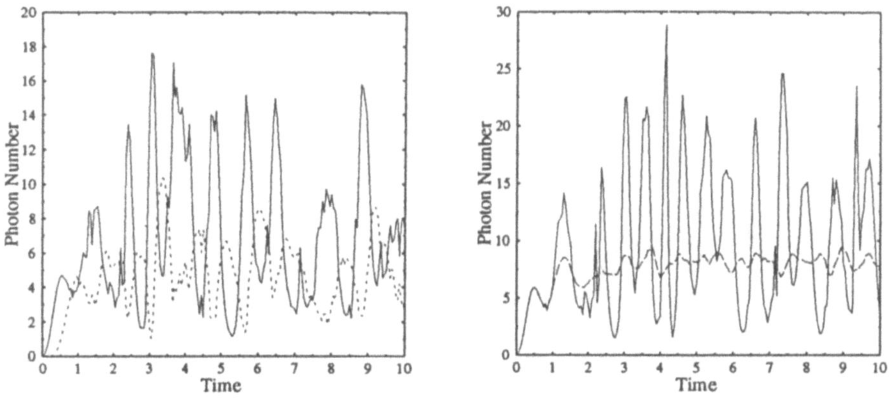

Figure 2: Second-harmonic generation: (a) Photon number $\langle\psi(t)|a_i^\dagger a_i|\psi(t)\rangle$ for a single quantum trajectory; solid line: fundamental mode, dotted line: second-harmonic mode; parameters: $\chi/\gamma = 2$, $\delta_{1,2}/\gamma = 1$, $F/\gamma = 6.0$. (b) Photon number for a single quantum trajectory (solid line) and mean photon number for an ensemble of 100 quantum trajectories (dashed line); parameters: $\chi/\gamma = 2$, $\delta_{1,2}/\gamma = 1$, $F/\gamma = 7.5$.

required basis becomes too large in the semiclassical limit and we are not able to make comparisons with results from the positive P-function [16], [18-20].

In the following we restrict ourselves to the numerical simulation of the stochastic Schrödinger equation in the quantum regime, i. e. we set $\chi/\gamma = 2$, $\delta_{1,2}/\gamma = 1$. For convenience we plot $\langle\psi(t)|a_i^\dagger a_i|\psi(t)\rangle$ rather than the quadrature components which are assumed to be measured. Results are shown in Fig. 2 for $F/\gamma = 6$ and $F/\gamma = 7.5$, respectively, corresponding to a limit cycle and chaotic behaviour of the classical dynamics. One clearly sees that there is no qualitative difference between the two graphs, indicating the purely fluctuation dominated quantal behaviour of the system. The evolution of the photon number for small times $t \leq 1.0$ shows the fundamental mode amplitude increasing linearly in time due to the driving force while the second-harmonic mode amplitude increases cubic in time due to the nonlinear coupling. However, the quantitative behaviour for times $t > 1.0$ strongly depends on the realizations of the Wiener processes. Indeed, the ensemble average over 100 single quantum trajectories also shown in Fig. 2b reveals the fact that the oscillations are strongly reduced by averaging and, in fact, the photon numbers seem to reach a stationary value. The convergence of the averaged solution, however, is poor: the size of the remaining stochastic fluctuations decreases with increasing number of elements in the ensemble only by a factor $N^{-1/2}$.

The ensemble average of the photon numbers can be compared with numerical solutions of the classical amplitude equations. For times $t \leq 2.0$ the quantal and classical solutions are similar while for times $t > 2.0$ they are totally different.

6 Conclusion

We have discussed how the stochastic results of continuous-in-time measurements of observables can be represented as quantum trajectories, i.e. as expectation values of stochastic wave functions governed by stochastic Schrödinger equations. Exact solutions for harmonic oscillators and parametric amplifiers have been presented. For models of a nonlinear absorber and of second harmonic generation we investigated numerically a special diffusion process corresponding to balanced heterodyne detection. In the case of nonlinear absorption we showed that the stochastic Schrödinger equation gives reliable solutions for the master equation, whereas an unphysical behaviour is predicted by the positive P representation. In the case of second-harmonic generation we showed that due to the large number of coupled differential equations required it is very difficult to obtain semiclassical solutions of the stochastic Schrödinger equation. However, the stochastic Schrödinger equation can be used to obtain solutions of the master equation in the range of purely quantal behaviour as shown by the example of Fig. 2.

Acknowledgement

This work was supported by the Deutsche Forschungsgemeinschaft through the Sonderforschungsbereich 237 "Unordnung und große Fluktuationen".

References

[1] N. Gisin, *Phys. Rev. Lett.* **52**, 1657 (1984); *Helv. Phys. Acta* **62**, 363 (1989)
[2] P. Pearle, *Phys. Rev. D* **13**, 857 (1976)
[3] L. Diósi, *Phys. Lett. A* **129**, 419 (1988)
[4] N. Gisin, I. C. Percival, *J. Phys. A* **25**, 5677 (1992); *Phys. Lett. A* **167**, 315 (1992)
[5] G. C. Ghirardi, P. Pearle, A. Rimini, *Phys. Rev. A* **42**, 78 (1990)
[6] A. Barchielli, V. P. Belavkin, *J. Phys. A* **24**, 1495 (1991)
[7] V. P. Belavkin, *J. Math. Phys.* **31**, 2930 (1990);
V. P. Belavkin, P. Staszewski, *Phys. Rev. A* **45**, 1347 (1992)
[8] H. J. Carmichael, *An Open Systems Approach to Quantum Optics*, Lecture Notes in Physics m18, Berlin, Springer 1993
[9] H. M. Wiseman, G. J. Milburn, *Phys. Rev. A* **47**, 642 (1993); *Phys. Rev. A* **47**, 1652 (1993)
[10] M. J. Collet, C. W. Gardiner, *Phys. Rev. A* **30**, 1386 (1984);
C. W. Gardiner, M. J. Collet, *Phys. Rev. A* **31**, 3761 (1985)
[11] P. Goetsch, R. Graham, *Ann. Physik* **2**, 706 (1993)
[12] M. Schlautmann, *Zeitlich kontinuierliche Messungen an chaotischen Quantensystemen*, Thesis, Universität Essen in preparation
[13] M. J. Collet, private communication

[14] R. Graham, in *Quantum Optics V*, ed. by J. D. Harvey, D. F. Walls, Berlin, Springer, p. 46 (1989)
[15] C. W. Gardiner: *Quantum Noise*, Berlin, Springer 1991
[16] R. Schack, A. Schenzle, *Phys. Rev. A* **44**, 682 (1991)
[17] C. M. Savage, D. F. Walls, *Optica Acta* **30**, 557 (1983)
[18] M. Dörfle, A. Schenzle, *Z. Phys. B* **65**, 113 (1986)
[19] M. Dörfle, R. Graham, in *Optical Instabilities*, ed. by R. W. Boyd, M. G. Raymer, L. M. Narducci, Cambridge University Press, Cambridge, p. 352 (1986)
[20] P. D. Drummond, K. J. McNeil, D. F. Walls, *Optica Acta* **28**, 211 (1981)
[21] I. J. D. Craig, K. K. McNeil, *Phys. Rev. A* **41**, 4009 (1990)
[22] P. D. Drummond, C. W. Gardiner, *J. Phys. A* **13**, 2353 (1980);
P. D. Drummond, I. K. Mortimer, *J. Comp. Phys.* **93**, 144 (1991);
A. M. Smith, C. W. Gardiner, *Phys. Rev. A* **39**, 3511 (1989)

Part V

Applications in Laser Physics

Intensity Coherence in Multimode Lasers

P. Mandel

Université Libre de Bruxelles, Campus Plaine CP 231,
B-1050 Bruxelles, Belgium

Abstract. We briefly review antiphase dynamics and then present analytic results explaining its occurrence in multimode Fabry-Perot lasers and in intracavity second harmonic generation.

1. Introduction

If two identical oscillators are coupled, there are usually two stable regimes of operation: either the oscillators are in phase or they are out of phase. If more than two oscillators are coupled, we can still classify the solutions as in phase and out of phase, though an exhaustive classification of the out of phase solutions is a challenging problem. In physics, the out of phase solutions are often called *antiphase* solutions.

Although antiphase dynamics (AD) can be traced back to the famous experiment of Huyghens who coupled two clocks, the recent interest in AD arose with the realization that this mode of operation persists when a large number of oscillators are coupled [1, 2]. Soon afterwards, this prediction was verified experimentally in a selfpulsing Nd:YAG laser with an intracavity frequency doubling crystal (ISHG) [3]. In the simplest case, the experiment showed N selfpulsing modes, each mode displaying the same periodic state but shifted by 1/N of a period from another mode. Recently an example of AD was reported in a selfpulsing Erbium-doped optical fiber laser [4]. The ISHG problem was analyzed theoretically [5] and an analytical description of the transition from the N-mode steady state to the selfpulsing states was obtained with the emergence of one in phase and (N - 1)! antiphase solutions explained.

An alternative way to induce oscillations in the lasing intensity is to modulate periodically the gain of the laser. This was analyzed numerically [6] for a solid-state laser in a Fabry-Perot configuration. The problems associated with attractor crowding as N increases were also considered in [6] and it was shown that injection seeding provides a simple and economical way to switch between antiphase attractors: injecting one period of another antiphase state is sufficient to switch the system and transients decay fast.

The concept of AD was generalized to transient states in [7] where it was shown that the relaxation towards the steady state of a multimode laser also displays "interference patterns" of the intensity phase. Here the typical situation is that each modal intensity displays in its power spectrum as many frequencies as there are modes (or clusters of modes for the fiber lasers [7] and [4]), while the total intensity displays only one frequency, the largest of the frequencies found in the separate modes. This was confirmed experimentally [7], [8] for a Nd-doped optical fiber laser, studied analytically and numerically in [9] and [10] for a solid-state laser and in [5] for ISHG.

An elaboration on transient AD is the noisy AD. It was shown experimentally with a miniature solid state $LiNdP_4O_{12}$ (LNP) laser [11] that the unavoidable small scale fluctuations in the output of a multimode laser can also display AD in the sense that the noise spectrum of each mode has as many frequencies as there are modes while the noise spectrum of the total intensity has only one frequency. This property was also found numerically in the laser rate equations [9] and in ISHG [5].

A further extension was the observation of AD in the chaotic regime. Here again, AD is generalized to imply some kind of destructive interference among the modal intensity phases. A first report was on ISHG [12] where histograms of the peak intensity in the chaotic regime were shown to be singular for the partial modes but fairly smooth (nearly gaussian) for the total intensity. The LNP laser was also made to operate in the chaotic regime. The power spectrum of the total intensity displayed only one peak (broadened by chaos) while the partial intensities displayed as many peaks as there were modes [13].

In order to study chaotic AD with the LNP laser, it was necessary to introduce a feedback to destabilize the laser [14]. The beam was modulated with N arbitrary frequencies and reinjected in the cavity. In this way, it was possible to analyze systematically the response, in the antiphase regime, of an N mode laser modulated with up to N frequencies. An interesting choice for the N frequencies is either the N natural oscillation frequencies or their nearest rational approximation. It was observed experimentally [15] that $2^N - 1$ different patterns could be generated in this way. They were analyzed numerically [16] and a classification of the patterns was proposed. In particular, the phenomenon of clustering (in which some modes are in phase and other are in antiphase) which had been introduced earlier [17] was shown to be a prominent feature.

A fascinating problem in AD is the analysis of the routes to chaos in the antiphase regime. The first report on this question is the experimental observation in a modulated Nd:YAG laser that while the total intensity displays the standard Feigenbaum route to chaos, partial mode intensities do not follow a regular sequence of subharmonic bifurcations [18]. When AD and passive Q-switching are combined, as in ISHG, we have found [19] that some of the modes can be chaotic while others remain periodic despite the global coupling of the modes.

From this brief review, it is clear that AD is a form of temporal selforganisation in multimode lasers. Therefore, we expect AD in Fabry-Perot (FP) lasers but not in ring lasers because a ring laser does not have a stable multimode regime. Thus an essential role is played by the spatial grating formed by the standing modes of the cavity fields. The grating affects the level of population inversion along the optical path. This spatial inhomogeneity induces an additional nonlinearity and coupling in the intensity evolution which is just sufficient to allow for multimode operation and AD. For ring cavities, an additional nonlinear element, such as the frequency doubling crystal in ISHG, is required. FP lasers do not necessarily display AD. Therefore, it will be necessary to find the domain of operating parameters in which this effect occurs. One negative criterion is that the mode-locking domain must be avoided.

2. Fabry-Perot lasers

The simplest and best documented model of laser physics which displays AD is the model of Tang, Statz and deMars [20] for a FP solid state laser:

$$\frac{dN_0}{dt} = w - N_0 - \sum_{k=1}^{N} \gamma_k \left(N_0 - \frac{N_k}{2}\right) I_k \,, \qquad (1\ a)$$

$$\frac{dN_m}{dt} = \gamma_m N_0 I_m - N_m \left(1 + \sum_{k=1}^{N} \gamma_k I_k\right) , \tag{1 b}$$

$$\frac{dI_m}{dt} = \kappa[\gamma_m(N_0 - \frac{N_m}{2}) - 1]I_m, \qquad m = 1, 2,..., N, \tag{1 c}$$

where N_0 is the space average of the population inversion $N(x,t)$, N_m is the space average of $2N(x,t)\cos(2k_m x)$, w is the pump parameter normalized to the laser first threshold, κ the photon decay rate which is assumed to be the same for all modes and γ_m is the gain of mode m normalized to the gain of the mode 1. The modes are labelled in order of increasing thresholds $1 \geq \gamma_2 \geq \gamma_3 \geq ... \geq \gamma_N$.

The steady state solution of eqs.(1) can be obtained explicitly in terms of a simple quadratic [9]:

$$w = N_0^0 + (N_0^0 S_1 - S_2)/[S_1 - (N - 1/2)N_0^0], \tag{2 a}$$

$$I_m^0 = (N_0^0 - 1/\gamma_m)/[\gamma_m S_1 - (N - 1/2)N_0^0 \gamma_m], \tag{2 b}$$

$$N_m^0 = 2(N_0^0 - 1/\gamma_m), \qquad S_p = \sum_k (1/\gamma_k)^p . \tag{2 c}$$

AD appears most clearly in the linear stability analysis of the steady solutions (2). We introduce the deviations from steady state

$$n_0 = \sqrt{\kappa}(N_0 - N_0^0), \quad n_m = \sqrt{\kappa}(N_m - N_m^0), \quad s_m = I_m - I_m^0 . \tag{3}$$

In terms of the scaled time $\tau = t/\delta$ with $\delta = 1/\sqrt{\kappa}$, one easily obtains from eqs.(1) the linearized evolution equations for the deviations

$$\frac{dn_0}{d\tau} = - \sum_k s_k - \delta\left(n_0 + \sum_k \gamma_k I_k^0 (n_0 - \frac{n_k}{2}) \right), \tag{4 a}$$

$$\frac{dn_m}{d\tau} = \gamma_m s_m N_0^0 - N_m^0 \sum_k \gamma_k s_k + \delta\left(\gamma_m n_0 I_m^0 - (1 + \sum_k \gamma_k I_k^0) n_m\right) , \tag{4 b}$$

$$\frac{ds_m}{d\tau} = \gamma_m I_m^0 (n_0 - \frac{n_m}{2}). \tag{4 c}$$

They depend on δ and the γ_m. In solid state lasers, κ varies typically between 10^3 and 10^6. Hence $\delta \ll 1$. Numerical and experimental results suggest that AD will occur when all the gains are close to unity. Let us therefore consider first the reference problem defined by $\delta = 0$ and $\gamma_m = 1$. The linear evolution equations become [10]

$$\frac{dn_0}{d\tau} = - \sum_k s_k, \quad \frac{dn_m}{d\tau} = s_m N_0^0 - \mathcal{N} \sum_k s_k, \quad \frac{ds_m}{d\tau} = \mathcal{I}(n_0 - \frac{n_m}{2}), \tag{5}$$

with $I_m^0 \equiv \mathcal{I}$ and $N_m^0 \equiv \mathcal{N} = 2(N_0^0 - 1)$. The total intensity $\mathbf{I} = \sum_k s_k$ verifies a closed equation $\mathbf{I}'' + (\mathcal{I}N + N_0^0 - 1)\mathbf{I} = 0$ which is characterized by a single frequency

$$\Omega_{R,0} = \sqrt{\mathcal{I}N + N_0^0 - 1} \,. \tag{6}$$

The analysis of the linear equations (5) indicates that there is one other frequency defined as

$$\Omega_{L,0} = \sqrt{\mathcal{I}N_0^0/2} \,. \tag{7}$$

This eigenvalue is $(N - 1)$-degenerate and is associated with the $N - 1$ eigenvectors $\{n_0 = 0,\ n_1 = 1,\ n_j = -1,\ s_1 = \pm i\Omega_{L,0}/N_0,\ s_j = \mp i\Omega_{L,0}/N_0\}$ with $j = 2, 3, \dots N$. The total intensity associated with these eigenvectors clearly vanish, implying AD. On the contrary, the eigenvectors associated with the high frequency (6) is of the form $\{n_0 = N,\ n_q = \beta,\ s_q = \pm i\Omega_{R,0};\ q = 1, \dots, N\}$. Thus in this reference problem, AD is clearly present.

If we keep $\delta = 0$ but allow for a dispersion of the gains, expressed by $\gamma_m = 1 - \varepsilon g_m$, $0 < \varepsilon \ll 1$, $g_m = \mathcal{O}(1)$, we find that the degeneracy of the low frequency root (7) is removed. The expansion $\Omega_L(p) = \Omega_{L,0} + \varepsilon\Omega_{L,1}(p) + \mathcal{O}(\varepsilon^2)$, $p = 1, 2, \dots, N -$

1, leads to an implicit equation for the ε corrections which has as many different roots as there are different gains.

On the contrary, if we keep all gains equal to unity but allow for a small δ, we find that the roots of the reference problem acquire an $\mathcal{O}(\delta)$ negative real part which is associated with damping. However, the N - 1 low frequency roots have the same real part $-\delta(1 + N\mathcal{I})/2$ which mean that they have the same linewidth. This linewidth scales like the total steady intensity $N\mathcal{I}$. The same scaling holds for the square of the high frequency (6).

A conjecture which remains without a proof and is in excellent agreement with the experimental results is the general form for the high frequency (6) which was proposed in [9]:

$$\Omega^2_{R,0} = (2 - \frac{N^0_0}{2})I^0_1 + (2 - N^0_0)\sum_{k=2}^{N} I^0_k \ . \qquad (8)$$

This expression is compatible with all analytical approximations of the high frequency derived as of now.

Thus in the limit of small photon life-time (in units of the population inversion life-time) and small dispersion of the gains, we expect to find AD characterized by one high frequency of oscillation (6) and N - 1 low frequencies weakly spread around (7). The transient evolution of the total intensity is characterized by the high frequency only whereas the modal intensities are characterized by all N frequencies. One way to display this property is to start with a laser in steady state and suddenly change the pump parameter. The transient relaxation towards the new steady state is recorded. Its power spectrum reveals the number of frequencies that characterize the dynamical variable. This was done experimentally and checked numerically using the rate equations (1). In the same way, we can add to eqs.(1) a source of noise. Using the stable steady state as initial condition, the stochastic solution is recorded and its power spectrum (or more precisely its noise spectrum) taken. We have found noisy AD in this way whenever there was deterministic transient AD.

3. Intracavity second harmonic generation (ISHG)

3.1. Steady and periodic states

The other model that has been studied for its AD properties is ISHG. The polishing of this model took years and the efforts of many groups. A recent account of this evolution is given in [5]. ISHG can be described by the rate equations for the modal intensities I_k and population inversions G_k

$$\tau_c \frac{dI_k}{dt} = (G_k - \alpha - g\varepsilon I_k - 2\varepsilon \sum_{j\neq k} \mu_{jk} I_j) I_k \ , \tag{9 a}$$

$$\tau_f \frac{dG_k}{dt} = \gamma - (1 + I_k + \beta \sum_{j\neq k} I_j) G_k \ , \tag{9 b}$$

where α is the cavity loss parameter, γ is the small signal gain which is related to the pump rate, β is the cross-saturation parameter and g is a geometrical factor whose value depends on the phase delays of the amplifying and doubling crystals and on the angle between the fast axes of these two crystals. We have assumed that α, β and γ are mode-independent, in good agreement with the experimental results [3]. τ_c and τ_f are the cavity round trip time and fluorescence lifetime respectively. The electric field modes can oscillate in one of two orthogonal polarizations. In eqs.(9), $\mu_{jk} = g$ if modes j and k have the same polarization while $\mu_{jk} = 1 - g$ if the modes have orthogonal polarizations. Let M and P be the number of modes in each of the two orthogonal polarizations, with M + P = N. Analytic results were obtained if either M = N or M = P. For arbitrary M and P, asymptotic results are available. Let us consider the simple case M = N where all modes oscillate in the same polarization. Then all modes have the same intensity and population inversion I_s and G_s, respectively. A linear stability of this steady solution indicates that it remains stable if

$$g < \frac{\tau_c}{\tau_f} \frac{\gamma}{\varepsilon I_s G_s} \ , \quad g < \frac{1 - \beta}{\varepsilon} \frac{G_s^2}{\gamma} \ . \tag{10}$$

Otherwise, an instability occurs. For the parameters of the

experiments reported in [3], $\beta < 1$ and $\varepsilon \ll 1$, so that the first bifurcation point is a Hopf bifurcation occurring at

$$I_H = \frac{\tau_c/\tau_f}{g\varepsilon - [1 + (M-1)\beta]\tau_c/\tau_f}, \tag{11}$$

provided that the geometrical factor verifies the condition

$$g > \frac{\tau_c}{\tau_f}\frac{1 + (M-1)\beta}{\varepsilon}. \tag{12}$$

In the vicinity of the Hopf bifurcation, we seek periodic solutions of (9) of the form

$$Z_k(\delta,T) = Z_H + \delta Z_{k1}(T) + \mathcal{O}(\delta^2), \qquad Z = I \text{ or } G, \tag{13}$$

$$\gamma\tau_f/\tau_c = \gamma_H \pm \delta^2, \qquad I_H(\gamma_H) = I_S(\gamma_H),$$

$$T = \omega\tau, \qquad \omega^2 = I_H G_H(1-\beta) - g\varepsilon I_H\gamma_H/G_H.$$

An essential property which derives directly from the evolution equations (9) is that the sums $S_I = \sum_k I_{k1}$ and $S_D = \sum_k D_{k1}$ vanish in the long time limit. It was shown in [5] that this vanishing implies for the modal intensities the form $I_{j1} = a_j\exp(iT) + \text{c.c}$ with the constraint $\sum_k a_k = 0$. Using the identity $\sum_{k=1}^{M} x^k = \frac{x(1-x^M)}{1-x}$ with $x = \exp(2\pi im/M)$, it is easy to verify that a solution of the constraint $\sum_k a_k = 0$ is $a_k = \exp(2\pi ikm/M)$ where $k = 1, 2, \ldots, M$ and $m = 0, 1, 2, \ldots, M-1$. Hence there is one in-phase state ($m = 0$) and $(M-1)!$ anti-phase states, each of which is shifted from another by $1/M$ of the period.

3.2. Passive Q-switching

As some control parameter is varied, the solution moves away from the Hopf bifurcation and it has been observed, experimentally and numerically, that there is a tendency of the periodic solution which is harmonic near the bifurcation point to become pulsed. When this form of passive Q-switching is coupled to AD, interesting new properties can occur as recently reported [19]. Because of the AD, each pulse may occur while all the other modes are practically off. In this situation, the coupling is mainly due to the overlap (if any) between the pulses and the requirements of continuity in time. In the case M > 1 and P = 1, we have observed that the M modes cooperate to produce a Feigenbaum sequence of subharmonic bifurcations to chaos followed by an inverse sequence, while the mode in the orthogonal polarization remains periodic. This is possible only because there is practically no overlap between the P mode and the M modes, while the M modes have a strong overlap among them. Such a situation offers the surprising case of a chaotic system with a synchronous internal clock.

Acknowledgments

This research was supported in part by the Fonds national de la Recherche Scientifique (Belgium) and the Interuniversity Attraction Pole of the Belgian government.

References

1. P.Hadley and M.R.Beasly, Appl. Phys. Lett. **50** (1987) 621.
2. K.Wiesenfeld and P.Hadley, Phys. Rev. Lett. **62** (1989) 1335; K.Y.Tsang and K.Wiesenfeld, Appl. Phys. Lett. **56** (1990) 495.
3. K.Wiesenfeld, C.Bracikowski, G.James and R.Roy, Phys. Rev. Lett. **65** (1990) 1749.
4. P.Le Boudec, C.Jaouen, P.L.François, J.-F.Bayon, F.Sanchez, P.Besnard and G.Stephan, Opt. Lett. **18** (1993) 1890 .
5. Wang Jingyi and P.Mandel, Phys. Rev. A **48** (1993) 671.
6. K.Otsuka, Phys. Rev. Lett. **67** (1991) 1090.
7. K.Otsuka, P.Mandel, S.Bielawski, D.Derozier and P.Glorieux Phys. Rev. A **46** (1992) 1692.
8. S.Bielawski, D. Derozier and P. Glorieux, Phys. Rev. A **46** (1992) 2811.

9. P.Mandel, M.Georgiou, K.Otsuka and D.Pieroux, Opt. Commun. **100** (1993) 341.
10. D.Pieroux and P.Mandel, Opt. Commun. (in press, 1994).
11. K.Otsuka, M.Georgiou and P.Mandel, Jpn. J. Appl. Phys. **31** (1992) L1250.
12. C.Bracikowski and R.Roy, Chaos **1** (1991) 49; Phys. Rev. A **43** (1991) 6455.
13. K.Otsuka, P.Mandel, M.Georgiou and C.Etrich, Jpn. J. Appl. Phys. **32** (1993) L318.
14. K.Otsuka, Jpn. J. Appl. Phys. **31** (1992) L1546.
15. K.Otsuka, D.Pieroux and P.Mandel, *Modulation dynamics and spatiotemporal pattern generation in a microchip multimode laser* (preprint).
16. D.Pieroux, P.Mandel and K.Otsuka, *Modulation dynamics in a multimode laser with feedback* (preprint).
17. K.Otsuka and J.-L.Chern, Phys. Rev. A **45** (1992) 8288.
18. N.B.Abraham, L.L.Everett, C.Iwata and M.B.Janicki, *Multimode dynamics of a sinusoidally modulated diode-pumped Nd:YAG laser* (preprint).
19. P.Mandel and Jingyi Wang, *Dynamical independence of pulsed antiphased modes* (preprint).
20. C.L.Tang, H.Statz and G.deMars, J. Appl. Phys. **34** (1963) 2289.

Electron-Photon Manipulation with Sub-Poissonian Semiconductor Junction Arrays

P.J. Edwards

Quantum Electronics Group, Faculty of Information Sciences and Engineering, University of Canberra, P.O. Box 1, Belconnen, ACT 2616, Australia

Abstract. Laboratory investigations have demonstrated the relative ease with which sub-Poissonian light may be generated by semiconductor junction laser and light-emitting diodes. A wide variety of electron - photon manipulations is possible with electrically and optically coupled arrays of sub-Poissonian semiconductor light emitters and receivers. We discuss systems capable of performing, inter alia, multiple sub-Poissonian noise-correlated beam generation; low noise photon number amplification, beam splitting and tapping; "QND-like" photon number measurement, low noise optoelectronic amplification; and sub-shot noise interferometry.

1. Introduction

The generation of sub-Poissonian light (SPL), that is, light having photon counting noise below the normal shot noise limit (SNL) with laser and light emitting diodes using the high impedance noise suppression method described by Yamamoto et alia [1], is now well established [2,3,4,5,6,7]. This technique offers a technically simple way of reducing quantum noise below the SNL in closely coupled semiconductor junction optoelectronic systems. The amount of noise suppression is limited mainly by the magnitude of the quantum transfer efficiency which can be realised between light emitter and detector [8]. Such a reduction in quantum noise is of practical interest. It has applications in shot noise - limited metrology, including gravitational wave detection, and in shot noise - limited information transfer, including optical computing and lightwave communications. Realisation of SPL based systems in the form of optoelectronic integrated circuits employing optically interconnected microcavity semiconductor laser and diode detector arrays appears likely to provide a range of novel electron - photon manipulations and applications.

2. Sub-Poissonian Light from Semiconductor Junctions

Sub-Poissonian light generation from open loop semiconductor light emitters was first demonstrated in 1987 by Tapster et alia [2] using infrared emitting diodes and by Machida et alia [3] using a laser diode. Infrared emitting diodes, cooled to increase their quantum transfer efficiency, η, to approximately 11%, exhibited small, up to 4%, noise reduction below the SNL. Machida et alia [3] measured a 7.3% noise reduction using a laser diode with $\eta = 0.3$ at room temperature This was followed by measurement of a 32% noise reduction for a laser diode cooled to reduce the threshold current and so permit strong pumping to reduce amplified spontaneous noise [4]. Because of incomplete electronic source noise suppression, none of these early measurements gave results in close accordance with the theoretical expectations (Equation (1)) of a fractional noise reduction below the SNL, numerically equal to the quantum transfer efficiency, η. However, Edwards et alia [5, 6, 7] measured intensity noise reductions of up to 30% with cooled infrared emitting diodes, which were in good agreement with the expression for the Fano factor [8],

$$F_o = \eta \cdot F_i + (1-\eta) \simeq 1-\eta, \quad (1)$$

for negligible source noise as specified by the Fano factor, F_i.

The applicability of the above equation has been subsequently confirmed by other workers [9, 10]. It has a simple semiclassical interpretation in terms of the random deletion of photons giving rise to the binomial number variance represented by the second term.

All the above demonstrations utilised the high impedance method of noise suppression [1].

2.1 High Impedance Current Noise Suppression

The photon number (lightwave intensity) fluctuations generated by a semiconductor source of unit quantum efficiency are determined solely by electron hole recombination current fluctuations in the source. Consequently, by raising the impedance of the photon generating current loop, low frequency electronic and photonic current fluctuations can be reduced to generate a near number state with $F_i << 1$. Referring to the conventional noise equivalent circuit [11] for a junction diode in Figure (1) we identify the noise current in the branch containing the differential resistance of the diode as the recombination current fluctuation, driven by the voltage generator, v_n, representing the internal stochastic processes (spontaneous emission, absorption, stimulated emission, and diffusion) and by Nyquist voltage generator, v_s, representing the external thermal pump noise.

This figure serves to illustrate the high impedance noise suppression principle for semiconductor junction light emitters generally, whether or not these exhibit population inversion and stimulated emission. At low frequencies, $\omega << 1/\tau$, the current through the "diffusion capacitance", C, may be neglected and the junction current and recombination current are consequently identical and easily shown to be negligible compared with the shot noise level providing

$$|Z| >> r \cong kT / e \cdot I \quad (2)$$

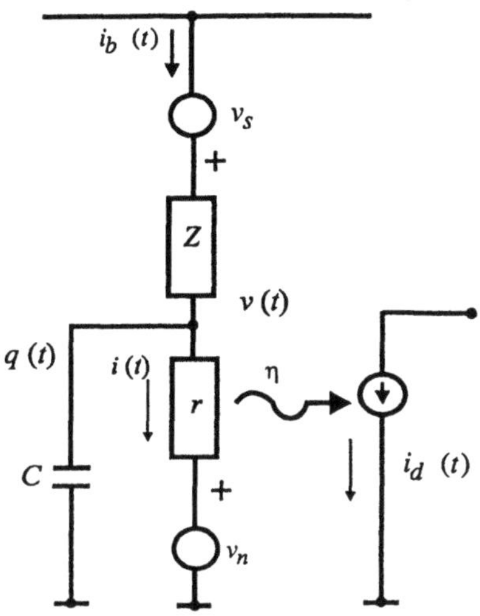

Figure 1

Ideal light-emitting semiconductor junction diode noise equivalent circuit. The overall quantum efficiency, η, is defined as the ratio of detector current $i_d(t)$ to junction current, $i(t)$.

We and others [6, 9, 10] have verified electronic noise levels of greater than 15 dB below the shot noise level with infrared emitting diodes driven by high impedance current sources. We have also carried out computer simulations of the systems which closely match the measurements and provide confidence in this simple model [12].

Although the capacitive current may be neglected at low frequencies, the capacitive element provides an important conceptual link with the recombination process. Thus for a light-emitting semiconductor junction, the carrier lifetime is easily shown to be the product of the differential resistance and capacitance. At high frequencies, or with a low impedance pump, the capacitor voltage, and therefore the charge carrier number in the active region of the device, the upper level population, is held constant. Consequently, we expect, and observe "classical" Poissonian photon emission. If, on the other hand, the junction voltage is allowed to fluctuate, which, by inspection of Figure (1) it will do at low frequencies provided $|Z| >> r$, the population number will also fluctuate. It will do so in a way which reflects the depletion of the population number by the stochastic emission processes. This can be represented as a classical negative feedback process and can be readily computer modelled [12].

2.2 Semi-Classical Model of SPL Generation

Referring to the light emitting diode equivalent circuit of Figure (1), the state equation for the junction voltage fluctuation, v, immediately yields the rate equation for the charge carrier population number fluctuation, $n(t)$ in the form,

$$dn/dt = -n(1/\tau + 1/RC) + f_n(t) \quad (3)$$

$$= [i_b(t)/e - s(t)] \, , \quad (4)$$

where the Langevin noise term, f_n, has covariance,

$$< f_n(t) \cdot f_n(t+t') > \; = \; (2 < N > /\tau \; + \; 4kT / e^2 R) \cdot \delta(t') \, . \quad (5)$$

At low frequencies, $\omega << 1/\tau$, $dn/dt = 0$, and the photon current fluctuation, $s(t)$, and electron current fluctuation, $i_b(t)/e$, are therefore equal. If, in addition, the condition $RC >> \tau$, is satisfied, then $n(t) = f_n(t) \cdot \tau$, and the spectral density of the charge number fluctuation, $S_n(\omega) = 2<N>\tau$, is as expected for a Poisson variable of mean, $<N>$. The spectral densities of the electron and photon noise currents then both vanish in the limit, $x=r/R=0$. Consideration of the rate equation shows that, as the pump impedance is raised, the thermal noise in the current pump eventually determines the low frequency quantum noise level For this diode model [11], which differs slightly from that discussed by Yamamoto and Machida [1], the photon Fano factor for unit quantum efficiency is equal to $x/(1+x)$.

An equivalent view, following Teich and Saleh, is to treat the electron recombination as a self excited point process exhibiting rate compensation by excitation feedback [8]. Low frequency stochastic fluctuations in the recombination rate are permitted to deplete and enhance the charge carrier population. It is this fluctuating population number (or alternatively, the fluctuating junction voltage) which mediates the negative feedback mechanism responsible for the quantum noise reduction on time scales long compared with the charge carrier lifetime [11].

The final quantum noise level at the detector is then calculable using equation (1). This equation lumps all photon number losses together in the form of a single quantum transfer efficiency, $\eta \leq 1$. This is the product of all the (statistically independent) probabilities which characterise the photon deletion processes in the chain from initial electron injection to final photoelectron excitation. Figure 3(a) shows a simplified representation of a sub-Poissonian light emitter.

3. Arrays of Correlated SPL Emitters

3.1 Generation of Multiple SPL Beams and Photon Number Amplification

It is apparent from the discussion above, that a series connected pair of semiconductor light emitters should exhibit positively correlated, sub-Poissonian photon number fluxes on long time scales. Consideration of classical electrical partition noise at a current branch indicates that shunt connected light emitters should provide negatively correlated sub SNL photocurrents. That such is the case has now been well verified [6, 10] and used to validate a partition model for the largest (8.3 dB) quantum noise reduction observed to date with semiconductor light emitters[13].

The low frequency noise equivalent circuit for a tandem array of diode emitters is shown in Figure (2). It is evident that a different noise reduction mechanism [14] is responsible for the SPL operation of such an array. This new mechanism is analogous to the dynamic pump noise suppression technique proposed for multilevel lasers [15]. It is well known in exponential queuing theory and leads to an N-Erlang recombination event waiting time distribution for any element in an array of N diode emitters. Fano factors and covariances are readily calculated [6]. Again, the physical mechanism responsible for SPL generation is evident in the occurrence of (correlated) low frequency fluctuations in population numbers, as for the high impedance noise suppression method discussed in the previous section.

The light from these individual beams may be separately detected or combined into a single composite beam and then detected. These configurations form the basis for low noise quantum taps (Fig. 3(b)), optoelectronic amplification, QND-like measurement and photon number amplification (Fig. 3(c)) [14, 16].

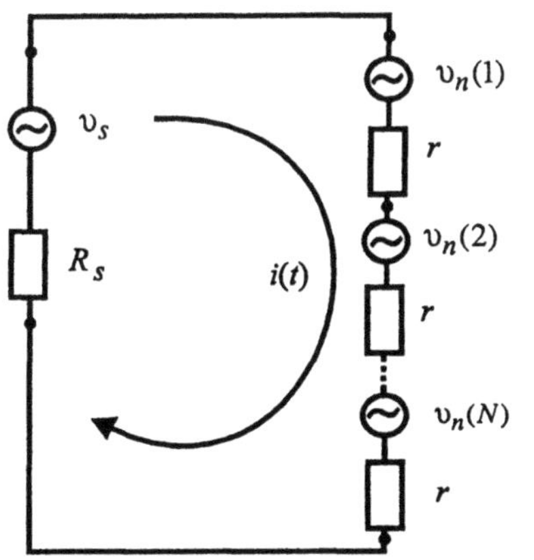

Figure 2

Low frequency noise equivalent circuit for a tandem array of N junction diode light-emitters connected to voltage source, υ_s.

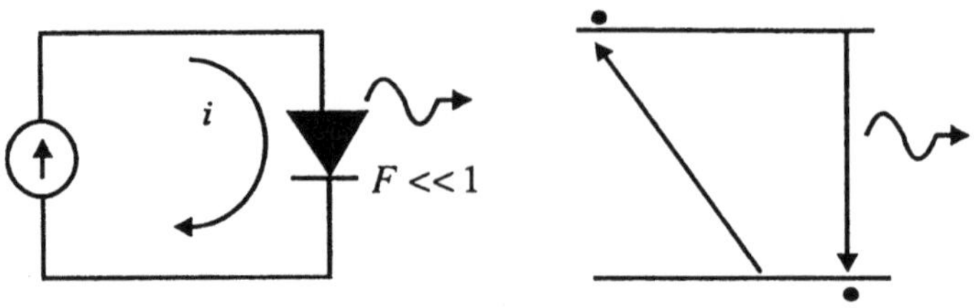

Figure 3(a)

Idealised representation of sub-Poissonian light-emitting semiconductor junction with photon number Fano factor, F.

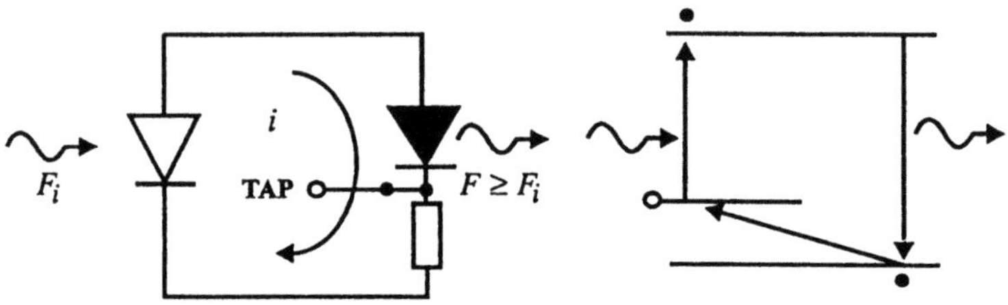

Figure 3(b)

Idealised representation of low noise photon number tap with input photon number Fano factor, F_i.

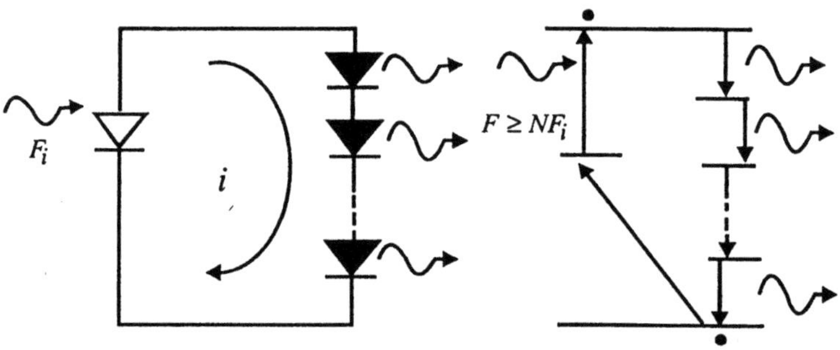

Figure 3(c)

Idealised representation of a photon number amplifier employing an array of sub-Poissonian light-emitting junctions.

It is evident from the preceding discussion, that a tandem array of SPL emitters provides the basis for photon number amplification. The tandem diode array consists of an array of N identical light emitters, each generating a sub-Poissonian beam with Fano factor, F_i. Both the noise and signal modulating currents are evidently the same in each element of the series connected array. Consequently the signal to noise ratio in the incoherently combined beams is the same as that in each of the individual beams while the optical power and the Fano factor are both increased N-fold. More generally, after attenuation and/or non-ideal detection, the Fano factor becomes

$$F_o = \eta \cdot N \cdot F_i + (1-\eta). \tag{6}$$

In the ideal case, $F_i = 0$, the combined beam remains in a number state. This arrangement therefore constitutes a low noise, high power modulated SPL generator. Efficiently coupled to a light detecting array, it constitutes a low noise optoelectronic amplifier. If driven by an ideal detector (Figure 3c) the array functions as a low noise photon number amplifier with a limiting noise figure of unity even for a number state input [14].

We have recently demonstrated the SPL operation of such an array in each of the above modalities and verified the utility of equation (6) for an SPL emitting array of eight infrared emitting diodes with a current gain of 2 and an optical power gain exceeding unity [17].

3.2 Low Noise Photon Number Replication

Yuen [21] has suggested the use of photon number preamplification of sub-Poissonian states as a means of partially compensating for their extreme fragility in a lossy environment. It is well known that an initially noiseless photon number state, with Fano factor, $F_i = 0$, rapidly degrades towards the Poissonian level with $F_i = 1$, through a succession of intermediate binomial states with binomial number variance equal to $\eta \cdot (1-\eta) \cdot < \eta_i >$ as the overall system quantum efficiency, η, decreases towards zero. Following preamplification, the direct detection signal to noise ratio is then, in terms of the input photon number, n_i, or detector photocurrent, I, measured over time interval, Δt,

$$\mathrm{SNR} = < I >^2 / < i^2 > = N \cdot < n_i > \cdot \eta \cdot / (1-\eta) \tag{7}$$

for mean photocurrent, $< I >$, and mean square photocurrent noise, $< i^2 >$.
This is a factor of $N/(1-\eta)$ higher than for a shot noise limited (Poissonian) signal with the same initial mean photon number, $< n_i >$ However the fragility remains in the form of the η / (1 - η) dependence of the signal to noise ratio.

The tandem diode array can be used to generate a robust form of light signal in which the signal to noise ratio remains enhanced by the factor ($1/F_i$) above that for coherent light with the same initial power as that of a single element, independently of attenuation and overall quantum transfer efficiency, η .

In order to achieve this, each element of the array is operated with Fano factor, $F_i < 1$, and the composite beam is arranged to be super-Poissonian with Fano factor, $N \cdot F_i >> 1$.

After attenuation (and/or non-ideal detection) with quantum transfer efficiency, η the signal to noise ratio becomes,

$$\begin{aligned} SNR &= \langle n_o \rangle / F_o \\ &= N \cdot \eta \cdot \langle n_i \rangle / [1 + \eta \cdot (N \cdot F_i - 1)] \\ &= \langle n_i \rangle / F_i . \end{aligned} \quad (8)$$

Thus, the signal to noise ratio is independent of η , and equal to that at one element of the transmitter, providing the condition $N \cdot F_i \cdot \eta >> 1$ is satisfied.

3.3 Low Noise Optoelectronic Amplification

Consider a tandem array of N high quantum efficiency light emitters, connected to a modulating signal source, each coupled with overall quantum efficiency, η, to a lightwave detector array. Referring to the noise equivalent circuit of Figure (2), it may be shown [14] that the excess system noise temperature,

$$T_s = \langle I \rangle \cdot e \cdot R \cdot (F_i + [1-\eta] / N \cdot \eta) / 2k. \quad (9)$$

Thus, for example, looking ahead to low threshold microlaser/LED array structures operating at room temperature with a quantum transfer efficiency of 90%, and a pump current of 1 mA, a power gain of 20 dB can be achieved in a 50 Ω system with an excess temperature of no more than 3 K.

3.4 Noiseless Photon Number Tapping

Figures (3b) and (4) shows how semiconductor junction detector and emitter arrays may be configured to effect a non destructive measurement of photon number. Evidently, non destructive measurement of the intensity of the incident beam can be achieved and the system can be used as a low noise quantum tap [22]. The necessary photon number to electron number conversion gain may also be achieved with a low noise electronic amplifier as described by Roch et alia [23] who show that necessary (although not sufficient) QND criteria are formally satisfied.

The photocurrent from a junction diode detector may also be used to pump a tandem light emitting diode array as in Figure (4). If each emitter is coupled with unit efficiency to a separate detector or information channel then noiseless N-fold replication of the detected signal is achieved. If the input detector is also of unit efficiency, then noiseless splitting of the photonic signal will have been effected If the emitting array is optically coupled to a single detector, array of detectors or single information channel as in Figure (3c), the configuration constitutes a photon number amplifier with a limiting noise figure of unity [11]. The summed detector array currents, may in turn be used to drive a single light emitter for onward transmission of the regenerated optical signal as shown in Figure (4). Goobar et

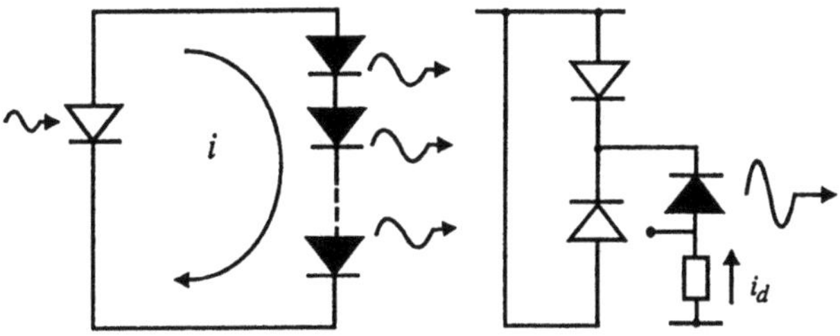

Figure 4
Idealised representation of a photon number amplifier/tap employing arrays of SPL emitters and detectors.

alia [16], following Yuen [24], describe the possible use of these configurations in transparent optical communication networks.

3.5 Sub-Poissonian Lightwave Interferometry

Holland and Burnett [18] have shown that the minimum level of phase noise permitted by the Heisenberg relation, $1/n$ radians, is reached in an optical interferometer in which both input ports are driven by number states containing equal photon numbers, n. The intensity difference noise and phase difference noise between the fields at the output ports of the first beam splitter in the interferometer are linked by the Heisenberg relation. Photon interference results in a number variance in the intensity difference (for large n) of $n^2/2$ [19] and hence a variance in the phase noise of $1/n^2$. The dual Fock state, $|nn>$, appears to provide advantages similar to those of the two mode squeezed state suggested by Caves [20], and offers potentially very great improvement in sensitivity over a shot noise limited system with coherent input to a single port. It is therefore appropriate to ask whether the dual number state, or some approximation to it, can be realised with "wired optics".

Strongly correlated, sub-Poissonian binomial states can readily be generated with series connected "tandem" arrays of light emitting diodes [6]. Although not yet demonstrated, one may expect this to apply also to arrays of laser diodes. The necessary requirement of first order coherence could be satisfied by a pair or an array of laser diodes enclosed within a common cavity or by a coupled cavity laser diode system.

4. Conclusions

The use of macroscopic electron currents to control, generate and manipulate sub-Poissonian photonic currents, foreshadows a variety of low noise applications.

The systems which have been demonstrated to date have utilised the relatively high electron to photon and photon to electron quantum conversion efficiencies available with semiconductor detectors, and to a lesser degree, with laser diodes and light emitting diodes at low temperatures.

The practicality of these wired optic systems will depend on the ability to achieve high quantum efficiencies at room temperature in arrays of low threshold, high spontaneous emission coefficient microcavity lasers fabricated in optoelectronic integrated circuit form.

References

1. Y. Yamamoto and S. Machida, Phys. Rev. A 35, 12, 5114 (1987).
2. P. R. Tapster, J. G. Rarity and J. S. Satchell, Europhys. Lett., 4 (3), 293 (1987).
3. S. Machida, Y. Yamamoto and Y. Itaya, Phys. Rev. Lett., 58, 10, 1000 (1987).
4. S. Machida and Y. Yamamoto, Phys. Rev. Lett., 60, 9, 792 (1988).
5. P. J. Edwards, Int. J. Optoelectronics, 6, 23 (1991).
6. P. J. Edwards and G. H. Pollard, Phys. Rev. Lett., 69, 12, 1757 (1992).
7. P. J. Edwards, G. N. French, P. Lyons, J. P. Rayner, in *Abstracts (2), 23rd General Assembly, URSI*, (URSI, Prague, 1990), p. 318.
8. M. C. Teich and B. E. A. Saleh, in *Progress in Optics* 26, (E.Wolf, Ed. Amsterdam: North Holland, 1988).
9. H. Bachor, P. Rottengatter and C. M. Savage, Appl. Phys. B55, 258 (1992).
10. E.Goobar, A. Karlsson, G. Bjork and P. J. Rigole, Phys. Rev. Lett., 29, 299 (1993).
11. P. J. Edwards, IEEE J. Quantum Electronics, 29, 3, 299 (1993).
12. P. J. Edwards, W. N. Cheung and B. Santosa, in *Abstracts, 24th General Assembly of the International Union of Radio Science*, (URSI, Kyoto, 1993), p.158.
13. W. H. Richardson, S. Machida and Y. Yamamoto, Phys. Rev. Lett., 66, 22, 2867 (1991).
14. P. J. Edwards, Electron. Lett., 29, 3, 299 (1993).
15. T. C. Ralph and C. M. Savage, Optics Lett., 16, 14, 1113 (1991).
16. E. Goobar, A. Karlsson and G. Bjork, in *Proceedings 19th European Conference on Optical Communication (ECOC '93)*, (SEE, Valbonne, 1993), p. 385.
17. P. Lynam and P. J. Edwards, presented at *Sixth International Symposium on Quantum Optics*, (University of Auckland, Rotorua, 1994).
18. M. J. Holland and K. Burnett, Phys. Rev. Lett., 71, 9, 1355, (1993).

19. R. A. Campos, B. E. A. Saleh and M. C. Teich, Phys. Rev. A 40, 1371 (1989).
20. C. M. Caves, Phys. Rev. D, 23, 8, 1693 (1981).
21. H. P. Yuen, Phys. Rev. Lett., 56, 20, 2176, (1986).
22. J. Ph. Poizat and P.Grangier, Phys. Rev. Lett., 70, 3, 271 (1993).
23. J. F. Roch, J. Ph. Poizat and P. Grangier, Phys. Rev. Lett., 71, 2006 (1993).
24. H. P. Yuen, Opt. Lett., 12, 789 (1987).

Near Dipole-Dipole Effects in Nonlinear and Quantum Optics with Applications to Piezophotonic Switching

C.M. Bowden[1], *J.P. Dowling*[2], *A.S. Manka*[2], *and M. Fleischhauer*[3]

[1]Weapons Sciences Directorate, AMSMI-RD-WS-ST Research, Development, and Engineering Center, U.S. Army Missile Command, Redstone Arsenal, AL 35898-5248, USA

[2]National Research Council Research Associates, AL 35898, USA

[3]Department of Physics, Ludwig-Maximilians-Universität, Munich, D-80333, Germany

Abstract. We summarize a new field-theoretical derivation of the near dipole-dipole or local field correction for dynamic electromagnetic fields interacting with a nonlinear medium. We present some of our recent results on a coherently-prepared collection of three-level atoms in a lasing without inversion scheme. In this system, we find local-field enchanced inversionless gain and absorptionless index, as well as a pressure sensitive or piezophotonic switching effect.

1. Introduction

It is well known that for static fields in a dielectric medium the local microscopic field $\mathbf{E}_L$, that couples to a point dipole in the medium, is linearly related to the macroscopic field $\mathbf{E}$ and the macroscopic volume polarization $\mathbf{P}$ by the Lorentz-Lorenz (LL) relation [1], that is,

$$\mathbf{E}_L = \mathbf{E} + \left[\frac{4\pi}{3} + s\right]\mathbf{P}. \tag{1.1}$$

Here, s is a structure factor that is zero for cubic or higher local symmetry.

The induced dipole moment $\mathbf{p}$ for an atom or molecule is related to the local field $\mathbf{E}_L$ by the polarizability α, where,

$$\mathbf{p} = \alpha\,\mathbf{E}_L. \tag{1.2}$$

The macroscopic volume polarization $\mathbf{P}$ is determined for an isotropic homogeneous medium in terms of the local field $\mathbf{E}_L$ and the polarizability α by the relation

$$\mathbf{P} = n\,\alpha\mathbf{E}_L, \tag{1.3}$$

where n is the volume density of microscopic dipoles. Instead of Eq. (1.1), we may equivalently write $\mathbf{E}_L = \mathbf{E} + 4\pi\mathbf{P}/3 + \mathbf{E}'$, where $\mathbf{E}'$ is the contribution to the field

Springer Proceedings in Physics, Vol. 77 Quantum Optics VI
Editors: D.F. Walls · J.D. Harvey

that couples to **p** via Eq. (1.2). This field arises from neighboring induced dipoles within a range of volume where discrete spacial properties are important, such as in a crystal lattice. For cubic or higher local symmetry, $E' = 0$ and thus $s = 0$. If we start with the local field $\mathbf{E}_L$ from Eq. (1.3), using Eq. (1.1) with $s = 0$, we can eliminate **P** using the constitutive relation

$$\mathbf{P} = \chi \mathbf{E}, \tag{1.4}$$

where χ is the electric susceptibility defined in terms of the dielectric function ε as,

$$\chi = \frac{\varepsilon - 1}{4\pi}. \tag{1.5}$$

Then, by using Eqs. (1.1) and (1.3) relating **P** and $\mathbf{E}_L$, we arrive at the important Clausius-Mossotti relation for the microscopic polarizability α and the macroscopic dielectric function ε, namely,

$$\alpha = \frac{3}{4\pi n}\left(\frac{\varepsilon - 1}{\varepsilon + 2}\right). \tag{1.6}$$

Thus, the microscopic polarizability α is a nonlinear function of the macroscopic dielectric function ε. Conversely, Eq. (1.4) can be used in conjunction with Eq. (1.1) and Eq. (1.3) to obtain

$$\chi = \frac{n\alpha}{\left[1 - \frac{4\pi}{3} n\alpha\right]}, \tag{1.7}$$

that gives the macroscopic susceptibility χ as a nonlinear function of the microscopic polarizability α.

We have derived an expression analogous to Eq. (1.1) for time-varying fields and, in particular, for the case where the medium is nonlinear in the local field and requires equations of motion for the atomic or molecular variables [2]. For isotropic homogeneous media, an expression identical to the LL relation, Eq. (1.1), is obtained with $s = 0$, but the result is nontrivial since one must explicitly account for causality and retardation in the case of nonstatic fields. For steady states, relations similar to Eqs. (1.6) and (1.7) follow; however, for nonlinear media we have in general that $\alpha = \alpha(E_L)$.

Section two is used to illustrate the effect of the local field on induced quantum coherence in a nonlinear optically dense medium composed of a collection of three-level atoms of the lambda energy level configuration. A novel, sensitive, density-dependent transition of the medium from strongly-enhanced absorption to strongly-enhanced gain is demonstrated. The pressure sensitive phenomenon, which we call the piezophotonic effect [3], is shown to be due entirely to local field effects combined with quantum coherence. Finally, the last section is used for a summary of results, conclusions, and a citation to further work.

2. Local Field Effects in Lasing Without Inversion: Piezophotonic Switching

It is well known that quantum coherence and interference in atomic systems can lead to interesting optical phenomena such as electromagnetically induced transparency [4], lasing without a population inversion [5], and generation of a large absorptionless index of refraction [6]. To our knowledge, in all of these works it is assumed that the density of atoms n is small (10^{13} atoms/cm^3 or less) compared to a cubic resonance wavelength. We wish to assay the role of local field or NDD contributions in systems where induced quantum coherence is important. Several recent studies by the present authors have shown that NDD effects may lead to a substantial enhancement of the inversionless gain and the absorptionless index of refraction [3, 7]. In an optically dense medium, the real and imaginary parts of the susceptibility — the dispersion and gain, respectively — have resonances of the Clausius-Mossotti sort, even in the low field limit. Hence, these quantities can be significantly enhanced due to the local field [7]. It was also pointed out in reference 7 that, for a certain threshold density, the dispersion curve flips sign due to NDD induced nonlinear effects. These results were obtained under the assumption of a linear atomic response and are therefore only valid for sufficiently small local fields. For the present discussion, we drop the approximation of the weak electric field limit and we study the influence of NDD interactions by taking into account the full nonlinear response of the atomic system. We predict an unusual and novel density-dependent switching effect that occurs for intermediate field strengths. In particular, the gain curve can be made to switch suddenly from high loss to high inversionless gain as a sensitive function of the atomic density.

To demonstrate the above-mentioned piezophotonic effect, we investigate a medium composed of three-level atoms in a Λ configuration, with the lower two levels, $|b\rangle$ and $|b'\rangle$, closely spaced and initially prepared in a coherent superposition state. Figure 1 displays the level diagram of the atoms of interest. These atoms are now injected into an interaction region of volume V at an injection rate r, and an external probe field E couples the upper level $|a\rangle$ to the coherent superposition of the two lower levels. The atomic preparation of the two lower levels can be accomplished by a coherent pulse, e.g., a microwave field, applied just before the atoms enter the interaction region. Using the rotating wave approximation and the Scully-Lamb technique of coarse-grained averaging [8] , we obtain the equations of motion for the density matrix elements:

$$\dot{\rho}_{aa} = r\rho^0_{aa} - \gamma_a\rho_{aa} - \frac{i}{\hbar}\left[(\mu\rho_{ba} + \mu'\rho_{b'a})E_L - c.c.\right], \tag{2.1a}$$

$$\dot{\rho}_{b'b'} = r\rho^0_{b'b'} - \gamma_{b'}\rho_{b'b'} - \frac{i}{\hbar}\left[\mu'\rho_{ab'}E^*_L - c.c.\right], \tag{2.1b}$$

$$\dot{\rho}_{bb} = r\rho^0_{bb} - \gamma_b\rho_{bb} - \frac{i}{\hbar}\left[\mu\rho_{ab}E^*_L - c.c.\right], \tag{2.1c}$$

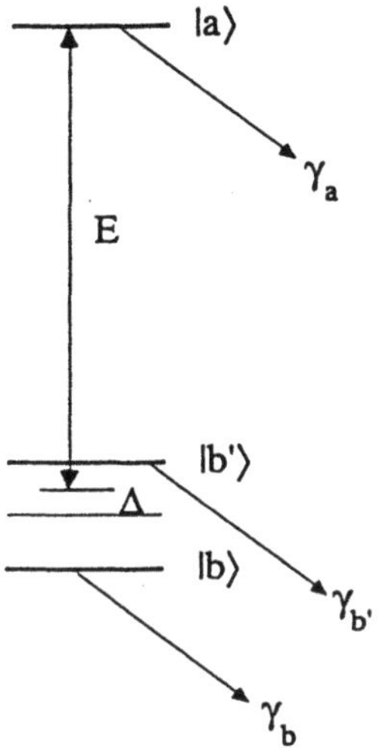

Figure 1. A schematic energy level diagram of the three-level system studied

$$\dot{\rho}_{ab'} = -(\gamma_{ab'} + i\Delta_{ab'})\rho_{ab'} - \frac{i}{\hbar}\mu'(\rho_{b'b'} - \rho_{aa})E_L - \frac{i}{\hbar}\mu\rho_{bb'}E_L, \tag{2.1d}$$

$$\dot{\rho}_{ab} = -(\gamma_{ab} + i\Delta_{ab})\rho_{ab} - \frac{i}{\hbar}\mu(\rho_{bb} - \rho_{aa})E_L - \frac{i}{\hbar}\mu'\rho_{b'b}E_L, \tag{2.1e}$$

$$\dot{\rho}_{b'b} = r\rho^0_{b'b} - (\gamma_{b'b} + i\omega_{b'b})\rho_{b'b} - \frac{i}{\hbar}\left(\mu'\rho_{ab}E^*_L - \mu\rho_{b'a}E_L\right), \tag{2.1f}$$

where ρ^0_{ij} are the elements of the initial density matrix ρ^0 of an injected atom that is of the form

$$\rho^0 = \begin{pmatrix} \rho^0_{aa} & 0 & 0 \\ 0 & \rho^0_{b'b'} & \rho^0_{b'b} \\ 0 & \rho^0_{bb'} & \rho^0_{bb} \end{pmatrix}.$$

Here μ' and μ are the dipole matrix elements (assumed real) of the $|a\rangle \rightarrow |b'\rangle$ and $|a\rangle \rightarrow |b\rangle$ transitions, respectively; $\Delta_{ij} \equiv \omega_{ij} - \nu$, where ν is the probe field frequency (i,j = a,b',b); and E_L is the local field. We also define the $\gamma_i = \gamma_i^{sp} + 1/\tau$ as the decay rates out of the various levels depicted in Fig. 1, where the rate of spontaneous decay to other levels is identified by γ_i^{sp}, and τ is the mean time-of-flight through the interaction region. The detuning from the center of the lower two energy levels is defined in terms of Δ_{ij} as $\Delta \equiv \Delta_{ab'} + \omega_{b'b}/2 = \Delta_{ab} - \omega_{b'b}/2$ and is illustrated in Fig. 1. The initial coherence, the crucial ingredient that gives rise to effects such as amplification without inversion and a large index of refraction at a point of vanishing absorption, is chosen as $\rho^0_{b'b} = -i|\rho^0_{b'b'}| = -i|\rho^0_{bb}|$ to maximize the gain in the small field regime. Furthermore, we assume $\gamma_{b'} = \gamma_b = \gamma$, $\omega_{b'b} = 1.5\gamma$, and $\gamma_a = 0.05\gamma$. The density-dependent NDD parameter C, that is a measure of the strength of the NDD interaction and which we use in our results as a control parameter, is defined as

$$C = \frac{\mu^2 n}{2\,\varepsilon_0 \hbar \gamma}, \tag{2.2}$$

where $n = \frac{r/\gamma}{V}$ is the effective atomic density in the interaction region. For values of C on the order of unity, if we in addition take the reasonable values of $\gamma \approx 10^8$ Hz and $\mu \approx 10^{-29}$ C-m, then the volume density required to see interesting behavior is about $n \approx 10^{15}$ atoms/cm^3.

To obtain the full nonlinear susceptibility that includes NDD contributions, we straightforwardly proceed by eliminating the local field using the LL relation, $E_L = E_L(E,P)$, in Eqs. (2.1). Then we can solve for the steady state solutions of the atomic polarization and construct the susceptibility. An analytic solution is difficult to obtain directly by this procedure, because the use of the LL relation in the density matrix equations immediately makes them nonlinear and somewhat intractable. An alternative but equivalent procedure, is to solve the linear Eqs. (2.1) analytically for the steady state solutions ρ_{ij} in terms of the (unknown) local field E_L, and then to use the LL relation to eliminate E_L self-consistently. We first construct the polarizability

$$\frac{n\alpha}{\varepsilon_0} = \chi_e' + i\chi_e'' = -\frac{\left(\mu\rho_{ab} + \mu'\rho_{ab'}\right)}{\varepsilon_0 V E_L}$$

$$= -\frac{\mu^2 r}{2\varepsilon_0 \hbar V}\,\frac{\left(A' + B'|E_L|^2 + C'|E_L|^4\right) + i\left(A'' + B''|E_L|^2 + C''|E_L|^4\right)}{a + b|E_L|^2 + c|E_L|^4 + d|E_L|^6}, \tag{2.3}$$

where we have made use of the results of Rathe, *et.al.* [9]. The quantities A', B', ... , a,b, ... are constant functions of the atomic parameters that do not contain the electric field and are defined in reference 9. In the last line of Eq. (2.3), we have assumed that $\mu' = \mu$. Since E_L itself is a function of the medium polarization, we must self-consistently eliminate the E_L in terms of the E. To carry this out, we use the LL relation together with the microscopic polarization; this yields

$$E_L = \frac{E}{1 - \frac{1}{3}\chi_e(|E_L|^2)}. \tag{2.4}$$

This expression, together with Eq. (2.3) for the polarizability, is used in the Clausius-Mossotti relation, Eq. (1.7), to self-consistently calculate the nonlinear susceptibility χ_{NDD}. Thus, χ_{NDD} has the form of Eq. (1.7) with Eq. (2.3) and Eq. (2.4) simultaneously used to replace α and E_L [10].

Equation (2.4) in general has multiple roots in $E_L(E)$. We must therefore examine the uniqueness and stability of our self-consistent solution. We introduce intensity and phase variables, implicitly defined by

$$E_L \equiv \sqrt{I_L}e^{i\phi_L}, \quad E \equiv \sqrt{I}e^{i\phi}, \quad 1-\frac{1}{3}\chi_e(I_L) \equiv \eta\ e^{i\theta}. \tag{2.5}$$

From Eqs. (2.4) and (2.5), we immediately find $\phi_L = \phi - \theta(I_L)$, which implies that the phases of the fields are uniquely defined once the intensity of the local field is fixed. In order to test the stability, we therefore have only to analyze the intensity relation

$$I_L = \frac{I}{\eta^2(I_L)}, \tag{2.6}$$

that is an odd-order polynomial in I_L. Hence, there exists at least one real-valued solution, provided η^2 is nonsingular. If $dI/dI_L \geq 0$, i.e., Eq. (2.6) is monotonic and therefore the solution of Eq. (2.6) found by self-consistent means must be unique and stable. For the parameters used in the results reported here, we have numerically calculated dI/dI_L and have found that the above stability criterion is satisfied.

We have studied the full nonlinear model in the high density regime for a wide range of electric field strengths and atomic densities. Since the main predictions of the linear model [7] — the enhanced non-inversion gain and absorptionless index of refraction — are due to a large increase of the local field relative to the macroscopic field ($|E_L|/|E| \gg 1$), it is important to analyze the impact of atomic nonlinearities on these effects. As the probe field strength E increases, the susceptibility of the medium becomes dependent on the intensity and can be expanded in the usual way as

$$\chi(\omega,E_L) = \chi^{(1)}(\omega) + \chi^{(3)}(\omega)\,|E_L|^2 + \cdots,$$

where the higher order terms lead to a saturation of the total polarization with increasing intensity I_L. As a result of this saturation, the polarization per atom p is always less than the dipole moment μ,

$$\left|\frac{p}{\mu}\right| = \left|\frac{\chi_{NDD}\,E}{\mu\, n}\right| \leq 1. \tag{2.7}$$

Equation (2.7) sets an upper bound for the magnitude of the susceptibility. Substituting for E in terms of the Rabi frequency, $\Omega = \mu E/\hbar$, scaling to γ, and solving for $|\chi_{NDD}|$ we find

$$|\chi_{NDD}| \leq \frac{2C}{\Omega/\gamma}. \tag{2.8}$$

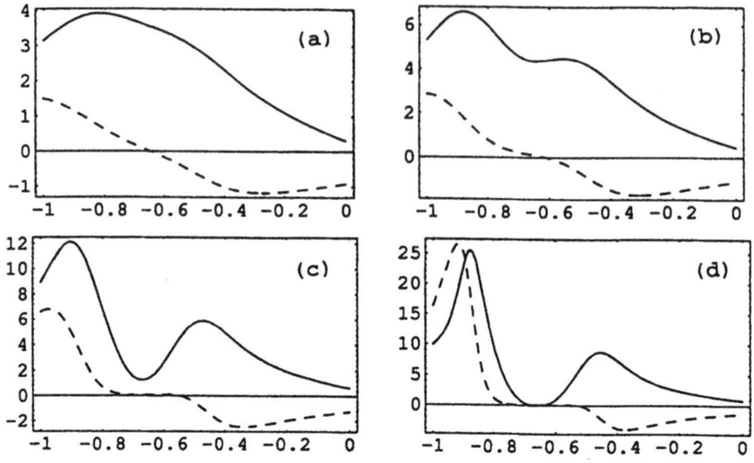

Figure 2. The effect of increasing density on χ_{NDD} (Δ/γ) showing the formation of a region of electromagnetically induced transparency [Fig. 2(d)]. $|\Omega| = 0.1\gamma$ and (a) C = 4, (b) C = 5, (c) C = 6, and (d) C = 7. The other parameters are $\rho^0_{aa} = 0.01$, and $\rho^0_{b'b'} = \rho^0_{bb} = 0.495$.

Equation (2.8) shows that the gain and index enhancements due to NDD effects are limited by atomic nonlinearities and are less pronounced for increasing probe field strength. Nevertheless, for sufficiently small values of E, an enhancement by more than two orders of magnitude is possible. An increase in inversionless gain has obvious benefits. As pointed out by Scully [6], the increase in the index of refraction could be used in a laser particle accelerator, in optical microscopy, or in a sensitive magnetometer. We display in Fig. 2 the result of increasing the atomic number density is on the dispersion (solid curve) and absorption (dashed curve) for different, intermediate, values of the macroscopic field strength. Note that in Fig. 2(d) there appears a self-induced transparency region where χ'_{NDD} and χ''_{NDD} and are simultaneously near zero (the numerical values are small but nonzero over this range).

We now present the primary result of this example — what we call the piezophotonic effect. As the density is further increased in the regime of moderate E, there is a threshold value of density where the absorption curve flips to become one of large gain, as we show in Fig. 3. This phenomenon is quite remarkable. By varying the density about the threshold value, this effect could be exploited in the form of an optical switch. This phenomenon is somewhat analogous to a piezoelectric crystal. Instead of squeezing a crystal to get a current, changing the density by a small amount will flip the system from an absorber to an amplifier which then results in an intense inversionless amplifying switch. We also predict the existence of a range of frequencies where the gain and the index of refraction are simultaneously very large after the flip [Fig. 3(b)].

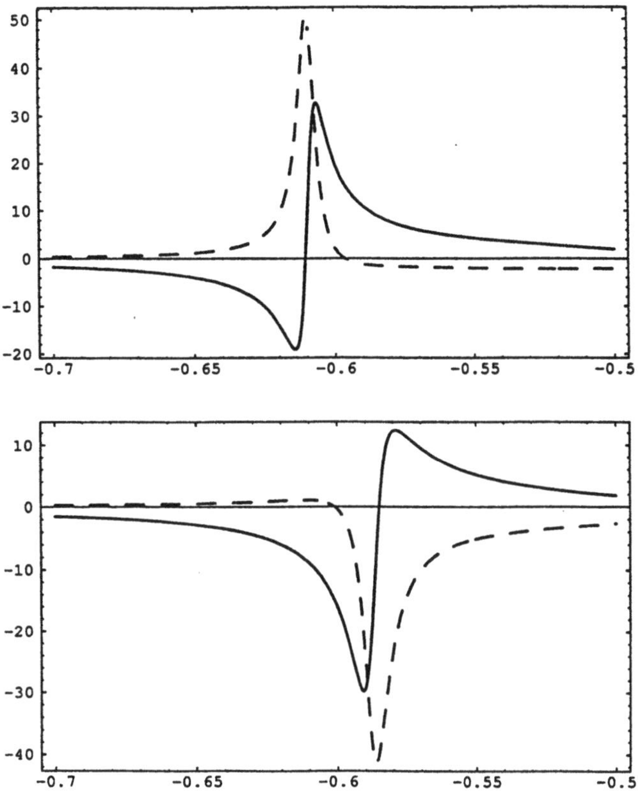

Figure 3. χ_{NDD} as a function of dimensionless detuning (Δ/γ) for parameters that lead to a flip in χ''_{NDD}. $|\Omega| = 0.1\gamma$ and C = 8.6 (top), C = 8.7 (bottom). The other parameters are the same as in Fig. 2.

3. Summary of Results

Historically, the Maxwell-Bloch equations are the fundamental relations used to describe light interacting with a collection of two-level atoms [11]. However, when the volume density of atoms becomes sufficiently dense such that there are — on the average — many atoms per cubic resonance wavelength of the field, then the usual Maxwell-Bloch equations are fundamentally incorrect and must be modified. The structure of the modification is formally the same as that of the LL relation that gives rise to the Clausius-Mossotti relation in the static field limit, as discussed in section one. However, it is not evident that the static relation, Eq. (1.1), should generalize as is to dynamic fields in a nonlinear media. The argument that this is indeed so is sketched from a field theoretic point of view in reference 12. In fact, this result constitutes a new proof of the static LL formula in the limit $\omega \rightarrow 0$. At this point we then, with confidence, use the now dynamic LL relation, Eq. (1.1), to modify the Maxwell-Bloch equations to be self-consistent for optically dense media. The resultant equations are nonlinear in the atomic variables, a fact that results in a host of new phenomena such as intrinsic optical bistability [13], spectral shifts [14], and fast optical switching [15].

In section two, as an example of our recent work on three-level systems, we study the effect of local fields in a coherently prepared three-level Λ model known to exhibit lasing without inversion. We confirm that the NDD or local field effects can enhance the inversionless gain and absorptionless index by two orders of magnitude. In particular, the index of refraction may be enhanced at a point of zero absorption or even gain. This is particularly useful for guided wave optics as well as other applications. Also of interest is the flip in the dispersion curve that is a signature of $\chi^{(3)}$ changing sign. This result could be exploited in the design of a new type optical switch in a photonic band material [16]. In addition, we observe a small region of near electromagnetically induced transparency where both the real and imaginary part of the susceptibility simultaneously approach zero. Finally, we predict a new type of density-dependent switching phenomenon — the piezophotonic switch. The switch operates when a small change in atomic density flips the dispersion curve so that a strong absorber suddenly changes into a powerful, local-field enhanced, inversionless amplifier. This type of sudden switching, similar to that seen in two-level systems [13,15], is a direct result of the nonlinearity of the Maxwell-Bloch or density matrix equations in the atomic variables.

References

1. J. D. Jackson, *Classical Electrodynamics*, 2nd Ed. (Wiley, New York, 1975), Chap. 4.
2. C. M. Bowden and J. P. Dowling, Phys. Rev. A **47**, 1247 (1993); *ibid* (erratum) **49**, 1514 (1994).
3. A. S. Manka, J. P. Dowling, C. M. Bowden, and M. Fleischhauer, "Piezophotonic Switching Due to Local Field Effects in a Coherently Prepared Medium of Three-Level Atoms," submitted for publication.
4. K. J. Boller, A. Imamoglu, S. E. Harris, Phys. Rev. Lett. **66**, 2593 (1991); J. E. Field, K. H. Hahn, and S. E. Harris, *ibid.* **67**, 3062 (1991).
5. O. A. Kocharovskaya and Ya. I. Khanin, Zh. Eksp. Teor. Fiz. **90**, 1610 (1986) [Sov. Phys. JETP **63**, 945 (1986)]; S. E. Harris, Phys. Rev. Lett., **62**, 1033 (1989); M. O. Scully, S.-Y. Zhu, and A. Gavrielides, *ibid.* **62**, 2813 (1989); L. M. Narducci, M. O. Scully, C. H. Keitel, S.-Y. Zhu, and H. M. Doss, Opt. Commun. **86**, 324 (1991); for reviews see: O. Kocharovskaya, Phys. Rep. **219**, 175 (1992); M. O. Scully, *ibid.* **219**,191 (1992).

6. M. O. Scully, Phys. Rev. Lett. **67**, 1855 (1991); M. Fleischhauer, C. H. Keitel, M. O. Scully, C. Su, B. T. Ulrich, and S.-Y. Zhu, Phys. Rev. A **46**, 1468 (1992); M. Fleischhauer, C. H. Keitel, M. O. Scully, and C. Su, Opt. Commun. **87**, 109 (1992); M. O. Scully and S.-Y. Zhu, *ibid.* **87**, 134 (1992).
7. J. P. Dowling and C. M. Bowden, Phys. Rev. Lett. **70**, 1421 (1993).
8. M. O. Scully and W. E. Lamb, Jr., Phys. Rev. **159**, 208 (1967); M. Sargent III, M. O. Scully, and W. E. Lamb, Jr., *Laser Physics*, (Addison-Wesley; Reading, MA 1974).
9. U. Rathe, M. Fleischhauer, S.-Y. Zhu, T. W. Hänsch, and M. O. Scully, Phys. Rev. A **47** 4994 (1993).
10. We use *Mathematica*, which is especially suited for this type of work, for the substitutions and concomitant plots of the real (χ'_{NDD}) and the imaginary (χ''_{NDD}) part of this susceptibility as a function of the dimensionless detuning Δ/g.
11. L. Allen and J. H. Eberly, *Optical Resonance and Two-Level Atoms* (Wiley, New York, 1975).
12. C. M. Bowden and J. P. Dowling, Phys. Rev. A **47**, 1247 (1993); erratum *ibid* **49**, 1514 (1994).
13. F. A. Hopf, C. M. Bowden, and W. H. Louisell, Phys. Rev. A **29**, 2591 (1984); Y. Ben-Aryeh, C. M. Bowden, and J. C. Englund, Phys. Rev. A **34**, 3917 (1986).
14. R. Friedberg, S. R. Hartmann, and J. T. Manasseh, Phys. Rep. **7**, 101 (1973); Phys. Rev. A **39**, 3444 (1989); *ibid.* **40**, 2446 (1989); J. J. Maki, M. S. Malcuit, J. E. Sipe, and R. W. Boyd, Phys. Rev. Lett. **67**, 972 (1991).
15. M. E. Crenshaw, M. Scalora, and C. M. Bowden, Phys. Rev. Lett. **68**, 911 (1992); M. E. Crenshaw and C. M. Bowden, Phys. Rev. Lett. **68**, 3475 (1992).
16. M. Scalora, J. P. Dowling, C. M. Bowden, and M. J. Bloemer, "Optical Limiting and Switching in Nonlinear Photonic Bandgap Materials", unpublished.

A Superradiant Laser

F. Haake[1], *M.I. Kolobov*[1], *C. Fabre*[2], *E. Giacobino*[2], *and S. Reynaud*[2]

[1]Fachbereich Physik, Universität GH Essen, D-45117 Essen, Germany
[2]Laboratoire de Spectroscopie Hertzienne de l'Ecole Normale Supérieure, Université Pierre et Marie Curie, 4 place Jussieu, F-75252 Paris Cedex 05, France

Abstract. We discuss stationary superradiance from a laser with N identical three-level atoms. Cooperative radiation is made possible by a cavity supporting two modes, each in resonance with an atomic transition. For large N, substantial squeezing results even in the presence of pump fluctuations and spontaneous emission.

We propose to investigate a laser [1] which has N identical three-level atoms driven by a coherent pump on one transition, $0 \leftrightarrow 2$, and coupled to two resonant modes on the other two transitions, $2 \leftrightarrow 1$ and $1 \leftrightarrow 0$ (see Fig. 1). For one of these modes, to be called the active one, the resonator should have a high finesse; the second, passive mode may be strongly damped. As long as spontaneous emissions into other, nonresonant modes are negligible, the dynamics of such a laser is fully collective, such that the atoms can be described in terms of global $S_{ij} = \sum_{\mu=1}^{N} S_{ij}^{\mu} = \sum_{\mu} |i\rangle_{\mu\,\mu}\langle j|$; these comprise three population operators S_{ii} and six polarizations S_{ij} with $i \neq j$. Denoting the creation and annihilation operators of photons in the active (passive) mode by a and $a^{\dagger}$ (b and $b^{\dagger}$) we have the interaction Hamiltonian

$$H = +i\hbar\Omega(S_{20} - S_{02}) + i\hbar g_a(aS_{21} - a^{\dagger}S_{12}) + i\hbar g_b(bS_{10} - b^{\dagger}S_{01}) \,. \quad (1)$$

Here Ω is the strength of the pump field, normalized so as to have the dimension and physical meaning of a Rabi frequency; similar is the meaning of the coupling constants g_a and g_b, except that these latter frequencies refer to the fictitious fields of a single photon in the active and passive mode, respectively. The Heisenberg equations of motion generated by the above Hamiltonian must be complemented by the cavity damping terms for the mode amplitudes [2, 3]

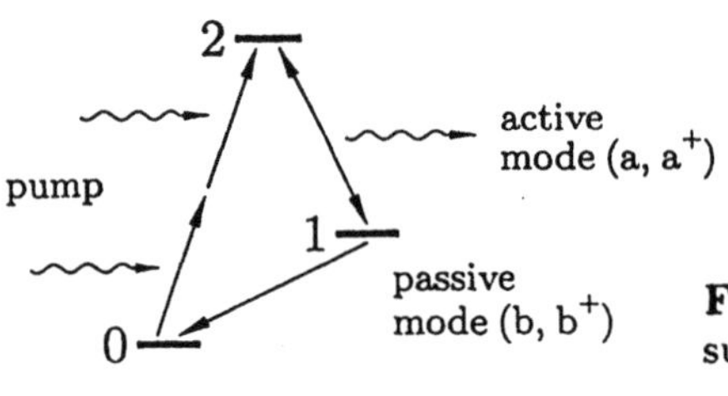

Fig. 1. Scheme of three-level superradiant laser.

Springer Proceedings in Physics, Vol. 77 Quantum Optics VI
Editors: D.F. Walls · J.D. Harvey

$$
\begin{aligned}
(\dot{a})_{irr} &= -\kappa_a a + \sqrt{2\kappa_a}\eta_a\,, \\
(\dot{b})_{irr} &= -\kappa_b b + \sqrt{2\kappa_a}\eta_b\,,
\end{aligned} \tag{2}
$$

where κ_a and κ_b are damping constants while η_a and η_b represent Gaussian white noise according to

$$
\begin{aligned}
[\eta_\alpha(t), \eta_\beta^\dagger(t')] = \left\langle \eta_\alpha(t)\eta_\beta^\dagger(t')\right\rangle = \delta_{\alpha\beta}\delta(t-t')\,, \\
\langle \eta_\alpha(t)\rangle = \left\langle \eta_\alpha^\dagger(t)\eta_\beta(t')\right\rangle = \langle \eta_\alpha(t)\eta_\beta(t')\rangle = 0\,.
\end{aligned} \tag{3}
$$

As already mentioned we take the passive mode to be so strongly damped that it may be adiabatically eliminated. The adiabatic equilibrium regime is then characterized by

$$
b(t) = \left(-g_b S_{01}(t) + \sqrt{2\kappa_b}\eta_b(t)\right)/\kappa_b\,. \tag{4}
$$

By inserting this into the Heisenberg equations we obtain the following set of evolution equations of our laser

$$
\dot{a} = -gS_{12} - \kappa a + \sqrt{2\kappa}\eta_a
$$

$$
\begin{aligned}
\dot{S}_{02} &= gaS_{01} - \Omega(P_2 - P_0) + \gamma S_{12}S_{01} - \sqrt{2\gamma}S_{12}\eta_b\,, \\
\dot{S}_{12} &= -ga(P_2 - P_1) + \Omega S_{10} - \gamma S_{10}S_{02} + \sqrt{2\gamma}\eta_b^\dagger S_{02}\,, \\
\dot{S}_{01} &= -ga^\dagger S_{02} - \Omega S_{21} + \gamma(P_1 - P_0)S_{01} - \sqrt{2\gamma}(P_1 - P_0)\eta_b\,, \\
\dot{P}_0 &= -\Omega(S_{02} + S_{20}) + 2\gamma S_{10}S_{01} - \sqrt{2\gamma}(S_{10}\eta_b + \eta_b^\dagger S_{01})\,, \\
\dot{P}_1 &= -g(aS_{21} + a^\dagger S_{12}) - 2\gamma S_{10}S_{01} + \sqrt{2\gamma}(S_{10}\eta_b + \eta_b^\dagger S_{01})\,, \\
\dot{P}_2 &= \Omega(S_{02} + S_{20}) + g(aS_{21} + a^\dagger S_{12})\,.
\end{aligned} \tag{5}
$$

We have simplified the notation a bit by introducing the rate constant $\gamma = g_b^2/\kappa_b$ and dropping the index a on g_a and κ_a.
Two conservation laws implied by our dynamics must be mentioned,

$$
\begin{aligned}
&\sum_i^{0,1,2} S_{ii} = N\,, \\
&\sum_{i,j}^{0,1,2} S_{ij}S_{ji} = N(N+2)\,.
\end{aligned} \tag{6}
$$

Obviously, the first of these conserved quantities is just the number of atoms. The second one generalizes the squared length of the familiar Bloch vector for N two-level atoms; for our three-level atoms the second conservation law represents a $SU(3)$ symmetry of the collective dynamics [4]; the numerical value $N(N+2)$ corresponds to maximum cooperativity as would arise, e.g., if all atoms were prepared in the ground state initially; the accessible atomic Hilbert space is then spanned by $\frac{1}{2}(N+1)(N+2)$ fully symmetric states.
In the semiclassical limit $N \gg 1$ all observables $a, a^\dagger, S_{ij}$ will execute small fluctuations $\delta X(t)$ around stationary mean values $\bar{X} \sim N$,

$$X(t) = \bar{X} + \delta X(t) \,. \tag{7}$$

We may determine the means $\bar{X}$ as stationary c-number solutions of the evolution equations (5) after neglecting the noise terms. The operator valued fluctuations $\delta X(t)$ can then be found by linearizing about the means $\bar{X}$ in (5). After appropriate rescalings all means and correlations can be expressed in terms of a dimensionless coupling strength c, a dimensionless pump strength p, and a time scale ratio ξ,

$$c = g^2/\kappa\gamma \,, \quad p = \frac{\Omega}{Ng}\sqrt{\frac{\kappa}{\gamma}} \,, \quad \xi = N\gamma/\kappa \,. \tag{8}$$

Simple analytic results are obtained in the good-cavity limit $\xi \gg 1$. We have found a unique nontrivial stationary solution in that case. It reads

$$\begin{aligned}
\bar{S}_{01}/N \equiv x &= \sqrt{\frac{c}{1+c}}\sqrt{p(1-p)} \,, \; \bar{S}_{02}/N = -x^2/(\sqrt{c}p) \,, \; \bar{S}_{12}/N = -x/\sqrt{c} \,, \\
\bar{P}_0/N &= \frac{c(1-p)}{1+c} \,, \; \bar{P}_1/N = p \,, \; \bar{P}_2/N = \frac{1-p}{1+c} \,, \\
\bar{a} &= \frac{N\gamma}{g}\sqrt{c}x \,.
\end{aligned} \tag{9}$$

Obviously, for this solution to exist the pump parameter must be restricted to the interval $0 \le p \le 1$. Stability yields the further restriction $(1-c)/2 < p < c/(1+2c)$. It is interesting to see that the mode amplitude $\bar{a}$ is maximal, $\partial\bar{a}/\partial p = 0$, for $p = 1/2$, a value just outside the stability range; maximum stable output is approached for large couplings, $c \gg 1$, and pumping $p \le 1/(2+1/c) \le 1/2$. The linear equations for the fluctuations $\delta X(t)$ fall into two separate blocks. A four dimensional one comprises the imaginary parts $\delta v = (\delta a - \delta a^\dagger/2i, \delta v_{ij} = (\delta S_{ij} - \delta S_{ij})/2i$ with $i \neq j$. Since the stationary means are all real this first block may be interpreted as describing phase fluctuations. A second block contains two population fluctuations δS_{00}, δS_{11} as well as the real parts $\delta u = (\delta a + \delta a^\dagger)/2$, etc. Evidently, the latter block contains the fluctuations of the photon number in the active mode.
After Fourier transforming we easily obtain the fluctuations of the mode amplitude $\delta u(\omega) + i\delta v(\omega)$. Of special interest are, of course, the fluctuations of output field $\delta a_{\text{out}} = \sqrt{2\kappa}\delta a - \eta_a$. From these we infer the line width

$$\Delta\nu = \frac{\kappa}{(\bar{a})^2}\frac{p^2(1+c)^2 + (1-p)^2(1-c)^2}{(-1+c+2p)^2} \tag{10}$$

which is obviously proportional to $1/N^2$. The correlation function of the amplitude fluctuations,

$$\begin{aligned}
\langle \delta u_{\text{out}}(\omega)\delta u_{\text{out}}(\omega')\rangle &\equiv \frac{1}{4}\delta(\omega+\omega')\, I(\omega,p,c) \,, \\
I(\omega,p,c) &= 1 - \frac{S_0(p,c)}{1+\omega^2\tau^2}
\end{aligned} \tag{11}$$

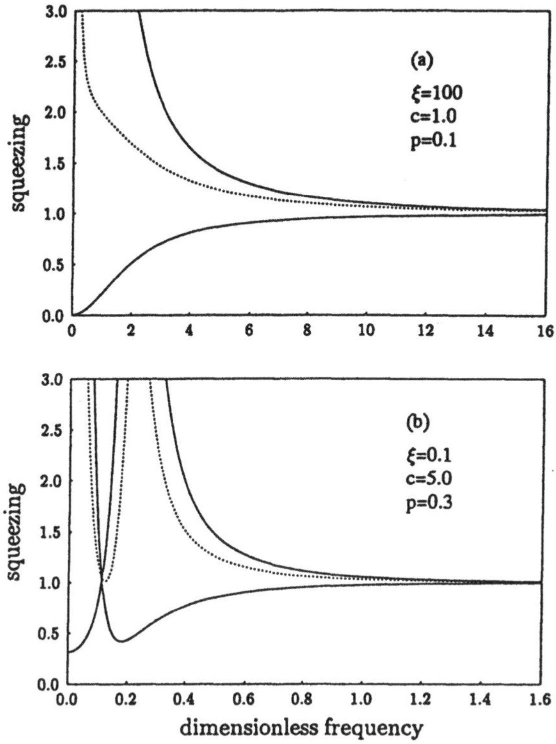

Fig. 2. Optimum squeezing and stretching spectra (I_Θ and $I_{\Theta+\pi/2}$) and their product for the superradiant laser without spontaneous emission: (a) good-cavity case, (b) bad-cavity case.

displays a Lorentzian dip at zero frequency of width

$$1/\tau = 4\kappa(1+c)(1-p)/(3+c-2p)\,. \tag{12}$$

Squeezing is present for positive values of

$$S_0(p,c) = \frac{1}{2} + \frac{2c}{(1+c)^2} - \frac{p^2}{2(1-p)^2}\,. \tag{13}$$

The ideal value $S_0 = 1$ is attained for $p = 0$, $c = 1$ which is, of course, not an interesting working point itself; however, the maximum of S_0 at that point is rather flat such that strong squeezing prevails in a sizeable region of p and c. For nonzero values of the frequency, optimal squeezing is not obtained in the amplitude quadrature of the output field but rather in some other quadrature $X_\theta = (ae^{-i\theta} + a^\dagger e^{i\theta})/2$. Fig. 2 displays, as functions of ω, the optimum squeezing and stretching spectra $I_\theta(\omega)$ and $I_{\theta+\pi/2}(\omega)$.
The product $I_\theta(\omega)I_{\theta+\pi/2}(\omega)$, also shown in Fig. 2, approaches the quantum limit unity for large frequencies as well as for certain finite values of ω. Fig. 2a

refers to the good-cavity limit $\xi \gg 1$ for which simple analytic results arise, while Fig. 2b was obtained for the bad-cavity limit $\xi \ll 1$. The latter case offers the especially interesting possibility of optimal squeezing, at some frequencies, in the phase quadrature.

We have checked the impressive potential for squeezing and line narrowing to be robust against pump fluctuations. More interesting is the effect of spontaneous emission since it tends to impair the collectivity of the dynamics. Allowing for spontaneous emission with rate γ_s on the lasing transition $2 \leftrightarrow 1$ we incur the incremental time rates of change, to be added to the r.h.s of (5),

$$\begin{aligned}
\delta\dot{S}_{02} &= -\gamma_s S_{02} + \sqrt{2\gamma_s}\sum_\mu S_{01}^\mu \eta_\mu\,, \\
\delta\dot{S}_{12} &= -\gamma_s S_{12} - \sqrt{2\gamma_s}\sum_\mu (P_2^\mu - P_1^\mu)\eta_\mu\,, \\
\delta\dot{S}_{01} &= -\sqrt{2\gamma_s}\sum_\mu \eta_\mu^\dagger S_{02}^\mu\,, \\
\delta\dot{P}_0 &= 0\,, \\
\delta\dot{P}_1 &= -\delta\dot{P}_2 = 2\gamma_s P_2 - \sqrt{2\gamma_s}\sum_\mu (\eta_\mu S_{21}^\mu + \eta_\mu^\dagger S_{12}^\mu)\,.
\end{aligned} \tag{14}$$

It is important to realize that the incoherent drift terms in (14) are linear in S_{ij}, as opposed to the collective drift terms in (5) which are quadratic. Moreover, the noise forces η_μ are independent from atom to atom while otherwise obeying (3). At any rate, the incoherent rates (14) are of order $1/N$ relative to the collective ones in (5). One might thus be tempted to argue that spontaneous emission is irrelevant in the semiclassical limit. But however small the incoherent terms may be, they do bring about a qualitative change of the dynamics by breaking collectivity, i.e., destroying the conservation of $\sum_{ij} S_{ij}S_{ji}$. The role of that conservation law is now taken over by one of the evolution equations (5),(14) which is no longer a linear combination of the remaining ones. For the stationary regime the new equation can be written, after a little rearranging, as

$$\gamma_s\left[p\sqrt{c}(p^2 - x^2)\,\bar{P}_2/N \pm (cx^2 + p^2)\sqrt{x^2 - \gamma_s\bar{P}_2/\gamma N^2}\right] = 0\,. \tag{15}$$

Obviously, now, in the fully cooperative case $\gamma_s = 0$ this consistency condition is trivially fulfilled. In the present case, the square bracket in (15) must vanish and that condition takes the role previously played by the conservation of $\sum_{ij} S_{ij}S_{ji}$ in nailing down all stationary means. At this point, i.e., after dividing (15) by γ_s, we may and should indeed let $\gamma_s \to 0$.

The inequivalence of $\gamma_s = 0$ and $\gamma_s \to 0$ is somewhat reminiscent of the symmetry breaking in, say, a Heisenberg magnet below the Curie temperature. The mean-field equations (which correspond to our semiclassical approximation) imply a vanishing magnetization in the strict absence of a symmetry breaking

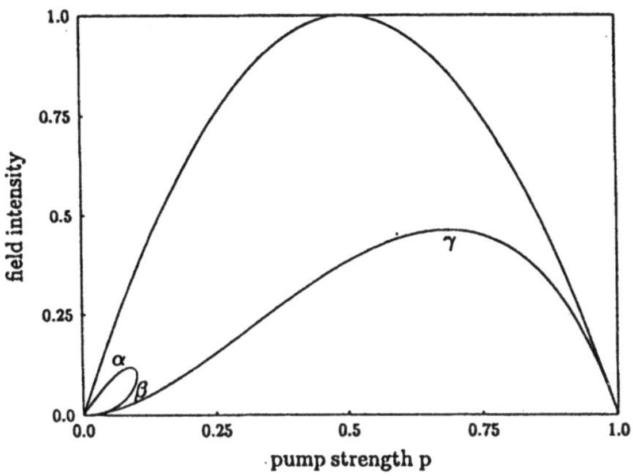

Fig. 3. Stationary intracavity intensity of the superradiant laser vs pump-strength p without (uppermost curve) and with (α, β and γ) spontaneous emission for the coupling strength $c = 0.1$. The solution α is unstable in the good-cavity limit.

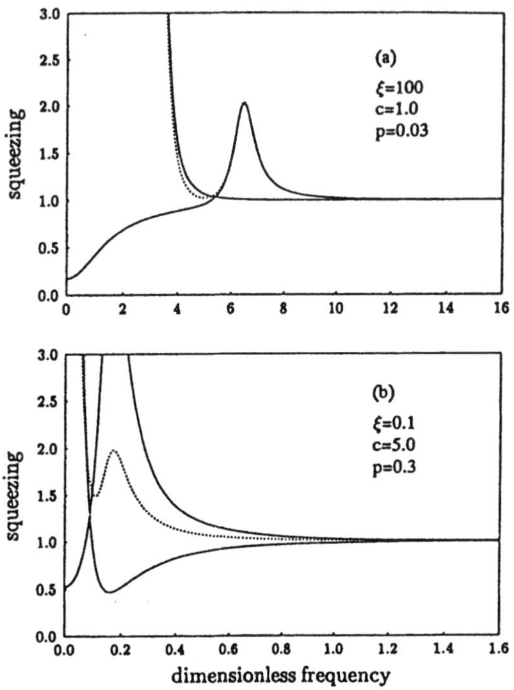

Fig. 4. Optimum squeezing and stretching spectra and their product for the superradiant laser with spontaneous emission on the lasing transition. Spectra refer to case γ in Fig. 3. (a) good-cavity case, (b) bad-cavity case.

magnetic field h. However, if $h \neq 0$ is allowed, the limit $h \rightarrow 0$ leaves a spontaneous magnetization. In our case $\gamma_s \neq 0$ breaks the SU(3) symmetry of the collective dynamics.
In the limit $\gamma_s \rightarrow 0$ all $\bar{S}_{ij}$ and $\bar{a}$ retain their proportionality to N. We now find three nontrivial stationary solutions. While in the bad-cavity case $\xi \gg 1$ bi- and even tristability is possible, the more easily tractable good-cavity case allows stability only for two of the three stationary solutions and these have non-overlapping stability ranges with respect to the parameters p and c. Fig. 3 depicts the p dependence of $\bar{a}$ at fixed c. Fig. 4 is analogous to Fig. 2 and shows the optimal squeezing and stretching spectra for one of the new stationary solutions. Good squeezing is still arises.

Acknowledgement

The collaboration in this project was funded in part by the École Normale Supérieure in Paris, the Sonderforschungsbereich "Unordnung und grosse Fluktuationen" der Deutschen Forschungsgemeinschaft, the NATO Fellowship Program, the U.S. Office of Naval Research, and the CEC grant ESPRIT; we express our gratitude for that support.

References

[1] F. Haake, M. Kolobov, C. Fabre, E. Giacobino, and S. Reynaud, Phys. Rev. Lett. **71**, 995 (1993) and references therein.

[2] H. Haken, *Handbuch der Physik* (Springer, Berlin, 1970),Vol. XXV/2c.

[3] C. Gardiner, *Quantum Noise* (Springer, Berlin, 1992).

[4] F. T. Hioe and J. H. Eberly, Phys. Rev. A **25**, 2168 (1982)

Laser-Induced Processes in Molecular Systems

S. Stenholm

Research Institute for Theoretical Physics and the Academy of Finland,
P.O. Box 9, Siltavuorenpenger 20 C, University of Helsinki,
SF-00014 Helsinki, Finland

Abstract. We formulate the problem of laser-induced dynamics on adiabatic molecular energy surfaces. The theory is applied to the recent experimental work on sodium dimers. For this we need to include spontaneous emission in a way that allows an interpretation in terms of a simulation. The results of the simplified model are found to agree qualitatively with the experimental observations.

1. Prolog

Femtosecond laser pulses can now be produced under well controlled conditions. This has initiated a lively experimental activity in time resolved molecular spectroscopy: see Ref.[1]. With modern computers, the time evolution of the Schrödinger equation can be followed in cases close to those achievable in the laboratory, see Ref.[2]. A particularly beautiful experiment was recently performed by Dunn et alii [3]. They were able to follow the spontaneous emission in the sodium dimer while its wave packet evolved on an excited electronic energy level. By choosing the detected frequency, they were able to probe the probability of the wave packet to occur at various positions along the vibrational coordinate.

The experiment in [3] combines many of the theoretical features of time resolved molecular dynamics. I choose this as a prototype problem to exhibit the various physical considerations required to build up the complete theory. By selecting the simplest possible situations, I can build up a toy model of the system, which allows me to calculate analytically the type of result expected in this experiment. My simplified treatment cannot, of course, be used to fit the observed data, but it turns out that the qualitative features of the experiment can be found.

In Sec.2, I build up the adiabatic approximation for the time-dependent case. The description of the vibrational states on the adiabatic energy levels is discussed in Sec.3. Finally, in Sec.4 I add the effect of spontaneous decay and present my simplified treatment of the experiment.

2. Time-Dependent Born-Oppenheimer

The ordinary Born-Oppenheimer approximation is usually presented as a method to determine energy eigenvalues, i.e. a time independent method, see e.g. Ref.[4]. It is obvious that the same approach can be used in a time dependent fashion. One may, however, have to reinvestigate the validity condition of the procedure because of the additional time parameter indicating the rate of change of the system. These discussions are postponed, and we present the theory and its applications without critical scrutiny of its validity.

The system under investigation consists of an assembly of nuclei, whose positions are collectively denoted by R, and their corresponding valence electrons with positions collectively denoted by r. The particles interact throught the Coulomb potential, and the whole system is coupled to the external laser field $\mathbf{E}$. The Hamiltonian of the system is then written as

Springer Proceedings in Physics, Vol. 77 Quantum Optics VI
Editors: D.F. Walls · J.D. Harvey

$$H = T_e + T_N + V_{ee}(r) + V_{NN}(R) + V_{eN}(r,R) - \mathbf{D}\cdot\mathbf{E}\ , \tag{1}$$

where T indicates the kinetic energy contributions and the subscripts $e(N)$ denote the electronic (nuclear) variables. The external laser electric field is $\mathbf{E}$, and it is assumed coupled linearly to some (dipole) operator $\mathbf{D}$ of the system.

In accordance with the adiabatic assumption, we can introduce the electronic states in terms of the frozen nuclear configuration by solving the eigenvalue problem

$$\Big(T_e + V_{ee}(r) + V_{NN}(R) + V_{eN}(r,R)\Big)\ |\varphi_n> = \mathcal{U}_n(R)\ |\varphi_n>\ . \tag{2}$$

The electronic eigenvalues $\mathcal{U}_n(R)$ and the eigenfunctions $< r|\varphi_n> = \varphi_n(r,R)$ depend parametrically on the nuclear configuration R.

Because the problem (2) is a Hermitean eigenvalue problem, the general time dependent state of the system can be expanded as

$$|\Psi(t)> = \sum_n \psi_n(R,t)\ |\varphi_n>\ . \tag{3}$$

This result is still exact. Using the Hamiltonian (1), we obtain the Schrödinger equation

$$H\ |\Psi(t)> = i\hbar \sum_n \dot{\psi}_n(R,t)\ |\varphi_n>\ . \tag{4}$$

We now project this equation on the state $|\varphi_m>$ and assume that the nuclear kinetic energy operator T_N acts only on the coefficient ψ_n in (3) (this is the Born-Oppenheimer assumption). Because the eigenvalue problem (2) is Hermitean, the eigenstates $|\varphi_n>$ are orthonormal for a fixed nuclear configuration, and the Schrödinger equation becomes

$$\Big(T_N + \mathcal{U}_m(R)\Big)\psi_m(R,t) - \sum_n < \varphi_m|\mathbf{D}\cdot\hat{\varepsilon}|\varphi_n > E(R,t)\psi_n(R,t) = i\hbar\frac{\partial}{\partial t}\psi_m(R,t)\ . \tag{5}$$

Here the electric field is taken to have the amplitude $E(R,t)$ and the polarization $\hat{\varepsilon}$. In the absence of an electric field coupling, this equation describes the time evolution of the nuclear wave functions $\psi_m(R,t)$ on the electronic energy surfaces $\mathcal{U}_m(R)$. These are coupled by the external field, whose variation is assumed well described by the coordinate R; this is the molecular dipole approximation.

If we consider only two levels, 1 and 2 say, and assume that the laser field is monochromatic at the frequency Ω, we can apply the rotating-wave-approximation and obtain the matrix problem

$$i\hbar\frac{\partial}{\partial t}\begin{bmatrix}\psi_1\\ \psi_2\end{bmatrix} = \begin{bmatrix} T_N + \mathcal{U}_1(R) & V \\ V & T_N + \mathcal{U}_2(R) - \hbar\Omega\end{bmatrix}\begin{bmatrix}\psi_1\\ \psi_2\end{bmatrix}\ . \tag{6}$$

Here the rotating-wave-approximation has shifted the upper level by $\hbar\Omega$, and the coupling

$$V = -< \varphi_1|\mathbf{D}\cdot\hat{\varepsilon}|\varphi_2 > E \equiv \mu_{12}E \tag{7}$$

is taken to be real.

Equation (5) is the main result of the time-dependent Born-Oppenheimer approximation. Assuming only one nuclear coordinate R, we obtain the form (6) as used in our earlier works [5]. This situation is realized in a diatomic molecule.

We can, however, develop the theory one step further. The molecular eigenstates on the electronic energy surfaces are the vibrational and rotational eigenstates of the molecule. A part of the kinetic energy operator T_N denotes rotation of the molecule as a whole; this part has the angular momentum eigenvalues $\hbar^2 L(L+1)$, which we do not consider in this work. Usually many such states are thermally excited, and involved averaging takes place; however, cooling or matrix isolation techniques can eliminate them completely.

The remaining eigenvalues describe adiabatic vibrational excitation on the electronic energy surfaces. These are given by the eigenvalue problem

$$\left(T_N + \mathcal{U}_m(R)\right)u_\nu^{(m)}(R) = \mathcal{E}_\nu^{(m)}u_\nu^{(m)}(R) . \tag{8}$$

The eigenvalue $\mathcal{E}_\nu^{(m)}$ is the νth vibrational state on the electronic potential surface m. For small ν, these are usually harmonic oscillator states. For a given m, the eigenstates in (8) give a complete orthonormal set and we write

$$\psi_m(R,t) = \sum_\nu C_\nu^{(m)}(t)u_\nu^{(m)}(R) . \tag{9}$$

Inserting this into the equation (5) and assuming that the matrix elements μ_{nm} (see Eq.(7)) do not depend on R, we obtain

$$i\hbar\frac{\partial}{\partial t}C_\xi^{(m)}(t) = \mathcal{E}_\xi^{(m)}C_\xi^{(m)}(t) + \sum_{\nu,n}\mu_{nm}E_0(u_\xi^{(m)}|u_\nu^{(n)})C_\nu^{(n)}(t) . \tag{10}$$

Here the electric field E_0 is evaluated at the average position of the molecule, and the coupling of the states $\{n,\nu\} \Rightarrow \{m,\xi\}$ is given by the matrix element

$$M(n,\nu;m,\xi) = \mu_{nm}(u_\xi^{(m)}|u_\nu^{(n)}) , \tag{11}$$

where the scalar product is the celebrated Franck-Condon factor giving the overlap between the initial and final vibrational wave functions. Solving the time dependent problems (5) or (10) gives the same behaviour. Using (10) in time dependent perturbation theory gives the conventional transition rates of molecular spectroscopy.

3. Harmonic Oscillator Models

In many cases one can obtain analytical results by assuming that all vibrational states are exactly described by harmonic oscillators, see Refs. [6]. In Fig.1.a we show the case where the excited state is only a displaced replica of the ground state potential. We have

$$\begin{aligned} H_1 &= \frac{P^2}{2M} + \frac{1}{2}M\Omega_1^2R^2 , \\ H_2 &= \frac{P^2}{2M} + \frac{1}{2}M\Omega_1^2(R-R_0)^2 . \end{aligned} \tag{12}$$

Using the harmonic oscillator creation operator $a^\dagger$ and introducing the unitary transformation

$$D(\xi) = e^{-\xi(a-a^\dagger)} = \exp\left(-i\frac{R_0P}{\hbar}\right) , \tag{13}$$

we find

$$D(\xi)RD^\dagger(\xi) = R - R_0 \tag{14}$$

when

$$\xi = R_0\sqrt{\frac{M\Omega_1}{2\hbar}} = \frac{R_0}{\sqrt{2}A_1} , \tag{15}$$

where A_i is the extension of the ground state wave function on level i.

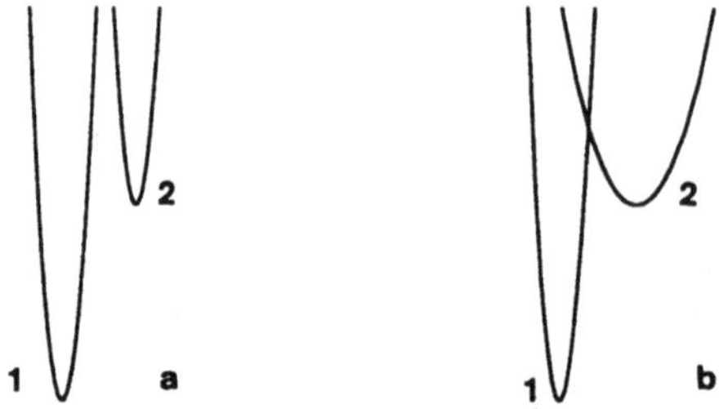

Fig.1. The model of two electronic energy surfaces. In (a) we have one oscillational frequency only; in (b) we have different frequencies.

Starting from the eigenvalue equation

$$H_1 \; |u_\nu^{(1)}> = \hbar\Omega_1\nu \; |u_\nu^{(1)}> , \tag{16}$$

we obtain the relations

$$\begin{aligned} D(\xi)H_1D^\dagger(\xi) \; |u_\nu^{(2)}> &= \hbar\Omega_1\nu|u_\nu^{(2)}> \\ H_2 \; |u_\nu^{(2)}> &= \hbar\Omega_2\nu \; |u_\nu^{(2)}> . \end{aligned} \tag{17}$$

The eigenfrequencies are the same, $\Omega_1 = \Omega_2$, and the eigenstates are related through

$$|u_\nu^{(2)}> = D(\xi) \; |u_\nu^{(1)}> . \tag{18}$$

The Franck-Condon factors are now

$$| < u_\nu^{(2)}|u_\mu^{(1)} > |^2 = | < u_\nu^{(1)}|D(\xi)|u_\mu^{(1)} > |^2 . \tag{19}$$

These states are expressible in terms of coherent states; if e.g. one state is the ground state we obtain the Poisson distribution

$$\begin{aligned} | < u_\nu^{(2)}|u_0^{(1)} > |^2 &= \\ | < \nu|e^{-\xi(a-a^\dagger)}|0 > |^2 &= \frac{(R_0/A_1)^{2\nu}}{2^\nu\nu!}\exp\left(-\frac{R_0^2}{2A_1^2}\right) . \end{aligned} \tag{20}$$

We now move to consider the more general model in Fig.1.b. Here the excited state oscillator has the frequency Ω_2 different from that of oscillator 1;

$$H_2 = \frac{P^2}{2M} + \frac{1}{2}M\Omega_2^2(R - R_0)^2 . \tag{21}$$

In order to describe this we introduce the squeeze operator

$$S(r) = \exp\left[-\frac{r}{2}\left(a^{\dagger 2} - a^2\right)\right] . \tag{22}$$

This transforms the dynamical variables according to

$$\begin{aligned} S(r)RS^\dagger(r) &= e^r R = sR , \\ S(r)PS^\dagger(r) &= e^{-r}P = s^{-1}P; \end{aligned} \tag{23}$$

here $s = e^r$.

From Eq.(16) we obtain directly

$$\begin{aligned} s^2 S(r) H_1 S^\dagger(r) S(r) |u_\nu^{(1)}> &= \hbar\Omega_1 s^2 \nu S(r) |u_\nu^{(1)}>, \\ \left(\frac{P^2}{2M} + \frac{1}{2} M \Omega_1^2 s^4 R^2\right) |\tilde{u}_\nu^{(2)}> &= \hbar\Omega_1 s^2 \nu \; |\tilde{u}_\nu^{(2)}> , \end{aligned} \tag{24}$$

which has got the same oscillational frequency as the Hamiltonian (21) if

$$r = \frac{1}{2} \log\left(\frac{\Omega_2}{\Omega_1}\right) . \tag{25}$$

We can now displace this Hamiltonian by the distance R_0 using the transformation (13). The new eigenfunctions are then given by

$$|u_\nu^{(2)} >= D(\xi) S(r) |u_\nu^{(1)} > . \tag{26}$$

Just like in Eq.(19), this can be used to evaluate the Franck-Condon factor. For example, from the ground state of the excited level

$$< R|u_0^{(2)} >= \left(\frac{s^2}{\pi A_2^2}\right) \exp\left[-\frac{s^2}{2}\left(\frac{R-R_0}{A_2}\right)^2\right] \tag{27}$$

the coupling to the various ground state levels is proportional to

$$| < u_\nu^{(1)} | u_0^{(2)} > |^2 = | < \nu | D(\xi) S(r) | 0 > |^2 = \tag{28}$$

$$\frac{2s}{s^2+1}\left(\frac{s^2-1}{s^2+1}\right)^2 \frac{1}{2^\nu \nu!} H_\nu^2\left[\frac{s^2}{s^4-1}\left(\frac{R_0}{A_2}\right)\right] \exp\left[-\frac{s^2}{s^2+1}\left(\frac{R_0}{A_2}\right)^2\right] ;$$

$H_\nu(x)$ is the νth Hermitean polynomial, see Ref.[7].

Using the results like (20) and (28), many conclusions about the coupling between the two levels can be drawn. In many cases, on the other hand, it is more advantageous to integrate the coupled channel Schrödinger equations (5) directly. A particularly simple case is treated in the next section.

4. Spontaneous Coupling between Levels

We are going to look at the situation where the upper level wave packet has been created by a short laser pulse. Subsequently the packet performs its oscillational motion unperturbed by any coherent coupling to the lower level. It will eventually decay spontaneously emitting one single photon. No more photon emissions are possible before the next excitation process. This corresponds closely to the physical situation described in Ref.[3].

Defining the coupling operator

$$A = |1 >< 2| \tag{29}$$

between the levels, we can describe the ensemble of systems undergoing emission by the density matrix Master Equation

$$\dot{\rho} = -\frac{i}{\hbar}[H, \rho] - \frac{\Gamma}{2}\left(A^\dagger A \rho + \rho A^\dagger A - 2 A \rho A^\dagger\right) . \tag{30}$$

According to the state vector Monte-Carlo method introduced by Carmichael and Mölmer et alii [8,9], we can describe the time evolution of the individual state vectors in the ensemble using the effective Hamiltonian

$$H_{eff} = H - i\frac{\Gamma}{2}A^\dagger A \,. \tag{31}$$

As long as no spontaneous emission has taken place, the state vector to be assigned to the system is

$$|\Psi(t)> = \frac{\psi_1(R,t)\ |1> + e^{-\Gamma t/2}\psi_2(R,t)\ |2>}{\sqrt{\|\ \psi_1(R,t)\ \|^2 + e^{-\Gamma t}\ \|\ \psi_2(R,t)\ \|^2}} \,, \tag{32}$$

where $\psi_i(R,t)$ is the time evolved initial wave packet on level $|i>$. For harmonic oscillator potentials, these are analytically calculable.

The survival probability of the state (32) is given by the distribution

$$P(t) = \Gamma e^{-\Gamma t} \,. \tag{33}$$

When a jump occurs to the lower level, the state after the event is (see e.g. Ref.[10])

$$|\Psi(t+)> = \frac{A|\Psi(t-)>}{\|\ A|\Psi(t-)>\|} = \frac{\psi_2(R,t)\ |1>}{\|\ \psi_2(R,t)\ \|} \,. \tag{34}$$

Thus the shape of the upper state vibrational wave packet is transferred undistorted to the lower level. This is in accordance with the Born-Oppenheimer assumption of slow nuclear adjustment to electronic variations.

Because the shape of the nuclear wave packet is preserved, the momentum and the kinetic energy are conserved in the decay process. This implies that the energy released in a decay process taking place at the position R carries away the energy difference

$$\hbar\omega(R) = \mathcal{U}_2(R) - \mathcal{U}_1(R) = \Delta E + \frac{1}{2}M\left[\Omega_2^2(R-R_0)^2 - \Omega_1^2 R^2\right] \,. \tag{35}$$

The frequency $\omega(R)$ is expected to be observed with the probability equal to that to find the particle in the upper state at the position R. During the time evolution of the wave packets, this will vary, and we find that

$$Prob\{\omega = \omega(R)|t\}\Delta t = \Gamma\ |\psi_2(R,t)|^2\Delta t \,. \tag{36}$$

This is the type of result expected in an experiment like that in Ref.[3]. The observation of a frequency varying in time is not in conflict with uncertainty here, because the resolution is not too close to the Fourier-limited case; if needed a suitable window-function must be used.

For a wave packet at the upper level, with the initial position R_0 and the initial width

$$<(R-R_0)^2> = b^2 \tag{37}$$

we find the solution

$$|\psi_2(R,t)|^2 = \frac{1}{\sqrt{\pi}\sigma(t)}\exp\left[-\frac{(R-R(t))^2}{2\sigma(t)^2}\right] \,, \tag{38}$$

where

$$\begin{aligned} R(t) &= R_0\cos\Omega_2 t \\ \sigma(t)^2 &= b^2\cos^2\Omega_2 t + \frac{\hbar^2}{4b^2M^2\Omega_2^2}\sin^2\Omega_2 t \,, \end{aligned} \tag{39}$$

see Ref.[11]. We can plot the probability density (38) as a function of R and t as shown in Fig.2. Here we have chosen the initial squeezing parameter of the wave packet

$$s^2 = \frac{2M\Omega_2 b^2}{\hbar} = \frac{1}{4} \,. \tag{40}$$

When Fig.2 is cut at a constant value of R, that is probed at a fixed frequency, we find the

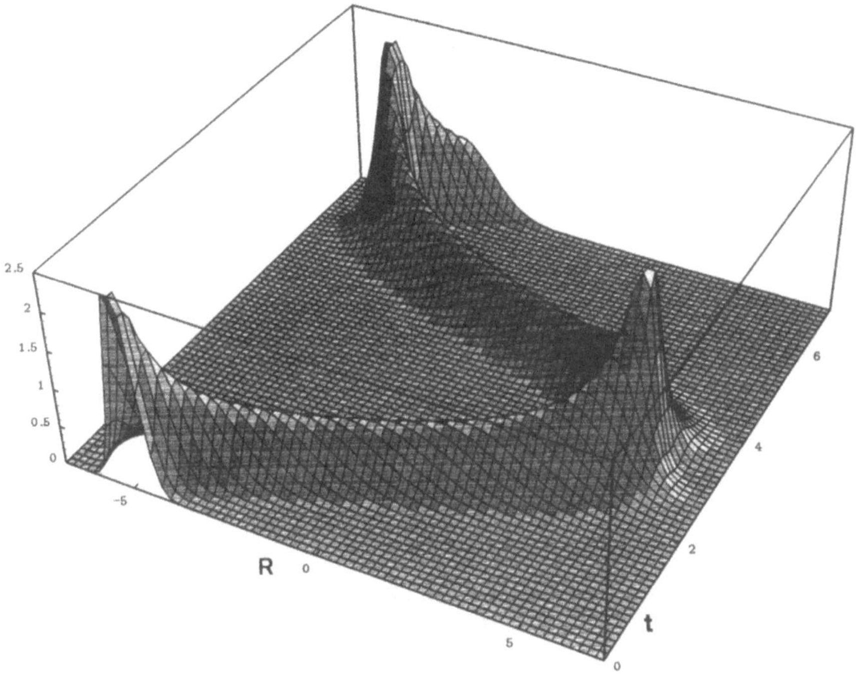

Fig.2. The space-time dependence of an oscillating wave packet on the excited state energy level.

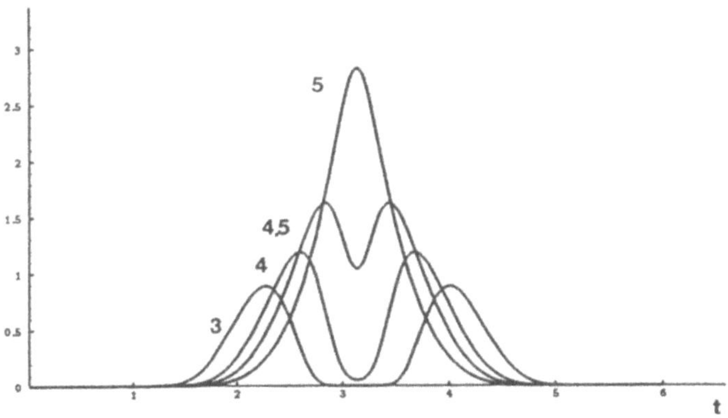

Fig.3 If we cut the function on the previous figure at constant R, we observe this type of time evolution. The curves are labelled by (R/A_2).

curves shown in Fig.3. Here we have looked at the time evolution of the probability in Fig.2 at the positions $(R/A_2) = 5$, 4.5, 4 and 3 respectively. At $(R/A_2) = 5$, we have a single peak, this is the turning of the motion. Gradually, when we move the probe away from this point, the spectrum splits up to show two well resolved peaks at the value $(R/A_2) = 3$. These results are in complete qualitative agreement with the results observed in the experiment [3]. To obtain a proper fit to the observations, one must consider the detailed shape of the potential surfaces and the influence of the parameters of the experiment. The principle of the calculation is, however, exactly like the one discussed in this lecture. Thus I have presented a theoretical framework for the treatment of a large class of problems in the laser-induced dynamical evolution of molecular systems.

References

[1] M. Gruebele and A.H. Zewail, Physics Today, p. 24, May 1990, and A.H. Zewail, Scientific American, **263**, p.40, December 1990

[2] B.M. Garraway, S. Stenholm and K-A. Suominen, Physics World, **6**, p.46, April, 1993

[3] T.J. Dunn, J.N. Sweetser and I.A. Walmsley, Phys. Rev. Lett., **70**, 3388, (1993)

[4] D.M. Hurst, *Potential Energy Surfaces*, (Taylor & Francis, London, 1985)

[5] B.M. Garraway and S. Stenholm, Opt. Commun., **83**, 349, (1991), and K-A. Suominen, B.M. Garraway and S. Stenholm, Phys. Rev., **A45**, 3060, (1992)

[6] J. Janszky and Y.Y. Yushin, Opt. Commun., **59**, 151, (1986), and A.V. Vinogradov and J. Janszky, Phys. Rev. Lett., **64**, 2771, (1990)

[7] W. Schleich and J.A. Wheeler, J. Opt. Soc. Am., **4**, 1715, (1987)

[8] H.J. Carmichael, *An Open Systems Approach to Quantum Optics*, (Springer-Verlag, Heidelberg, 1993)

[9] Y. Castin, K. Mölmer and J. Dalibard, J. Opt. Soc. Am. **B10**, 524, (1993)

[10] Wai Lai and S. Stenholm, Opt. Commun., in press.

[11] S. Stenholm, *Quantum Measurements in Optics*, p. 15, Eds. P. Tombesi and D.F. Walls (Plenum, New York, 1992)

Coherent Pulse Propagation in Mode-Locked Laser Systems

J.D. Harvey[1], *J.M. Dudley*[1], *D.B. Hirst*[1], *S.J. Carter*[2], *and P.D. Drummond*[2]

[1]Physics Department, University of Auckland, Private Bag 92019, Auckland, New Zealand
[2]Physics Department, University of Queensland, St. Lucia 4072, Brisbane, Australia

Coherent pulse propagation effects are shown to manifest themselves in both solid state and gas laser systems. An extension of this semiclassical model to include the effects of spontaneous emission in the laser cavity leads to an explanation of the noise characteristics of the laser system.

1. Introduction

Five years ago it was realised that coherent pulse propagation effects were important in explaining the behaviour of a modelocked argon laser system. In fact, the first extensive results of investigations into these effects and a comparison of theoretical simulations with experimental observations was published in the last volume of the proceedings of this symposium [1]. A more detailed study of the behaviour of the laser has been published recently [2]. While the effects seen in the argon laser system are dramatic, it might well be assumed that this is a pathological case and that coherent interactions are unimportant in other laser systems. This paper is concerned with two questions which arise from the observations of coherent effects in the argon laser. Firstly it is important to know if the coherent pulse propagation effects are apparent in, or whether the theory is relevant to, other laser systems. A second question concerns whether the theory, when extended to include the effects of spontaneous emission, can lead to a deeper understanding of the operation of the modelocked argon laser system. The next section concerns a brief review of results for the argon laser, while sections 3 and 4 are concerned with the questions raised above.

2. Coherent Effects in the Mode Locked Argon Laser

The results of experiments and numerical simulations [1-4] have shown that the operation of a modelocked argon laser system can only be fully understood on the basis of a model for pulse propagation in the laser system based on the fully coherent Maxwell-Bloch equations. The Maxwell-Bloch equations governing the propagation of a pulse in an inhomogeneously broadened medium are given by:

Springer Proceedings in Physics, Vol. 77 Quantum Optics VI
Editors: D.F. Walls · J.D. Harvey

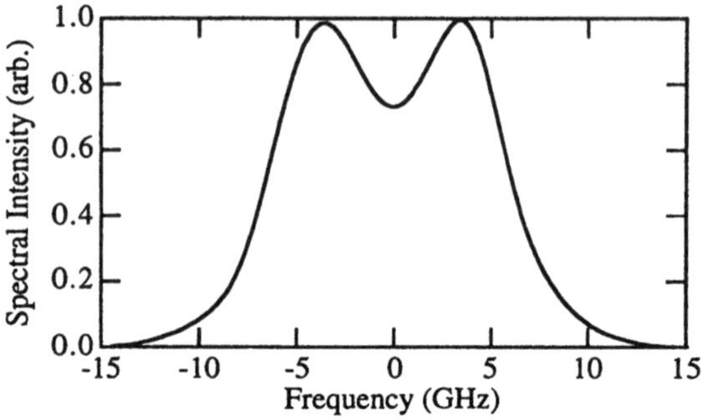

Figure 1 Experimental spectrum of the modelocked argon laser. The pulses corresponding to this spectrum had duration (FWHM) of 40 ps.

$$\frac{\partial E}{\partial t} + c\frac{\partial E}{\partial z} = i\frac{\omega_0}{2\varepsilon_0}\int_\Delta \rho(\Delta)\, P_\Delta \, d\Delta\,,$$

$$\frac{\partial P_\Delta}{\partial t} = -\; P_\Delta\left[\frac{1}{T_2} + i\Delta\right] + |\mu|^2\,\frac{ED_\Delta}{i\hbar}\,,$$

$$\frac{\partial D_\Delta}{\partial t} = -\frac{1}{T_1}(D_\Delta - D_0) - \frac{1}{2i\hbar}(EP_\Delta^{\,*} - E^*P_\Delta)\,,$$

where E is the amplitude of the incident field and P_Δ and D_Δ represent the induced polarisation and the population inversion in the medium at the detuned frequency Δ. Here, μ is the atomic dipole matrix element of the transition, T_1 is the relaxation time of the population inversion and T_2 is the homogeneous relaxation time of the atomic polarisation. D_0 is the equilibrium value of the inversion in the absence of any applied field.

In the argon laser, where the dominant broadening mechanism is inhomogeneous, the atoms are broken into N Doppler classes and the equations are solved separately. The numerical solution is much simpler in the case of a homogeneous system but gives qualitatively similar results. The key features of the predictions of this model are the following:

(i) The appearance of low amplitude trailing pulses behind the main pulse. These afterpulses are generally described as "coherent ringing" or "edge echoes".

(ii) The broadening of the power spectrum of the pulse train which develops a characteristic double peaked shape.

Both of these predicted features have been seen in experimental work using an argon laser. The ringing and the spectral broadening (accompanied by a correspondingly narrower main pulse) become more pronounced as the intra cavity peak power is increased. A typical power spectrum of the pulse train in a high Q cavity is shown in figure 1. It is important to realise that in this system the pulse duration is determined primarily by the intra cavity peak power, not by the inhomogeneously broadened linewidth of the transition. Indeed the pulse duration can be estimated quite reasonably from the peak Rabi frequency of the dipoles excited by the circulating pulse. If the pulse is thought of as a damped Rabi oscillation excited by the high peak power of the intracavity pulse, the pulse duration can be estimated as one third the Rabi period. This estimate is within 20% of the measured pulse width for mean intracavity powers varying by up to an order of magnitude (from 40 ps at 4.71 W to 150 ps at 0.4 W).

A further interesting feature of the experiment concerns the detection of coherent ringing at all levels of intracavity power. At low power levels, the longest pulses corresponding to a stable mode locked pulse stream of 160 ps have a duration longer than the inverse inhomogeneous linewidth and yet still exhibit coherent ringing. This is a consequence of the long homogeneous dephasing time T_2 (600 ps) associated with the argon laser system. Coherent ringing would not be expected in a laser based on a homogeneously broadened transition of the same linewidth, for pulses of this duration.

3. Coherent Pulse Propagation Effects in other Laser Systems

Since the significant effects seen in the argon laser cannot be explained without invoking the optical Bloch equations, it is important to consider whether these or related effects can be seen in other laser systems. It is generally accepted that laser systems are well described by the rate equation limit of the optical Bloch equations and that coherent pulse propagation effects are only seen when high power laser beams are used as probes of narrow linewidth simple systems [5]. The first experimental observation which prompted the argon laser work, however, was the discovery of unexpectedly narrow pulses emitted from a novel argon laser cavity. Recently there has been considerable technological progress in developing laser systems to produce shorter and shorter pulses in a variety of gain media. A survey of some recent results is given in Table I. It can be seen that in most cases now the minimum observed pulse durations are comparable to or even shorter than that which can be sustained by the transition linewidth (ie shorter than would be expected on the basis of a rate equation model). Coherent pulse propagation effects might therefore be anticipated in many current laser systems. The key reason why these effects have not been detected, however, is because, with the exception of the krypton laser, all the active gain media in Table I are solid state where strong non linear optical effects, such as self-phase modulation (SPM) coupled with group velocity dispersion (GVD), are the dominating influences on the pulse shape. Indeed, the presence of these nonlinear effects can be used to explain the observed increase in bandwidth of the laser pulses and hence account for a pulse width shorter than would be expected from a simple rate equation model.

The first question of whether the effects can be seen in any other laser system at all, can be answered by reference to figure 2 where coherent ringing effects have been successfully observed on the pulse train from a modelocked cavity dumped krypton laser operating at 647 nm. This result was certainly expected on the basis of the argon laser work and shows that the effects can be expected universally in gas laser systems. In the case of solid state laser systems, however, the effects of SPM and

Table I Modelocked lasers which may exhibit coherent pulse propagation effects. The expected pulse durations are calculated from the transition linewidth assuming a gaussian pulse shape.

Laser Type	λ (nm)	Fluorescent Linewidth	Expected Pulse Width	Minimum Reported Pulse Width
Argon	514	4 GHz	110 ps	37 ps [1]
Krypton	647	3.8 GHz	116 ps	61 ps [6]
Nd:Glass	1064	5.24 THz	84 fs	88 fs [7]
Nd:Fibre	1064	8.5 THz	52 fs	38 fs [8]
Nd:YAG	1064	180 GHz	5.5 ps	1.7 ps [9]
Nd:YLF	1047	360 GHz	2.7 ps	2.3 ps [10]
Er:Fibre	1590	5.3 THz	83 fs	50 fs [11]
Ti:sapphire	790	100 THz	4.4 fs	11 fs [12,13]

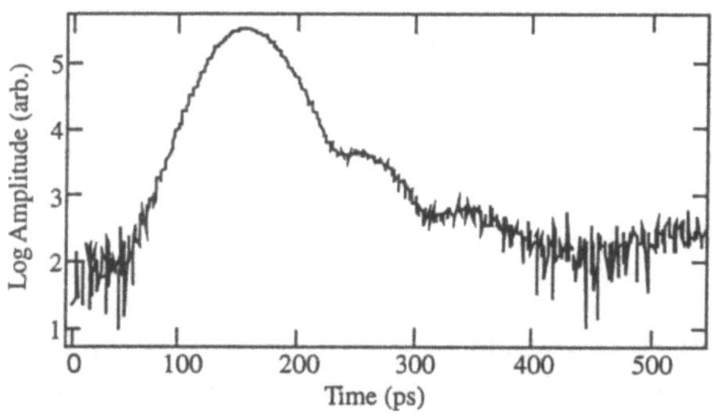

Figure 2 Logarithmic streak camera display of a 60 ps pulse from the modelocked cavity dumped krypton laser

GVD in most cases far outweigh the small perturbations induced by the coherent interaction. Recently, however, exceptionally broad spectra have been recorded in a self mode locked Ti:sapphire laser system operating at 840 nm where the effects of second and third order dispersion have been carefully compensated, and the effects of even higher order dispersion are negligible [12]. The experimental spectrum (shown in Figure 3) is strikingly similar to that shown in Figure 1 apart from being three orders of magnitude broader. The double peak structure is characteristic of a main pulse followed by low amplitude coherent ringing (physically, it is the change in

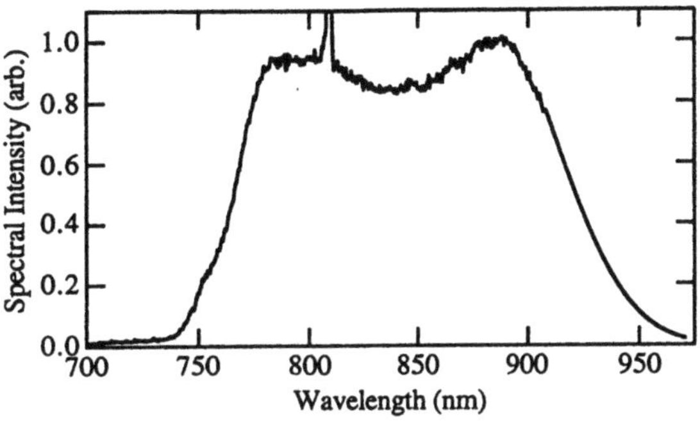

Figure 3 Spectrum of the self-modelocked Ti:sapphire laser operated in the regime of vanishing third-order dispersion (from reference [12]).

sign of the envelope of the electric field which generates the central dip in the power spectrum).

Numerical simulations showed that the experimental autocorrelation functions were also very well fitted by the coherent ringing model [14] with afterpulses spaced at intervals of 18 fs. The Ti:sapphire laser, however, is a homogeneously broadened system with a T_2 of only 3.2 fs and genuine Rabi oscillations having a period of the order of 30 fs would not be sustained, due to dipole dephasing. In other words, the laser must be operating in the "overdamped" regime. Estimates of the Rabi frequency based in the spot-size in the crystal and the known intracavity peak power give results of ≈10THz which are inconsistent with the apparent ringing frequency of ≈30THz. Indeed, true Rabi oscillation as seen in the argon laser system would not be expected unless the intracavity peak power were increased by over two orders of magnitude. This is unlikely to be achieved with current Ti:sapphire laser systems. In this context, it is important to note, however, that the spectrum shown in Figure 3 was obtained in a laser configuration where independent wavelength control meant that the peak of the pulse spectrum was significantly detuned from line centre. Under these conditions the weaker interaction means that the inversion is not significantly altered by the passage of the pulse (in contrast, for the argon laser in a high Q cavity the pulse approaches a π pulse which completely inverts the population of the two participating levels).

With this simplifying assumption, the detuned interaction in the homogeneously broadened Ti:sapphire laser can be described using only two of the Maxwell-Bloch equations:

$$\frac{\partial E}{\partial t} + c\frac{\partial E}{\partial z} = i\frac{\omega_0}{2\varepsilon_0}P,$$

$$\frac{\partial P}{\partial t} = -P\left[\frac{1}{T_2} + i\Delta\right] + |\mu|^2\frac{ED}{i\hbar},$$

which can be solved to yield the perturbation $\Delta E(t)$ generated by the passage of the pulse through the laser medium:

$$\Delta E(t) \sim \exp(-t/T_2)\exp[-i\Delta t\,],$$

where Δ is the detuning between the peak of the pulse spectrum and the peak of the transition gain. This perturbation is heavily damped for the Ti:sapphire laser system operating with a pulse duration of $\approx$10 fs, but the perturbation can accumulate over many round trips in a cavity configuration where the main pulse duration is controlled by the fundamental soliton shaping behaviour of the cavity (in which SPM and GVD are dominant influences).

The important qualitative feature of this perturbation is the oscillatory term at a frequency given by the detuning from line centre even when the system is heavily damped by the dipole dephasing time T_2. In these experiments the detuning was 60 nm, corresponding to an oscillatory frequency of 28 THz in exact agreement with that required to generate the double peaked spectrum shown in figure 3 and, furthermore, to give an excellent fit to the experimentally obtained interferometric autocorrelation functions, including the third and fourth sets of weak subsidiary oscillations, and the phase changes between these sets of oscillations [14].

We conclude therefore that this delayed coherent response of the atomic polarisation is responsible for the experimentally observed data, and that coherent coupling is important in this system as well as in the argon laser system. As peak powers are further increased and pulse durations reduced, it is likely that coherent effects will become important in more and more laser systems.

4. Phase Waves

The semi-classical model of the modelocked argon laser is also described as a "superfluorescent" model since the theory used to describe the laser is essentially that developed more than a decade earlier to describe the swept gain superfluorescent system. In the superfluorescent swept gain model (SSGM) the atomic media is prepared in the inverted state just ahead of the laser pulse which is the only pulse able to initiate and propagate. In the argon laser, the modelocker only allows one pulse to circulate which also "sees" a medium which is always in the inverted state and hence the systems are equivalent.

The SSGM has been developed to include the effects of spontaneous emission, and this has been found to have large scale effects on the pulse evolution Microscopic fluctuations in phase in the leading edge of the pulse have been shown to lead to macroscopic effects on the main pulse, including pulse collapse and revival associated with the passage of a "front" through the pulse where the phase of the underlying electric field changes by a large amount. These "phase waves" move slower than the main pulse envelope and thus progress through the main pulse after being initiated in the weak leading edge, where the spontaneous emission noise background is comparable in amplitude to that of the coherent pulse. The net effect is that the SSGM leads to a very unstable propagating pulse.

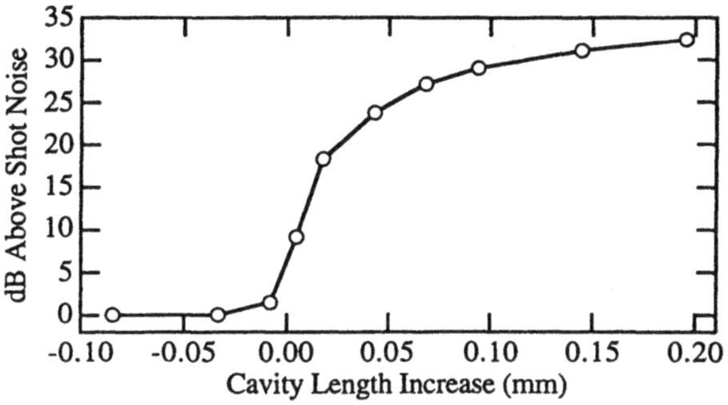

Figure 4 Amplitude noise of the modelocked argon laser (at 1.4 MHz) measured as function of cavity length detuning.

If this model is a correct description of the argon laser, the question which arises therefore, is how the laser can ever be made useful? Alternatively, how can these phase waves be suppressed to yield a stable pulse train? Experimentally the modelocked argon laser is a useful system, and if these phase waves exist then they must be a characteristic of an unstable operating regime normally avoided by experimentalists. Furthermore, the initiation of these phase wave events is stochastic so their effect would be merely to add noise to the pulse train. In searching for evidence of the existence of the predicted phase waves, our investigations have centred on the noise characteristics as a function of length detuning of the cavity. While no absolute length measurements are possible, it is known that the laser operates stably over a small range of lengths, assumed to correspond to a matching of the round trip time of the pulses to the period of the active mode locker in the cavity. If the cavity length is longer than a certain threshold, then the pulse train becomes noisy and the pulses cannot be synchronised to a sampling oscilloscope but the spectral width is maintained.

We have measured the amplitude noise on the pulse train from the modelocked argon laser as a function of cavity length and the results are shown in figure 4. The behaviour as a function of cavity length shows a dramatic onset of noise at longer cavity lengths, while this noise is totally suppressed as the cavity length is reduced. Indeed, in this quiet regime associated with short cavity lengths, we have measured the laser to be shot noise limited in the low frequency regime below 5 MHz.

It is tempting to suppose that this threshold length corresponds to the onset of phase waves which are totally suppressed for shorter cavities. Numerical simulations support this hypothesis since they also show these effects when the equations are modified to include the effects of spontaneous emission. The simulations generate the phase waves, as predicted by Hopf and Overman [15], but their frequency decreases as the cavity length is shortened until they are totally suppressed. This effect is generated by the action of the mode locker which of course has no analogue in the true swept gain superfluorescent system, and can be understood on the basis of two effects. Firstly, in a short cavity, the pulse arrives at the mode locker at a time of decreasing loss, and the leading edge is attenuated more than the trailing edge. This leads to a small shift in the peak of the pulse or an effective pulse retardation due to loss shaping. Secondly the spontaneous emission noise ahead of the pulse is strongly attenuated and this reduces the gain/loss ratio in the regime of phase wave initiation (which is 2-3 pulse widths ahead of the main pulse). This in turn reduces the rate of initiation of phase waves and ultimately suppresses them completely. A detailed investigation of these effects is currently in progress and will be published elsewhere.

A further verification of our interpretation of the noise characteristics has been obtained in a different experiment. The initiation of the phase waves is generated by the interaction of the quantum noise background ahead of the pulse and the coherent field of the mode locked pulse. A recently introduced "coherent seeding" technique has been shown to suppress the noise of a dye laser and lead to transform limited pulses in a system which was previously operating as a "noise burst" [16]. This technique involves inserting a weak coherent replica of the main pulse ahead of the main pulse. This has the effect of swamping the spontaneous emission noise. In our system we would expect that this weak coherent field would also suppress the initiation of phase waves and indeed experimentally the use of a feedback plate has been found to lead to shot noise limited operation of the laser at cavity lengths where the laser was previously extremely unstable [17]. Noise suppression of over 30dB have been achieved. It is important to note that the feedback plate does not introduce a significant loss into the cavity and only has to be aligned to specularly reflect a very weak fraction from the edge of the beam in the resonator back into the main mode. The plate is normally adjusted to be several beam radii away from the centre of the mode and no interferometric stability is required. Under these circumstances the laser produces significantly more peak power than when it is length detuned for stable operation.

5. Conclusion

This paper has been concerned with extending previous work on coherent pulse propagation in laser systems. Recent results have shown that the coherent interaction of the light field and the atomic dipoles of the gain medium are important in both solid state and gas laser systems, and will need to be considered in an increasing number of systems as the pulse duration decreases and the peak power increases. In addition, the noise characteristics of the modelocked gas laser, and potentially of other laser systems, can only be understood on the basis of a model which includes the spontaneous emission noise background surrounding the propagating pulse. This noise has its origin in a macroscopic manifestation of the quantum noise ahead of the pulse. In this sense, while many of the effects discussed here can be understood on the basis of a semiclassical model, the study of the noise on the laser system is clearly in the domain of quantum optics.

Acknowledgments

The study of coherent propagation effects in the Ti:sapphire laser is in collaboration with Ch. Spielmann, P. F. Curley and F. Krausz from Technische Universität Wien, Austria. We are particularly grateful for their permission to include experimental results from reference [12] in this manuscript.

References

1. J. D. Harvey, J. Dudley, R. Leonhardt, S. Carter and P. D. Drummond, in Quantum Optics V, J. D Harvey and D. F. Walls eds. (Springer-Verlag, Berlin, 1989), 214.
2. J. M. Dudley, J. D. Harvey and R. Leonhardt, J. Opt. Soc. Am. B, **10**, 840 (1993).
3. J. D. Harvey, R. Leonhardt, P. D. Drummond and S. Carter, Phys. Rev. A **40**, 4789-4792 (1989).
4. J. M. Dudley, J. D. Harvey and R. Leonhardt, Opt. Lett. **15**, 335-337 (1990).
5. A. E. Siegman, *Lasers*, (University Science Books, Mill Valley, Ca.., 1986), 113.
6. S. Murdoch, MSc Thesis University of Auckland, *in preparation,* (1994).
7. Ch. Spielmann, F. Krausz, T. Brabec, E. Wintner and A. J. Schmidt, Appl. Phys. Lett. **58**, 2470 (1991).
8. M. Hofer, M.H. Ober, F.Haberl, and M.E. Fermann, IEEE J. Quant. Electron. **28**, 720 (1992).
9. J. Goodberlet, J. Jacobson, J.G. Fujimoto, P.A. Schulz, and T.Y. Tan, Opt. Lett. **15**, 504 (1990).
10. M. Ramaswamy, A.S. Gouveia-Neto, D.K. Negus, J.A. Izatt, and J.G. Fujimoto Opt. Lett. **18**, 1825 (1993).
11. D.J. Richardson, V.V. Afanasjev, A.B. Grudinin, and D.N. Payne, Opt. Lett. **17**, 1596 (1992)
12. C.H. Spielman, P.F. Curley, T. Brabec, and F. Krausz : "Ultrabroad-band femtosecond lasers", IEEE J. Quant. Electron. *in press*, (1994).
13. M. Asaki, C. Huang, D. Garvey, J. Zhou, H. C. Kapteyn and M. M. Murnane, Opt. Lett. **18**, 977 (1993).
14. J. D. Harvey, J. M. Dudley, C.H. Spielman, P.F. Curley and F. Krausz, "Coherent Effects in a Self-Modelocked Ti:sapphire Laser", Opt. Lett., *in press,* (1994).
15. F. A. Hopf and E. A. Overman, Phys. Rev. A **19**, 1180 (1979).
16. D. S. Peter, P. Beaud, W. Hodel and H. P. Weber, Optics Letter **16**, 405 (1991).
17. J. D. Harvey, N. Christensen, J. M. Dudley and D. B. Hirst, Opt. Comm. **97**, 219 (1993).

Long-Range Atom-Dielectric Interactions

M. Ducloy

Laboratoire de Physique des Lasers, URA 282 du C.N.R.S.,
Institut Galilée, Université Paris-Nord, F-93430 Villetaneuse, France

Abstract : Recent advances in the monitoring of long-range interactions between vapor-phase atoms and dielectric surfaces via selective reflection spectroscopy, are reviewed. Particular emphasis is given to the influence of the dispersivity of the dielectric medium, and the angular properties of the surface potential. One discusses the possibility that dielectric resonances induce giant van der Waals attraction or repulsion of excited-state atoms.

1 - Introduction

The interactions between isolated particles and solid-state surfaces cover a large domain of modern physics, and have important implications in many fields : collision and ion physics, nuclear physics, surface physics, chemistry (catalysis, chemiadsorption) etc... [1]. At short distances, the collisional character of the interaction and the surface adsorption mechanisms are predominant. For complex particles like atomic systems, their internal properties are deeply altered, in a way which strongly depends on the microscopic structure of the surface.

On the other hand, at larger distances, this detailed structure of the surface gets ineffective, and the atoms retain most of their free-space characteristics. The modifications of the atomic structure are generally mediated by the electromagnectic (e.m.) field, in particular via the alteration of the e.m. vacuum modes which is imposed by the boundary conditions associated to the presence of a reflecting surface (cavity quantum electrodynamics, QED) [2]. For instance, for a perfectly conducting plane, the long range attraction of a neutral atom can be viewed as the interaction between the fluctuating atomic dipole and its electric image induced in the surface. In the non-retarded regime (short-range limit), the perfect correlation between the atomic dipole and its instantaneous image leads to the well-known van der Waals - London attraction potential scaling like z^{-3} with the atom-surface distance, z [3] . In the long-range limit, the dipole correlations tend to vanish. Casimir and Polder have shown that this retarded regime is thus characterized by a z^{-4} distance law, for a ground state atom [4].

The first experimental investigations of these long-range forces have been based on the observation of mechanical deflection of atomic beams by metallic surfaces [5, 6]. Since 1989, the Villetaneuse group has started an optical investigation of atom-surface interactions, by devising new high-resolution spectroscopic techniques at vapor-dielectric interfaces . This allows one to monitor the modification (shift, asymmetry...) of atomic spectra close to the interface [7-10]. This experimental approach has raised new questions : what is the influence of the dielectric response (which is in general different from that of a perfect reflector), what is the influence of the state of atomic excitation ?

Springer Proceedings in Physics, Vol. 77 Quantum Optics VI
Editors: D.F. Walls · J.D. Harvey

2 - Frequency-modulated selective reflection spectroscopy

The spectroscopy of gas-phase atoms in the vicinity of a dielectric surface can be performed through selective reflection (SR) techniques [11-14]. SR spectroscopy covers the field in which one monitors the spectral dependence of the optical reflection coefficient $R(\omega)$ at an interface between a dielectric window and a vapor cell. This reflection coefficient exhibits resonant variations whenever the frequency ω of the incident light beam gets into resonance with the atoms of the vapor (see fig. 1). Indeed, the optical dipoles resonantly induced in the vapor by the incident irradiation coherently reradiate an e.m. field $\Delta E_r(\omega)$ which adds to the non-resonant reflected field E_r. Under normal incidence irradiation, one gets [15]

$$\Delta E_r(\omega)=\frac{ik}{(n+1)\varepsilon_0}\int_0^{+\infty} p(z)\,e^{2ikz}\,dz \tag{1}$$

where p(z) is the complex amplitude of the induced oscillating dipole and n the dielectric refractive index. In eq. (1), e^{2ikz} is a atom-surface round-trip phase factor ($k = \omega/c$) appearing when evaluating the interference between the field radiated by the induced dipoles. This phase delay is responsible for the fact that the dominant contribution in the SR signal originates in a *typical vapor layer* $z<\lambda/2\pi$.

In numerous cases, the SR signal can be evaluated through a local index model (i. e. $p(z) = \varepsilon_0\,\chi E(z)$) so that SR spectroscopy becomes a way of monitoring the refractive index variations. However, at the interface with a low-pressure gas medium, the thermal motion generally invalidates the prediction of a local index theory, and vapor *spatial dispersion* must be taken into account. Indeed, when analyzing the SR lineshape, one notices a symmetric sub-Doppler contribution superimposed to the expected Doppler-broadened dispersive profile [12-13]. This narrow structure is a signature of the asymmetric behavior between arriving atoms ($v_z < 0$) resonantly driven by the incident field and already in a steady state, and the departing atoms ($v_z > 0$) in a transient evolution just following desorption.

To single out this narrow contribution, a simple frequency modulation (FM) technique can be used. Demodulation of the SR signal, which yields the frequency derivative of the original SR lineshape, provides a purely Doppler-free dispersion signal (in the large Doppler approximation) [10]. This signal originates in atoms *moving parallel to the surface* ($v_z = 0$), and which remain inside the $\lambda/2\pi$ vapor layer during the dipole lifetime. Atoms moving quickly with respect to the interface, do not stay long enough in this vapor layer to bring a sizeable contribution. Thanks to this selection of $v_z = 0$ atoms in a $\lambda/2\pi$ layer, FM selective reflection at normal incidence allows one to perform *Doppler-free, linear,* dispersion spectroscopy of a kind of "two-dimensional atomic beam", confined inside a wavelength-deep plane.

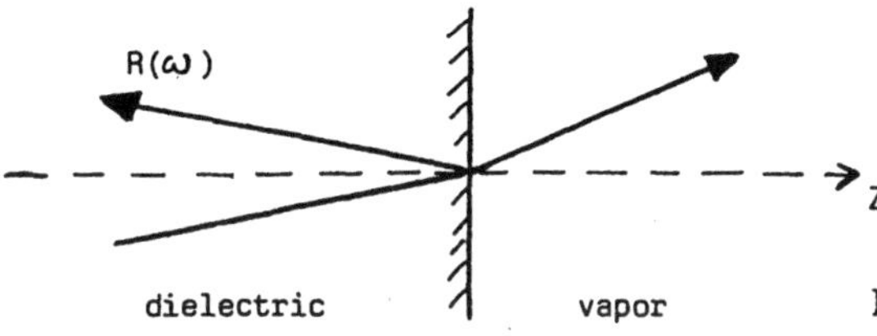

Fig. 1 *Principle of SR Spectroscopy.*

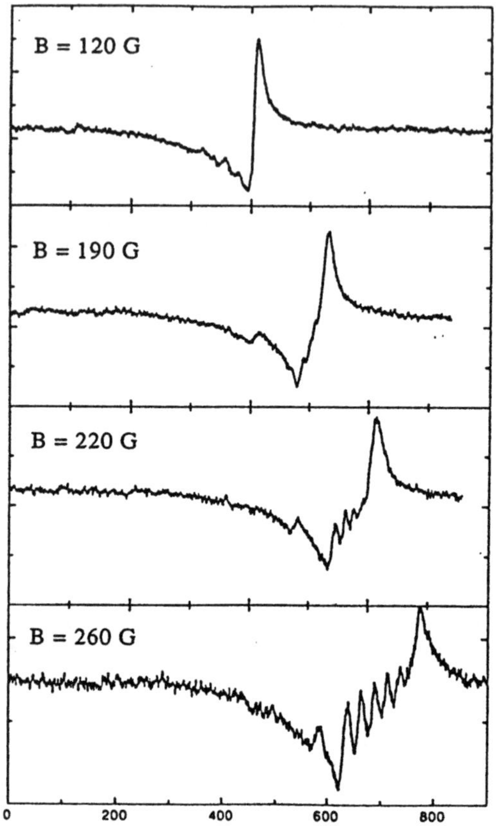

Fig. 2 *FM-SR Zeeman spectrum of the Cs D_2 line, monitored with σ^+ - polarized light (from Ref [17]).*

This spectroscopic approach gives direct access to Doppler-free diagnostics of optically dense vapors and has been applied to various alkalis (Cs, Rb, K). This has been used to analyze :

(i) collisional effects in dense vapors (pressure shift and broadening [9,16]).

(ii) Doppler-free Zeeman spectra of alkali vapors.
Such a spectrum on the Cs resonance line is given in Fig. 2 [17].

(iii) long-range van der Waal (vW)-type atom-dielectric surface interactions, which induce line shifts and distortions [7-10].

In the latter case, one takes advantage of the great sensitivity of low-velocity atoms to frequency shifts induced by the surface potential. A general theory of FM-SR spectroscopy for two-level atoms moving into an inhomogeneous potential has been developed [10]. Specific applications of this theory to a z^{-3} scaling law for the atomic resonance frequency

$$\omega_o(z)=\omega_o-C/z^3 \quad (2)$$

has allowed one to interpret all the experimentally observed spectra, the lineshape of which depends only on a single dimensionless parameter $A=2Ck^3/\gamma$ (γ:optical linewidth). A gives the vW transition shift in $\hbar\gamma$ units at a distance $\lambda/2\pi$. Two distinct regimes are observable [9] : (i) the weak vW regime ($A<<1$), which is characterized by a relatively weak red shift, as expected for an attraction process, and a small line assymetry, is generally observed on the first resonance line. (ii) In the strong vW regime ($A>>1$), quite anomalous lineshapes are observed, including absorption-like or even inverted dispersion-like red-shifted lineshapes, which reflect the strong spatial inhomogeneous broadening of the line. This regime appears for highly excited atomic levels (e. g., on the second Cs resonance line, $6S_{1/2}$ - $7P_{3/2}$ [8]).

Accurate fitting of the experiments has allowed us to measure the strength of the vW attraction on various Cs transitions, with a relatively good precision. For instance, the vW strength is C = 1.37 (0.4) kHz μm^3 for the Cs $6S_{1/2}$ - $6P_{3/2}$ resonance line, in the case of a quartz window. For the $6S_{1/2}$ (F=4) - $7P_{3/2}$ (F=5) line, the vW strength measured at a Cs - sapphire interface is quite large, 20 (3) kHz μm^3. On both resonance lines, the vW strength is found to be independent of the hyperfine component, within the experimental accuracy.

3 - Atom-surface interaction potential

The fitting of the experimental lineshape with the predictions of a z^{-3} interaction potential clearly demonstrates that the nonretarded part of the atom-surface interaction is predominant in the wall-induced modifications of the atomic response. This observation is consistent with the fact that the average atom-surface distance probed in SR spectroscopy is about 100 nm, while the characteristic atomic transition wavelengths are in the mid-infrared range.

If one supposes a perfectly reflecting surface, the non-retarded dipole-image-dipole interaction Hamiltonian is

$$H^o_{vw}=-\left(\vec{\mu}^2+\mu_z^2\right)/16z^3 \quad (3)$$

where $\vec{\mu}$ is the atomic dipole operator. The vW-induced energy shift of state $|j>$ is the diagonal matrix element of H^o_{vw}, and is given by a sum over all the virtual dipole couplings

$$\left\langle j\middle|H^o_{vw}\middle|j\right\rangle=-\sum_{j'}\left|\mu_{jj'}\right|^2/12z^3 \quad (4)$$

with $\mu_{jj'}=\langle j|\mu|j'\rangle$. In Eq.4, one has only considered the scalar shift i.e. $\langle\mu_z^2\rangle=\langle\mu^2\rangle/3$.

For a non-dispersive dielectric medium, with dielectric permittivity ε, there is a dielectric "reflection coefficient" $(\varepsilon-1)/(\varepsilon+1)$, appearing as a multiplicative factor in

Eqs. (3-4), which describes the dielectric response of the surface. However some of the atomic-transition frequencies, $\omega_{jj'}$, may lie inside the dielectric absorption range. In that case, the frequency-dependence of ε (ω), i. e. the dispersion of the dielectric response must be taken into account.

4 - Dispersive dielectric surfaces

4-1 Ground state atoms

The analysis of ground state atoms interacting with an homogeneous dispersive dielectric has been performed by several authors [18, 19]. One can show [20] that the dielectric response can be accounted for by introducing in Eq. 4 a *frequency-dependent* dielectric reflectivity, which is simply given by an integral over the imaginary frequency axis :

$$r(\omega > o) = \frac{2}{\pi} \int_o^{+\infty} \frac{\varepsilon(iu) - 1}{\varepsilon(iu) + 1} \frac{\omega \, du}{\omega^2 + u^2} \tag{5}$$

The ground state energy shift is thus given by

$$\langle g | H_{vw} | g \rangle = - \sum_j |\mu_{gj}|^2 r(\omega_{gj}) / 12z^3 \tag{6}$$

in which a different dielectric image coefficient appears for each virtual coupling $|g> - |j>$. Due to the well-known properties of ε (iu), r (ω) is a positive function, smaller than one, and monotonically decreasing with ω. As an example, Fig. 3a shows the sapphire reflectivity (for the ordinary index).

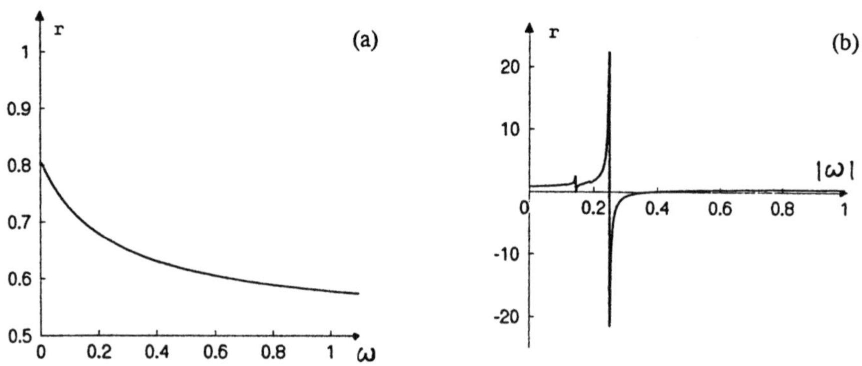

Fig. 3 *Dielectric image reflection coefficient* r(ω) *versus frequency* ω *(in* 10^{14} *Hz) for the ordinary mode of sapphire. (a) absorption, (b) emission. The sapphire vibration bands induce image resonances at* $\lambda = 12\mu m$ and $21\mu m$ *(from Ref [20].*

4-2 Excited-state atoms

In most cases, excited-state atoms interacting with dielectrics may still be described by Eqs. 5-6. This is in particular the case when the virtual dipole couplings of Eq. 6 appear in *absorption* $(\omega_{gj} > 0)$, e. g. for a metastable state. This is no longer true for excited state atoms exhibiting dipole coupling in *emission* (i.e. to lower lying level). This case has been studied in detail by Wylie and Sipe [21]. It can be shown that Eq.6 still applies, but with a dielectric reflectivity defined for negative frequencies by [20]

$$r(\omega < 0) = \frac{2}{\pi}\int_0^{+\infty} \frac{\varepsilon(iu) - 1}{\varepsilon(iu) + 1} \frac{\omega du}{\omega^2 + u^2} + 2\,\mathrm{Re}\frac{\varepsilon(|\omega|) - 1}{\varepsilon(|\omega|) + 1} \qquad (7)$$

Contrary to the first term of the right hand side, the second term, thanks to its dependence on real frequencies, closely reflects the dielectric dispersion at the atomic emission frequency. This implies that the dielectric resonances deeply alter the surface image coefficient, and, in particular, can enhance the dielectric response by more than one order of magnitude, as compared with a perfect reflector $(r=1)$. This behaviour, which can be viewed as a resonant virtual exchange of excitation between the atom and the dielectric medium, is shown on Fig. 3b for the ordinary mode of sapphire. Therefore, *giant* van der Waals *attraction* or *repulsion* should be observable for excited atoms whose main emission coupling would be in the 11.5-12.5 μm region. It should be outlined that, simultaneously to this enhanced shift, resonant surface-enhanced level broadening is also predicted [21, 22].

From the above analysis, one calculates the predicted vW frequency shift to be 15 kHz $(\mu m)^3$ for the Cs $6S_{1/2}$ - $7P_{3/2}$ transition, in relative good agreement with the experimental findings. One should note that 60 % of this shift originates in the 7P - 6D dipole coupling which lies in the mid-infrared range $(\lambda = 15 \mu m)$[20].

5 - Anisotropy of the van der Waals interaction

The symmetry properties of the atom-surface interaction potential are contained in the angular dependence of $(\mu^2 + \mu_z^2)$. This Hamiltonian has a cylindrical symmetry around the normal to the surface. In addition to the scalar component $4\mu^2/3$, there is a quadrupolar component.

$$T_0^{(2)} = \mu_z^2 - \mu^2/3 \qquad (8)$$

which is responsible for a M-sublevel energy shift, which depends on the absolute value of the magnetic quantum number, $|M|$. The influence of the surface is similar to the effect of a fictitious inhomogeneous electric field, orthogonal to the surface. The z-dependent surface attraction distorts the atom's electronic cloud in a way that it gets an ellipticity along the surface normal.

This wall-induced resonant Kerr effect is responsible for a vapor anisotropy in the vicinity of the surface. The first vapor layers exhibit a birefringence whose axis is parallel to the surface normal, which, in particular, implies that the atom-field interaction should depend on the state of light polarization. An important feature is that this wall-induced

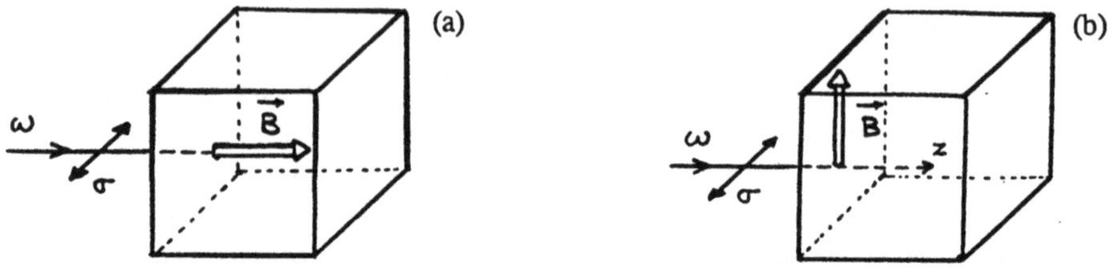

Fig. 4 *Field configurations for the monitoring of FM-SR Zeeman spectra with linearly polarized light, and with an applied magnetic field either normal (a), or parallel (b) to the surface. The two geometries are equivalent in the absence of surface interaction, or when the latter one is purely scalar.*

birefringence cannot be observed at *normal* light incidence, where the SR technique is the most suitable for selective detection of atoms close to the wall. Indeed, in this geometry, the atomic response does not depend on the light polarization, because of the cylindrical symmetry. However, in presence of an hyperfine structure, the quadrupolar interaction (Eq. 8) should break the internal atomic symmetry and introduce a dependence of the vW shift on the hyperfine level, as well as coupling between different hyperfine levels. However, due to the dominant role played by P-D couplings in the 6P and 7P level shift, one can show that the vW interaction is scalar within 5 to 10 %, and the hyperfine symmetry breaking is very small. This interprets the experimentally observed independence of the shift with respect to the various hyperfine components.

There are several approaches to improve the sensitivity of the reflection signal to the quadrupolar component of the surface interaction.

(i) The off-diagonal coupling should be responsible for the appearance of additional (free space forbidden) hyperfine lines in SR spectra, like $\Delta F = \pm 2$ components. Theoretical calculations show that, in spite of a considerable (*spatial*) inhomogeneous broadening (and shift) of such transitions (whose probability varies as z^{-6}), the lineshape of these forbidden components should remain within the Doppler width of a typical hyperfine structure, and that their maximum amplitude should bring an observable extra-distortion to standard SR lineshapes.

(ii) Another approach is to force the wall-induced vapor birefringence to rotate, in order to make it observable. This is, for instance, possible through the Larmor precession in an external magnetic field. More precisely, the comparison of Zeeman spectra recorded with linear σ-polarization, in two different geometries - i. e. with a magnetic field either parallel or orthogonal to the surface - should be a sensitive probe of anisotropic interactions (see Fig.4). Indeed, if the atom surface interaction is scalar, the two geometries are fully equivalent. For large enough magnetic fields, it is thus possible to explore the quadrupolar contribution on each spectrally resolved Zeeman components. Such experiments are in progress in Cesium vapor. It should be noted that, for the resonance level, in addition to vW shift anisotropy, there is also an important anisotropy in the wall-induced modification of the spontaneous emission rate, which can be enhanced or inhibited according to the dipole orientation and the nature of the surface [23, 24]. We thus expect that both amplitude and linewidth of the Zeeman resonances should be M-dependent.

6 - Conclusion

Selective reflection spectroscopy provides a sensitive way of monitoring velocity-selected atoms in a wavelength-confined 2D space. In addition to Doppler-free spectroscopy of optically dense vapors, this should have applications in devising compact, absolute frequency stabilization schemes. This is also a way of performing quantum optics of resonant systems in open, dielectric-bound microcavity, and, as such, gives access to the spectroscopy of long range atom-dielectric complex. In this type of micron-sized quantum optics, the excited atom acts as a selective detector of the quantum noise of the dielectric-vacuum interface, at a set of discrete atomic frequencies. This is to compare with usual cavity QED, in which the resonant cavity is used to filter, and eventually suppress, the photon noise at the atomic resonance frequency.

For weak atom-dielectric couplings, the atom keeps its main free-space characteristics, simply exhibiting surface-induced level shift and broadening, and symmetry breaking. Resonant enhancement of the van der Waals interaction could give access to strongly coupled excited-atom-dielectric systems, and their novel potential characteristics : long-range bound states, repulsive states, radiative properties etc...

References

[1] See e.g. ZANGWILL A, "Physics at Surfaces" (Cambridge University Press, Cambridge, 1988).

[2] See e.g. HAROCHE S. in "Fundamental Systems in Quantum Optics" (J. Dalibard, J-M. Raimond and J. Zinn-Justin, Eds, North-Holland, 1992) p. 771.

[3] LENNARD-JONES J.E., Trans. Faraday Soc., 28 (1932) 334.

[4] CASIMIR H.B.G. and POLDER D., Phys. Rev., 73 (1948) 360.

[5] RASKIN D. and KUSCH P., Phys. Rev., 179 (1969) 172 ; SHIH A., RASKIN D. and KUSCH P., Phys. Rev. A 9 (1974) 652 and 1507.

[6] ANDERSON A., HAROCHE S., HINDS E.A., JHE W. and MESCHEDE D., Phys. Rev. A 37 (1988) 3594 ;SUKENIK C.I., BOSHIER M.G., CHO D., SANDOGHDAR V. and HINDS E.A., Phys. Rev. Lett. 70 (1992) 560.

[7] ORIA M., CHEVROLLIER M., BLOCH D., FICHET M. and DUCLOY M. Europhys. Lett. 14 (1991) 527.

[8] CHEVROLLIER M., BLOCH D., RAHMAT G. and DUCLOY M. Opt. Lett. 16, (1991) 1879.

[9] CHEVROLLIER M., FICHET M., ORIA M., RAHMAT G., BLOCH D. and DUCLOY M., J. Phys. II (France) 2 (1992) 631.

[10] DUCLOY M. and FICHET M., J. Phys. II (France) 1 (1991) 1429.

[11] WOOD R.W., Phil. Mag. 18 (1909) 187.

[12] COJAN J.L., Ann. Phys. (France) 9 (1954) 385.

[13] SCHUURMANS M.F., J. Phys. (Paris) 37 (1976) 469.

[14] AKUL'SHIN A.M., VELICHANSKII V.L., ZIBROV A.S., NIKITIN V.V., SAUTENKOV V.V., YURKIN E.K. and SENKOV N.V., Pis'ma Zh. Eksp. Teor. Fiz 36 (1982) 247 [J.E.T.P. Lett. 36 (1982) 303].

[15] NIENHUIS G., SCHULLER F. and DUCLOY M., Phys. Rev. A 38 (1988) 5195.

[16] PAPAGEORGIOU N., FICHET M., SAUTENKOV V., BLOCH D. and DUCLOY M., "Laser Physics", in press (1994).

[17] PAPAGEORGIOU N., WEIS A., SAUTENKOV V., BLOCH D. and DUCLOY M., "Applied Physics B", in press (1994).

[18] MAVROYANNIS C., Mol. Phys., 6 (1963) 593.

[19] McLACHLAN A.D., Proc. Roy. Soc (London), A 271 (1963) 387.

[20] SCHULLER F., FICHET M., BLOCH D. and DUCLOY M., to be published.

[21] WYLIE J.M. and SIPE J.E., Phys. Rev., A 30 (1984) 1185 ; A 32 (1985) 2030.

[22] CHANCE R.R., PROCK A. and SILBEY R., J. Chem. Phys., 62 (1975) 2245.

[23] JHE W., ANDERSON A., HINDS E.A., MESCHEDE D., MOI L. and HAROCHE S., Phys. Rev. Lett. 58 (1987) 666.

[24] LUKOSZ W. and KUNZ R.E., Optics Commun. 20 (1977) 195 ; J. Opt. Soc. Am. 67 (1977) 1607 and 1615.

Index of Contributors

Springer-Verlag and the Environment

We at Springer-Verlag firmly believe that an international science publisher has a special obligation to the environment, and our corporate policies consistently reflect this conviction.

We also expect our business partners – paper mills, printers, packaging manufacturers, etc. – to commit themselves to using environmentally friendly materials and production processes.

The paper in this book is made from low- or no-chlorine pulp and is acid free, in conformance with international standards for paper permanency.

MIX
Papier aus verantwortungsvollen Quellen
Paper from responsible sources
FSC® C105338

If you have any concerns about our products,
you can contact us on
ProductSafety@springernature.com

In case Publisher is established outside the EU,
the EU authorized representative is:
Springer Nature Customer Service Center GmbH
Europaplatz 3, 69115 Heidelberg, Germany

Printed by Libri Plureos GmbH
in Hamburg, Germany